STRATHCLYDE UNIVERSITY LIBRA
30125 00378893 1
KV-119-900

ANDERSONIAN LIBR
WITHDRAWN
FROM
LIBRARY STOCK
UNIVERSITY OF STRATHCLYDE

WATER AND FOOD QUALITY

WATER AND FOOD QUALITY

Edited by

THELMA M. HARDMAN

Department of Chemistry, University of Reading, UK

ELSEVIER APPLIED SCIENCE
LONDON and NEW YORK

ELSEVIER SCIENCE PUBLISHERS LTD
Crown House, Linton Road, Barking, Essex IG11 8JU, England

Sole Distributor in the USA and Canada
ELSEVIER SCIENCE PUBLISHING CO., INC.
655 Avenue of the Americas, New York, NY 10010, USA

WITH 29 TABLES AND 55 ILLUSTRATIONS

British Library Cataloguing in Publication Data

Water and food quality.
1. Food. Preservation & processing. Effects of water
I. Hardman, Thelma M.
664'.02

Library of Congress Cataloging-in-Publication Data

Water and food quality.

Includes bibliographies.
Contents: Water activity and food quality/ J.A. Troller—Dehydration of foodstuffs/ J.G. Brennan—Interpreting the behavior of low-moisture foods/H. Levine and L. Slade—[etc.]
1. Food—Water activity. 2. Food—Quality.
I. Hardman, Thelma M.
TX553.W3W364 1989 664'.07 88-33390

ISBN 1-85166-306-1

Printed by The Universities Press (Belfast) Ltd.

PREFACE

Moisture, its control, equilibration and mobility are of paramount concern to both the food scientist and technologist. The ambrosial delight given by a crisp, fragile meringue and the disgust produced by a limp cookie are both predetermined by the moisture content and its distribution within a foodstuff. The earliest civilisations found empirically that water content is a critical factor in determining the spoilage and nutritional qualities of certain foods, many foods were preserved by sun-drying and salting. In modern food preservation the prevention of moisture exchange and the controlled removal of water have become important quality control techniques, extending storage life. Chemical changes, many undesirable, and microbial growth, both increase with moisture content.

Many food materials in uncontrolled conditions will change their moisture content by adsorbing or desorbing water. The amount and rate of water exchange will depend not only on the concentration of water vapour present in the atmosphere to which the sample is exposed but also on the nature and concentration of water soluble substances present; indeed this is a function of the water activity.

For many foods there is an optimum value of the water activity at which the maximum stability is achieved and deterioration minimised. For example, condensation of water in a sugar-based jam raises the water activity at the surface and moulds subsequently proliferate. Many food systems exhibit hysteresis in their interactions with water so that the same activity is not attained by simply adsorbing or desorbing moisture.

As the factors that determine water activity have become better understood, so new food products have been developed. Many are marketed in metastable states and rely for palatability on a kinetic barrier to the equilibration of water in all phases and the continuing drive towards minimal use of food preservatives, the attainment of food stability by controlling water activity is increasingly pursued. Although many degradation processes can be accounted for by increase of molecular mobility and reactivity with unbound water content, some solute-related phenomena associated with bacterial growth remain perplexing. Microbiologists are exploring the phenotypic adaptation and physiological requisities for propagation at low water activities. Of the basic food components fats make least contribution to temporal changes in the textural qualities of foodstuffs under isothermal conditions; the isomorphic state, be it liquid or particulate, is retained. Amphiphilic lipids interact with water producing the mesomorphic phases necessary for batter stability in the production of baked products, and interact with proteins to influence emulsion and foam stability. It is the carbohydrates and proteins that bind with water; hydrophilic properties of both and hydrophobic interactions of the latter are major determinants of the three-dimensional structure and texture.

Polysaccharides form liquid crystals and glasses. Proteins bind water and influence the water holding capacity of the food; interaction with lipids alters the conformation and bonding properties. Emulsification and production of unstable systems of large surface area is reliant on their amphiphilic nature. Food textures may be completely determined by protein structure as in the denaturing of protein to form meringue; a trace of fat prevents the foaming of the egg white. In this volume, the more general ambience of the effects of water on food quality, in the control of emulsion stability, gel structure and food texture, and in the provision of a medium for the distribution and subsequent degradation of components, is considered.

In the initial chapter John Troller presents an overview of the static and dynamic aspects of water activity and how it affects the purely aesthetic aspects of food presentation, the nutritional quality in degradation of amino acids and vitamins, as well as the sinister aspect of microbial growth. Next James Brennan considers the practical aspects of water removal from foodstuffs; modern methods of industrial food dehydration are examined and trends in research and development in this area scrutinised. Assignment of a value for the water activity does not give information about the nature of the binding of water molecules in the foodstuff or knowledge of the kinetics of binding processes. A contentious question is 'Can the amount of 'bound water' be identified?' This can be defined as the 'unfreezable water content' and is determined using differential calorimetric techniques. Harry Levine and Louise Slade, experts in this area, apply a polymer science approach to low-moisture food systems, plasticised by water.

The interaction between proteins and water determines many functional properties of proteins in foods. Marilynn Schnepf's contribution describes the types of binding between water and protein molecules and the methods used to determine this interaction. James McKay concentrates on the relationship between moisture content and enzymic activity. The action of some enzymes during storage may be beneficial, however most are detrimental to food quality on storage.

Some of the ways in which packaged food is found on the supermarket shelf rely on the skill of the food technologist in producing stable emulsions and gels. Only a limited number of synthetic surfactants are permitted in the food industry, so the emulsifying properties of proteins must be fully exploited. Douglas Dalgleish outlines the science behind the art of protein emulsification and indicates the limits of current awareness in this area.

The trend towards continuous production methods in the food industry, particularly using extrusion techniques, depends for its success on the rheological properties of the food system at various processing stages. Alan Bell explains the intricacies of obtaining meaningful rheological parameters and outlines current knowledge of the efficacy of different gelling agents.

Finally three areas are examined in detail. Karl Honikel describes how water movement in the mechanical processing and freezing of meat determines the eating quality. For example water content and water binding affects the firmness, toughness and juiciness of meat and water retention is influenced by early chilling or freezing. In the processing of meat products salt, fat and other additives are added to improve the water holding capacity.

Barbara Brockway surveys the connexion between water content, shelf-life and mouthfeel of confectionery products. Stability of a one-phase or homogeneous confectionery such as a boiled sweet or fudge is ensured by suitable packaging, but for two distinct components, such as toffee and biscuit, either mutual equilibrium or a barrier to water mobility is necessary. The visco-elasticity of flour dough, determined by the interaction of carbohydrate and protein with water, sets the character of the final baked product.

In the last chapter Janice Ryley considers the very important aspect of nutritional quality. Vitamins, as necessary as fats, carbohydrates and proteins to the balanced diet, brook that the consumption of ever increasing quantities of convenience food must go hand-in-hand with stringent nutritional control. Shelf-life and labelling claims must be commensurate. Preservation of vitamins by improved processing techniques is important as within the food structure they are more stable than in forms added for food fortification.

In a book of ten chapters it is impossible to cover all areas; foams and vegetables had regretfully to be omitted. It is hoped that the reader will make use of the extensive bibliography and citations of newer concepts at the end of each chapter to explore each particular area in greater detail.

THELMA M. HARDMAN

CONTENTS

LIST OF CONTRIBUTORS

A. E. BELL
Department of Food Science and Technology, University of Reading, Whiteknights, PO Box 226, Reading, RG6 2AP, UK

J. G. BRENNAN
Department of Food Science and Technology, University of Reading, Whiteknights, PO Box 226, Reading, RG6 2AP, UK

B. BROCKWAY
Department of Food Science and Technology, University of Reading, Whiteknights, PO Box 226, Reading RG6 2AP, UK

D. G. DALGLEISH
Physical Chemistry Group, Hannah Research Institute, Ayr, Scotland KA6 5HL, UK

K. O. HONIKEL
Bundesanstalt für Fleischforschung, E.-C.-Baumann—Strasse 20, 8650 Kulmbach, FRG

HARRY LEVINE
Nabisco Brands, Inc., Corporate Technology Group, PO 1943, East Hanover, New Jersey 07936–1943, USA

J. E. MCKAY
Procter Department of Food Science, University of Leeds, Leeds LS2 9JT, UK

JANICE RYLEY
Procter Department of Food Science, University of Leeds, Leeds LS2 9JT, UK

MARILYNN SCHNEPF
Department of Human Nutrition and Foods, Virginia Polytechnic Institute and State University, Blacksburg, Virginia 24061, USA

LOUISE SLADE
Nabisco Brands, Inc., Corporate Technology Group, PO 1943, East Hanover, New Jersey 07936–1943, USA

JOHN A. TROLLER
The Procter & Gamble Company, Winton Hill Technical Center, 6071 Center Hill Road, Cincinnati, Ohio 45224–1703, USA

Chapter 1

WATER ACTIVITY AND FOOD QUALITY

JOHN A. TROLLER
The Procter & Gamble Company,
Cincinnati, Ohio, USA

SYMBOLS AND ABBREVIATIONS

a_w	Water activity
ERH	Equilibrium relative humidity
ln	Natural logarithm
m	Molal concentration
M	Molar mass of water in kg mol^{-1}
n_1	Number of moles of solvent
n_2	Number of moles of solute
p	Vapor pressure of a solution
p_0	Vapor pressure of pure water
p^{equ}	Partial vapor pressure of water in equilibrium with a solution
p^0	Vapor pressure of pure water at the same temperature and pressure as the solution
R	Gas constant
T	Absolute temperature
$\bar{V}$	Partial molal volume of water
ν	Ions generated per molecule
ϕ	Molal osmotic coefficient

INTRODUCTION

During the middle and end of the 19th century, those scientists who applied what was then modern technology to the chemistry and physics of foods began to discover that a direct relationship often existed between the presence and amount of water in a food and its relative tendency to spoil. These learnings eventually were interpreted in terms of reductions in water vapor pressure which had been observed to occur during the drying of foods and other materials. It was, then, not a distant extrapolation to consider the relationship between vapor pressure and stability and, in fact, those who made this extrapolation were 'rewarded' with what proved to be an enlightened approach to the effects of water availability or binding as opposed to water content. In effect, they began to realize that the 'condition' of water could be much more important to the stability of food than the amount that was present.

It was left to W. J. Scott in 1953, a microbiologist working in Australia on the growth of bacteria on beef carcasses, to note that the relative humidity of the refrigerated chamber in which the meat was stored seemed to exert considerable influence on the rate at which microbial growth on the meat surfaces developed. Although relative humidity was, and still is, standard terminology for the characterization of ambient atmospheres, it was an inadequate descriptor for the condition of moisture extant in, for example, food products. As a result, Scott searched for a better term to circumscribe the events that occur as a result of the binding of water.

There is some doubt about the original application of the term water activity to food. Mossel and Westerdijk had used the term in 1949 to denote one of the factors that controlled the microbial stability of foods. Scott (1957) later described water activity as a 'fundamental property of aqueous solutions' and defined it as the ratio of vapor pressures of pure water (p_0) and a solution (p):

$$a_w = \frac{p}{p_0}$$

A stricter, thermodynamically correct definition is (Hardman, 1976):

$$a_w = \frac{p^{equ}}{p^0}$$

where p^{equ} is the partial vapor pressure of water in equilibrium with the solution and p^0 is the vapor pressure of pure water at the same temperature and pressure as the solution.

Since Raoult's law states that the lowering of the vapor pressure of a solution is equal to the mole fraction of its solute, it follows that the above ratio of vapor pressure (i.e. a_w) also is dependent on the number of moles of solute (n_1) and solvent (n_2). We then have according to Raoult's law:

$$a_w = \frac{p}{p_0} = \frac{n_1}{n_1 + n_2}$$

The degree to which various solutes reduce a_w is an intrinsic property of the solute itself, determined by its chemical constituents, its dissociation, solubility, extent of intramolecular binding, etc. These are characterized for each solute by the molal osmotic coefficient (ϕ) which relates to a_w as:

$$\ln a_w = -\nu m M \phi$$

where ν is the number of ions generated per molecule, m is the molal concentration and M is the molar mass of water in kg mol^{-1}.

A convenient way of obtaining these data for saturated electrolytes is to refer to the prepared tables of Stokes and Robinson (1949) or Greenspan (1977) or any number of similar published tables which have appeared in the literature.

For practical purposes a quantity called the equilibrium relative humidity (ERH) is defined by

$$\mathrm{ERH} = (p^{\mathrm{equ}}/p^{\mathrm{sat}})_{Tp=1\,\mathrm{atm}}$$

where p^{equ} is the partial pressure of water vapour in equilibrium with the sample in air at one atmosphere total pressure and temperature T, p^{sat} is the saturation partial pressure of water in air at a total pressure of one atmosphere and temperature T.

Another parameter that has been used to characterize and describe the 'condition' of water is osmotic pressure, a term favored by biologists which relates to a_w:

$$\text{Osmotic pressure} = \frac{-RT}{\bar{V}} \ln a_w$$

where $\bar{V}$ is the partial molal volume of water, R is the gas constant and T is the absolute temperature. Water potential, a term frequently used by plant physiologists, is closely related to osmotic pressure.

A search for an alternative term to define moisture condition has some merit if for no other reason than that there are numerous situations in which the a_w term seems to be deficient. Mossel (1975), who, as noted above, may have been one of the originators of the term, himself has expressed some doubt concerning its appropriateness, suggesting substitution of the term p_w or relative water vapor pressure. The well known solute effects described by microbiologists in which a microbial species may grow at lower a_w in the presence of one solute than in the presence of another is an example. Marshall *et al.* (1971), Corry (1974), Troller and Stinson (1978) and numerous other workers have reported data of these kinds. Perhaps, though, we miss the point. To a degree what we describe as water activity really may be irrelevant providing that we define our terms in an agreeable and consistent fashion. Certainly a definition of this term in which fugacity, or relative vapor pressure is an important element could be useful in explaining many things. Perhaps we could come to grips with water and the way that it participates in chemical and physical reactions more realistically if we could include in our definition a parameter relating to water mobility, for it is the ability to mobilize, to move one reactant into reactive juxtaposition to another, that might best explain the effects of water in foods and other biological materials. Nuclear magnetic resonance (NMR) may offer the most efficient means of characterizing this capability (Duckworth, 1981) while a closely related technique, electron spin resonance (ESR), has been described by Simatos and her coworkers (1981) as a similarly appropriate method to help characterize water mobility.

At some time in the future, we may, indeed, be able to define a parameter that is more closely related to water mobility than a_w. In the interim, however, we cannot ignore the fact that this term is extremely useful in explaining many observations relating to foods. Some of these will be discussed in this chapter. W. J. Scott (Troller and Christian, 1978) himself, probably put this situation in best perspective when discussing the role of water in experiments:

> 'We [should] be able to fill in some of the obvious gaps in our knowledge and will soon discover whether a_w can be replaced by something more meaningful or whether, like pH, it will be something to measure and consider for perhaps another 100 years or so.'

METHODOLOGY AND CONTROL OF a_w

The most commonly used instruments for the measurement of a_w are those which measure the equilibrium relative humidity (ERH) of a closed space in equilibrium with a food. ERH can be measured indirectly by using a crystalline material as a resistor. The amount of current passed through the crystal is dependent on its degree of hydration and is measured by a simple galvanometer which has been previously calibrated to specific a_w levels. One of these hygrometers is shown in Fig. 1. The use and calibration of this type of instrument has been reviewed by Troller (1977). Its principal disadvantages are that it is indirect (must be calibrated) and that it may be subject to contamination (Pollio *et al.*, 1986). Several companies manufacture a simple and inexpensive hygrometer which is a variant of the horsehair hygrometer (Rodel *et al.*, 1979). Current models of these devices employ hygroscopic polyamide filaments which stretch when subjected to humid environments. These filaments are attached to sensitive springs housed in the lid of a canister connected to an indicator needle. When food is placed in the canister and the lid/dial is put in place, equilibrium is permitted to be reached in the closed canister, and the final ERH is read directly from the dial (Rodel and Leistner, 1971). Other, more sophisticated, types of instruments to measure a_w have been described in the literature. Among these are capacitance

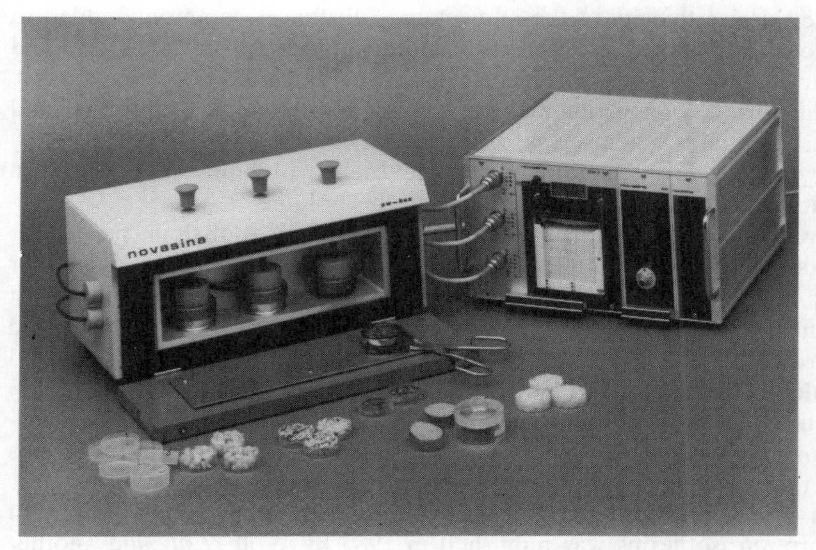

FIG. 1. Novasina hygrometer. (Photo courtesy, Novasina AG.)

TABLE 1
Some criteria for humectants to be used in foods

Criteria
Safe
Approved by regulatory agencies
Effective at reasonable concentrations
Compatible with the nature of the food
Flavorless at use concentrations
Colorless and/or imparts no color changes in the food

types (Favetto *et al.*, 1983) that are based on changes in capacitance which occur in a thin film capacitor when it absorbs moisture from an atmosphere in equilibrium with a food. Direct measurements of a_w via vapor pressure determinations (see equation above) also have been obtained (Taylor, 1961; Labuza *et al.*, 1982), however the manometers employed for this purpose are cumbersome and very difficult to use. Troller (1983*a*) recently described a capacitance manometer which effectively eliminated many of these shortcomings. These and other instruments and methods have been reviewed by Prior (1979), Rodel *et al.* (1979) and Troller (1983*b*).

The adjustment of a_w in foods by means of added humectants is necessarily limited by many factors (Table 1), not the least of which is the relative toxicity of the humectant. Substances commonly added to foods to reduce a_w are glycerol, sodium chloride, propylene glycol and various sugars such as fructose, sucrose, and modified corn syrups. The primary constituents of foods, for example, proteins and carbohydrates, themselves can affect a_w when water is removed by drying. In this situation, these materials will bind some water, a subject that is discussed elsewhere in this volume.

Other solutes may be employed for investigational purposes; for example, to determine the effects of a_w on microorganisms. A number of investigators have described the water binding capabilities of various solutes including Solomon (1951) (sulfuric acid); Anand and Brown (1968) (polyethylene glycol); Greenspan (1977) (saturated salts); Chirife *et al.* (1980) (amino acids); Ferro Fontan *et al.* (1981) (sucrose, glucose); and Resnik *et al.* (1984) (saturated salts). The relationship between the water content and a_w of these and other solutions is termed a sorption isotherm which can be constructed graphically (Fig. 2) or derived mathematically. A volume containing a bibliography of sorption isotherms was published in 1985 by Wolf *et al.* and another book by Iglesias and Chirife (1982) contains more than 500 different

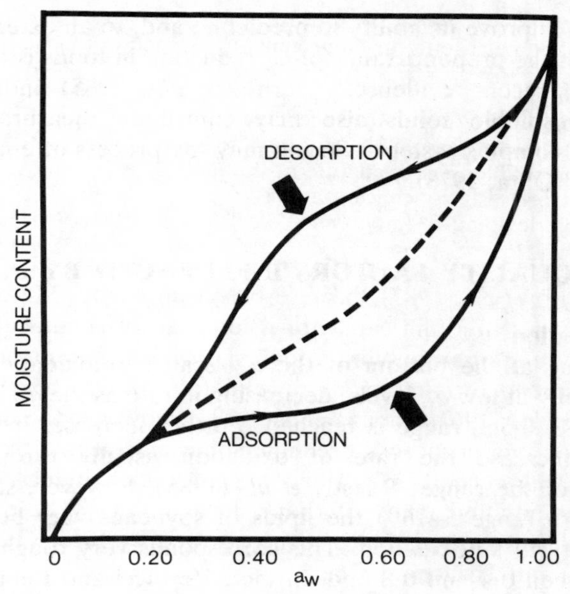

FIG. 2. Sorption isotherm.

food isotherms as well as a brief description of techniques for the use and calculation of isotherms. In addition, the chapter by Labuza (1985) contains an excellent discussion of water binding and the chemical and physical factors which affect it. The reader should consult these sources for further information on this very important aspect of the water relations of foods.

Control of a_w also can be attained by combinations of solutes. The Ross equation (Ross, 1975) is one of the most widely used attempts to predict the effect of various solutes in foods on the 'total' a_w of multicomponent systems. This equation:

$$a_w = (a_w^0)(a_w^0)_2(a_w^0)_3 \cdots$$

states that the a_w of a food is the product of a_w values for simple solutions of each solute measured at the same concentration as in the food. This equation provides a reasonably accurate estimate providing the food possesses a fairly high a_w. At lower a_w levels (<0·60), however, significant deviations occur. Others (Seow and Teng, 1981; Chirife *et al.*, 1985; Favetto and Chirife, 1985) have modified the Ross equation to improve its ability to predict a_w and, to an extent, simplify it. Although the preponderance of a_w reduction in foods is contributed by solutes, recent evidence (Chirife *et al.*, 1985) indicates that relatively insoluble solids also may contribute measurably to a_w lowering in complex systems. So too may the process of emulsification (Koiwa and Ohta, 1978).

QUALITY FACTORS INFLUENCED BY a_w

Lipid Oxidation

Commencing at the bottom of the a_w 'scale', autoxidation of lipids occurs rapidly at low a_w levels, decreasing in rate as the a_w is increased until the 0·3–0·5a_w range is reached. Further increases in a_w beyond 0·5 then increase the rate of oxidation usually throughout the remainder of the range. Priestly *et al.* (1985), however, state that in the moisture range 5–16% the lipids of soybeans vary little in their susceptibility to autoxidation. This corresponds very roughly to an a_w range between 0·4 and 0·8 and, in fact, Vertucci and Leopold (1984) state that at 5% (about 0·4a_w) moisture content, all of the water molecules in soybeans are tightly bound and presumably unavailable for the reactions that occur during autoxidation.

In addition to the lipids contained in soybeans, the oxidative rancidity of a number of materials has been investigated. Safflower oil at a_w levels of 0·20–0·91 was studied by Verma and Prabhaker (1982), who found that a_w was a factor in autoxidation when the surface area of the test sample was great. The formation of secondary oxidation products, was strongly influenced by a_w levels as indicated by increased thiobarbituric acid values. Secondary reactions of walnut oil also have been studied by Prabhakar and Amla (1978). At low levels of relative oxidation, changes in water activity appeared to have relatively little influence, however at peroxide values >100, total carbonyl content and quantity and composition of classes of carbonyl compounds were influenced by a_w level.

The most extensive basic research on the influences of a_w on autoxidation has come from the laboratories of Karel and his coworkers at the Massachusetts Institute of Technology (MIT) (Karel, 1985). Working primarily with model systems, these workers have proposed a number of hypotheses for the mechanisms of a_w effects on lipid oxidation. These are summarized in Table 2. Generally, free

TABLE 2
Mechanisms proposed for the suppression of lipid oxidation by water

Hypothesis	*References*
1. Antioxidants produced during nonenzymatic browning suppress oxidation	Eichner (1981)
2. Hydration of metal ions prevent their participation as 'catalysts'	Karel (1985)
3. Promotes formation of cross-links between proteins via oxidized lipids which act as cross-linking agents	Funes and Karel (1981)

radicals, the critical elements in most examples of lipid oxidation, are quenched and undergo decay as the water content of model systems is increased. The MIT studies have been extended to investigations of the effects of a_w on the interactions between protein and other food components and oxidizing lipids (Funes and Karel, 1981). In the case of interactions with proteins in which the transfer of free radicals from oxidizing lipids to proteins was studied, it was revealed that as free radicals disappeared with increasing a_w, more radicals were initiated (higher peroxide values). The general effect, however, was to reduce radical concentration in these combined systems. Karel (1975) proposes that the most likely mechanisms for these reactions are radical recombination and/or cross-linking and radical termination through donation of a proton by water.

Nonenzymic Browning

Browning reactions in foods may occur as a result of heating, dehydrating or concentrating food constituents. Therefore, the ability to influence these reactions assumes considerable practical importance because not only is the color of food influenced by browning, but also nutritional content and flavor also may be affected.

Hodge and Osman (1976) have suggested that browning reactions should be delineated into oxidative and nonoxidative types by virtue of the initiative reactions involved in the process since enzymatic catalysis may occur in both reaction sequences. Sugars must initially be present and then proceed through a series of reactions involving enolization, fragmentation and dehydration to produce a series of carbonyl intermediates. While these intermediates are capable of reacting to form polymers and flavors associated with browning products, the presence of amino compounds greatly enhances this reaction. Maillard reactions, commonly associated with browning in foods, require that amino compounds condense with sugars followed by enolization and dehydration. These reactions are influenced by the types of reactant sugars and amines, pH, temperature, and a_w.

The role of bound and unbound moisture in the reactions that occur during browning has been investigated for many years in a number of laboratories. The work of Chung and Toyomizu (1976) has shown a maximal browning rate in the a_w range 0·40–0·67 in model freeze-dried foods. Browning rates in whey powders were found by Labuza and Saltmarch (1981) to be greatest at 0·44a_w whereas Petriella *et al.* (1985) investigated browning rates in the a_w range within which bacteria grow (0·90–0·95a_w) and found that changes in a_w levels of the systems had relatively less effect on browning rates than did alterations in temperature and pH. At very high water contents (>0·95a_w) moisture strongly inhibits browning.

There appears to be a maximum browning rate in many foods in the 0·40–0·60a_w range. Reduced browning rates which are observed at a_w levels below this, according to Toribio *et al.* (1984), result from an increase in viscosity whereas retardation of rates at elevated a_w levels are due to the dilution of reactants. The type of humectant used to achieve the adjustment of a_w also was found by Warmbier *et al.* (1976) to have an influence on browning rates. Maximal browning rates in dehydrated foods are observed usually in the 0·65–0·75 range. If, however, glycerol is employed to reduce a_w, the range for maximal browning shifts to the 0·40–0·50 range. This drop is attributed to the fact that glycerol, being liquid, has water-like properties which increases reactant mobility, and/or solubility, slowing browning rates accordingly.

Eichner and his coworkers at the University of Munster (FRG) have intensively investigated the progression of browning reactions during the dehydration of vegetables. They have found (Eichner *et al.*, 1985) that it is possible to optimize critical parameters during drying through manipulation of temperature and a_w. The quantity of Amadori reactants (Maillard browning intermediates) was found to be minimized and product quality relative to browning could be improved by reducing the a_w and, more importantly, the temperature during the final stages of drying.

In summary, most rapid browning can be expected to occur at intermediate a_w levels in the 0·40–0·60 range. Whether or not it is maximized at the lower or upper portion of this range depends significantly on the specific solutes used to poise a_w, the nature of the food (especially amino compounds and simple sugars that might be present) as well as the pH and a_w of the product. It is somewhat paradoxical that at a_w levels which minimize browning, autoxidation of lipids is maximized.

Nutritional Quality

To at least some extent the effect of a_w on the nutritional quality of foods parallels that of nonenzymatic browning. This is not unexpected inasmuch as nutritionally essential amino acids such as lysine may be important reactants in the browning sequence. Vitamin degradation too has been studied extensively in foods and model systems with early emphasis on the degradation of ascorbic acid (Troller and Christian, 1978), whereas attention has turned, within the past five years, to the effect of a_w on other vitamins such as riboflavin, thiamin and α-tocopherol.

Protein Quality

Early studies (Goldblith and Tannenbaum, 1966) had indicated that lysine loss was highly correlated with the browning reaction although more recent work has shown that while browning can be associated with a decrease in lysine content, the latter can occur in the absence of the development of browning (Labuza and Saltmarch, 1981). Burvall *et al.* (1978) also noted this in their studies on lactose—hydrolyzed dried milk citing the fact that up to 50% available lysine losses occurred without visible evidence of browning. They further noted that unless this product was dried to, and maintained at, a_w levels in the 0·11 range, substantial losses (20–40%) in lysine content could be expected to occur during storage. Chirife and coworkers (1979) examined the effect of extended storage (60 days) on the loss of available lysine in an intermediate moisture meat product. At the a_w of this product, 0·83, no loss of lysine, determined chemically, could be found. A pasta product was examined by Labuza *et al.* (1982) who found that browning rate and lysine loss increased with increased temperature and a_w. Mathematical models derived to predict the loss of lysine produced variable results. Obviously, the nutritional quality of proteins suffers as a result of lysine loss during storage at elevated a_w levels. This is due primarily to the vulnerability of the α-amino group. The effect of storage at various a_w levels on other essential amino acids has not been thoroughly investigated.

Vitamins

Until recently, studies on ascorbic acid have formed the basis for most of our knowledge of the fate of vitamins in foods poised at various a_w levels. Generally, these data have shown that the rate of ascorbic acid destruction increases as a_w is increased. This effect appears to be relatively independent of temperature (Jensen, 1967), although other authors (Kirk *et al.*, 1977) disagree on this point. It also is highly oxygen dependent at high a_w levels, but not at low a_w (Mishkin *et al.*, 1982).

Ascorbic acid normally is subject to rapid deterioration in the presence of transition metals such as copper and iron. The effect of a_w on this deterioration was found by Dennison and Kirk (1982) to be rather profound. No catalysis was observed at a_w levels of 0·10 and 0·40 due to the lack of metal ion mobility at these low a_w levels. However, in systems poised at 0·65 (and presumably above), a 2- to 4-fold increase in rate of ascorbic acid destruction was observed. This probably was due to the requirement for complete hydration of the metal to attain mobility which would only occur in the free water of the isotherm's capillary region.

Two forms of commercially used thiamin, the hydrochloride and mononitrate, were evaluated for relative stability at various a_w and temperature levels by Labuza and Kamman (1982). They found that the mononitrate was less stable than the hydrochloride below 95°C due to differences in their respective activation energies. Above this temperature the reverse was true. The stability of both forms was slightly less at 0·86a_w than at 0·58a_w. Their data generally indicate that temperature is more critical in determining relative degradation of both the mononitrate and the hydrochloride than a_w level. The effect of a_w and other factors on the photodegradation of another B-complex vitamin, riboflavin, in macaroni was studied by Woodcock *et al.* (1982). These authors found no significant differences in the reaction rates at either 0·33 and 0·44, however, this narrow range precluded speculation on the effect at other a_w levels. Furuya *et al.* (1984) later extended this range to 0·11–0·75 and found that riboflavin photodegradation increased at high (0·75)a_w levels although significant differences in percentage retention could not be observed in the a_w range 0·11–0·52. These studies were done in a pasta which normally has an a_w of 0·20–0·54, hence it was concluded that riboflavin should be stable under usual storage conditions for pasta-type products.

Much remains to be learned about the stability of a wide variety of vitamins and amino acids poised at elevated a_w levels. Based on the studies which have been reviewed for this report , the stability of some vitamins and amino acids is reduced at high a_w. For ascorbic acid and lysine (in some systems) the effects may be significant in terms of nutrient depletion.

Texture and Rheology

As with many other factors that affect food and food processing, the influence of a_w on texture and rheology also has been investigated recently. One of the principal objectives governing the practical application of water activity-related factors to the development of new food products is to maintain textural properties as close as possible to those of the original food or, failing in this, the texture must be such that it is readily acceptable to the consumer. Kapsalis (1975) in his review of work on this topic alluded to several factors which might explain the behavior of mechanical textural properties as they relate to a_w. Generally, in meat products, tenderness forces increased with a_w to $a_w = 0{\cdot}85$ but decreased with further hydration. Some publications (Reidy and Heldman, 1972), however, indicate that meat texture values (toughness) are highest in the 0·40–0·60a_w range.

Dry cereals, snacks and cookies have been studied more recently. Zabik *et al.* (1979) studied the effect of relative humidity on the crispness and tenderness of cookies. Two ranges of relative humidities were studied, 11–93% and 52–79%. These researchers found that variations in relative humidity values of 5–10% did not significantly affect textures. As might be expected, high humidities produced the greatest reductions in crispness and breaking strengths. Work by Quast and Karel (1972) on potato chips showed that the critical water activity for optimal chip crispness was about 0·40. This was confirmed by Katz and Labuza (1981) who in a study on crispness as a function of a_w, found that snack products were unacceptable in the 0·35–0·50 range. These workers speculated that water plasticizes and softens the starch/protein matrix present in most snack foods which results in a change in mechanical breaking strength.

Intermediate moisture cheese products manufactured to an a_w level of 0·81 or 0·86 were found by Kreisman and Labuza (1978) to be quite tough. Cheeses of higher a_w (0·90–0·94) possessed a more acceptable texture. Sodium chloride, when added to certain vegetable and meat products may produce rather profound chemical changes which alter the texture of these foods. Because these effects are not primarily related to a_w adjustment (concentrations of NaCl are low, therefore effect on a_w is minimal), they will not be discussed in this chapter.

Enzymes

It is often desirable to control and, even, to limit the progress of enzymatic reactions during the manufacture and storage of foods. The sources of these enzymes can be diverse. Microorganisms growing in a food may secrete extracellular enzymes, enzymes may be intentionally added to a food during its manufacture or, as is the case frequently, the enzymes may be endogenous. In the latter situation, they may be relatively inactive unless released from cellular structures by tissue destruction or by mixing which would permit substrates to contact the catalyst.

Water acts in a number of ways to promote enzymatic reactivity:

1. As a mobilizing medium thereby promoting diffusion.
2. To stabilize the structure and configuration of the enzyme.
3. As a reagent during hydrolysis.
4. Disrupts hydrogen bonds between polar groups.
5. Releases reaction products from the reactive complex sphere.

In some instances more than one of these mechanisms may be concurrently operative.

As noted above, microorganisms can function as important sources of enzymes. One of the first reports relating to the effect of water limitation on enzyme production by microorganisms was coauthored by Nashif and Nelson (1953) who found that the synthesis of a lipase by *Pseudomonas fragi,* a common psychrotrophic food spoilage organism, was inhibited by as little as 1% NaCl. A concentration of 4% NaCl (approximately $0{\cdot}975a_w$) resulted in total inhibition of lipase production. Much later, Troller and Stinson (1978) observed similar inhibition of lipase synthesis in cultures of two enterotoxigenic strains of *Staphylococcus aureus*. Neither trioleinase nor tributyrinase production occurred at $0{\cdot}91a_w$ or below.

The ability of *Pseudomonas fluorescens* to metabolize glucose when grown at various a_w values (adjusted with NaCl, sucrose, potassium glutamate, glycerol and polyethylene glycol) was studied by Prior and Kenyon (1980). These workers noted significant differences in enzymatically catalyzed metabolic pathways when the organism was grown at reduced a_w levels, depending on the specific solute used to alter the a_w. This effect could have been due to the capacity of the solute to penetrate the cell membrane. In a later publication Prior *et al.* (1987) reported that the intracellular accumulation of solutes was dependent on the solute used to adjust a_w. If NaCl was the adjusting solute, a 23-fold increase in intracellular levels of glutamine was observed. If sorbitol was used to adjust the a_w of the medium, this solute accumulated within the cell to a concentration twice that existing in the medium. This suggests that an alternative, active (energy requiring) transport system exists for sorbitol since accumulation occurred against a concentration gradient.

The synthesis of pectolytic enzymes by various plant pathogens has been studied by several workers. Using a variety of organisms and substrates most researchers have observed a consistent decrease in production of enzymes within the range of 0·96–0·99a_w. One hypothesis proposed to explain this was that lower relative humidities concentrated nutrients at the surface of carrots used in this study which led to increased enzyme production. The work of Mildenhall *et al.* (1981), however, was done in a microbiological medium hence the source of this stimulation at high a_w may be other than that related to a specific product. These authors indicated that the results observed were highly solute dependent noting that in systems poised with NaCl to 0·98a_w there was an increased production of extracellular pectic acid lyase whereas similar reductions obtained with mannose or lactose resulted in a significant decrease in synthesis. The reason for this difference is discussed with some speculation that ionic effects might stimulate the release of the enzyme although previous work (Gould and Measures, 1977) had indicated no increase in cellular permeability with reduction in a_w. On the other hand, permeability may not necessarily be a prerequisite for release of the enzyme.

The production of a number of extracellular enzymes by *S. aureus* was followed by Troller and Stinson (1978). These authors found that the synthesis of lipase, catalase, coagulase and deoxyribonuclease was inhibited by reductions in a_w which permitted growth. Two enzymes, acid and alkaline phosphatase were not as sensitive to a_w reductions as a number of other enzymes studied.

Enzymatic reactions normally accelerate as the a_w level increases (Potthast, 1978). As noted earlier, this increase in reaction rate may be attributable to the increased mobility of enzymes as a function of a_w although water at the capillary level also may be required for stoichiometric reasons and may even be essential in its role as a reactant. If, however, the reactants are liquid, adequate mobility (and reactivity) can, in some circumstances, be observed even below monolayer levels of water.

In an effort to predict the behavior of enzyme-containing systems at different a_w levels, Drapron (1972) studied the threshold of hydration at which enzyme activity occurs as well as the effect of increasing levels of hydration on enzymatic activity. Working with a number of substrates, this author established that incremental increases in a_w resulted in elevations in the amount of enzyme activity. He also found that the increase is usually not linear, accelerating most rapidly in the lowest a_w range. It was concluded that the rate of reaction at each active enzyme site is much faster than the diffusion of the substrate or products, to or away from the reactive site, hence mobility must be the rate-limiting factor.

The situation with regard to lipolysis, however, is quite different. In this case, in which a liquid substrate is involved, the mobility provided specifically by water is not necessary since the substrate, being liquid, has the capacity to contact the enzyme. As a consequence, lipolysis may occur at extremely low water activities i.e. in the 0·025–0·25a_w range. Within this low range, the rate of lipolysis may be linear. Usually activity reaches a peak in the 0·75–0·95 range and diminishes above this level, probably as a result of dilution effects which would tend to reduce the concentrations of substrate and enzyme. A similar pattern was reported for soybean lipoxygenase reactions which were studied by Brockman and Acker (1977), however, Bone (1987) reports that cereal lipase reaction rates in a baked product are minimal in the 0·22–0·33 range and that a formulation to 0·33a_w or lower is required to attain stability.

Despite their critical importance in causing enzymatic browning of fruits and vegetables, there is relatively little known about the moisture relations of polyphenoloxidases (PPO). In one of the few references available, Halim and Montgomery (1978) found that PPO from pears was inhibited slightly (12%) by 10 mM of sodium chloride. This corresponds to an $a_w > 0·98$ at 25°C. Surprisingly, references are not available on the effect of sucrose and corn syrups on PPO.

Reports on the stabilization of enzymes by humectants have been reviewed by Hahn-Hagerdahl (1986). This author found that the stabilizing effect of sugar appears to depend on the mean number of equatorial hydroxyl groups per sugar molecule. In the presence of polyhydric alcohols, on the other hand, similar correlations could not be found.

Based on the above reports we can state that many enzymes will function minimally in the 0·65–0·70 range, increasing their reactivity as a_w levels are increased. These enzymes include amylases, phospholipases, phytases and many types of oxidases. Lipases, however, have much lower minimal a_w levels for functionality. As Drapron (1985) has stated, classical reaction kinetics appear to be followed even at a_w levels approaching the minimal range, however, it is the thermodynamic 'state' of water that must be considered at low a_w. Traditionally, these effects have been considered in terms of molecular mobility, however, more recent work has also begun to focus on the effects of a_w limitation on the conformation and structural integrity of the enzyme molecule itself.

Microbial Activity

There can be several outcomes relating to the growth of microorganisms in foods. A beneficial outcome occurs when a desirable change is brought about in an edible product to increase its value as a food. The ripening of cheese is an example. Undesirable or nonbeneficial changes occur if a food spoils as a result of microbial growth or if the food becomes a potential disease threat because it contains pathogenic microorganisms.

Microorganisms respond to circumstances of reduced water activity in a number of ways. In the most manifestly 'visible' sense, we observe that microbial growth fails to occur at reduced a_w levels; but this can be a symptom of many things. Figure 3 shows a generalized diagram of the effect of a_w reduction on the hypothetical growth curve of a bacterial culture. The effect is usually characterized by an extension of the lag phase, a suppression of the log phase and, in general, a reduction in the total number of viable microorganisms. At least one of these responses occurs in all microorganisms subjected to conditions of limiting a_w. Concomitant with these effects, other changes may occur in the relative presence and amounts of metabolites, both primary and secondary, which are out of all proportion to the effects on growth.

The minimal water activity levels for growth and toxin production of a number of food-borne pathogens are shown in Table 3. These data suggest that foods poised at a_w levels below 0·78–0·80 should not be a disease threat. They also indicate that mold species are generally more osmotolerant than bacterial species both in their ability to grow and to produce toxins at low a_w levels.

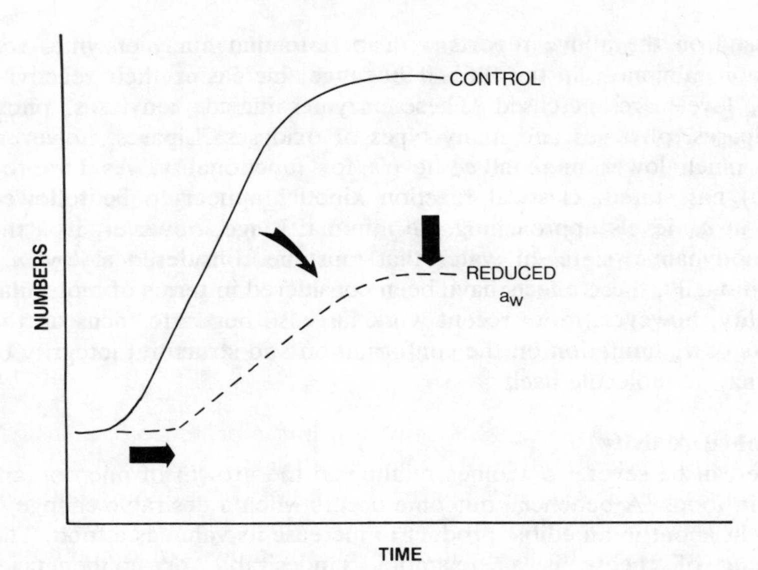

FIG. 3. Effect of a_w reduction on growth curve of microorganisms

Staphylococcus aureus is the bacterial species capable of growth at the lowest a_w, 0·86, although toxin synthesis may cease at levels well above this. The capacity to prevent the growth of disease-producing microorganisms through enlightened manipulation of a_w has been exploited in some food regulations in which a_w limits figure

TABLE 3
Effect of a_w on growth and toxin production by some food-borne pathogens

Organism	*Minimum a_w*	
	Growth	*Toxin production*
Salmonella species	0·94	—
Clostridium botulinum—Types A and B	0·93	0·84
Clostridium perfringens	0·95	Unknown
Bacillus cereus	0·93	—
Aspergillus flavus	0·83	0·78
Vibrio parahemolyticus	0·94	—
Staphylococcus aureus	0·86	0·89
Listeria monocytogenes	0·93	—

prominently. Examples of this are the requirement for an a_w of $<$0·85 for the transportation of dried foods in the Federal Republic of Germany or the US Good Manufacturing Practices provisions (Ref. US Code of Federal Regulations Title 21, Part 113) which require a thermal process for low acid foods (pH $>$ 4·6) with an a_w level $>$0·85. Another example is the sanitary directive of the Conseil des Communautes Europeenes No. 77/99/LEE which permits nonrefrigerated transport and storage of meat-based foods whose a_w is $\leq$0·95 and pH $\leq$5·2. At present, regulations relating to a_w control are relatively rare although their numbers are beginning to increase as awareness of this factor and its importance to food safety increases.

Solute Effects

Frequently a given species will grow well in the presence of a specific solute, for example glycerol, whereas if NaCl is substituted for glycerol, at identical a_w levels, growth will not occur. An example of this phenomenon is shown in Table 4, in which Jakobsen and Trolle (1979) determined the minimal a_w levels for growth of a number of *Clostridium* species. With these different humectants, it can be seen that the response within members of this single genus was quite variable and that glycerol appears to permit growth at lower a_w levels than either glucose or NaCl. This is the norm in most situations, NaCl tends to exert the greatest solute effect whereas glycerol and other polyols tend to exert the least with sugars in between. An exception is the important food-borne pathogen, *S. aureus* (Marshall *et al.*, 1971) which at identical a_w levels is inhibited to a greater extent by glycerol

TABLE 4
Growth of *Clostridium* species in media poised at various a_w levels with three humectants (Jakobsen and Trolle, 1979)

	NaCl			*Glucose*			*Glycerol*		
	0·965	0·945	0·930	0·965	0·945	0·930	0·965	0·945	0·930
C. butyricum	−	−	−	+	−	−	+	+	−
C. fallax	−	−	−	−	−	−	+	+	−
C. propionicum	+	−	−	+	−	−	+	+	+
C. perfringens	+	−	−	+	−	−	+	+	+
C. novyi	+	−	−	+	−	−	+	+	+
C. difficile	+	−	−	+	−	−	+	+	+
C. sporogenes	+	+	−	+	−	−	+	+	+

than by NaCl. It also should be noted that molds exhibit similar solute-related effects.

The reason for these effects is unknown at present although their presence should not be unexpected given the diversity of physiological pathways among various microorganisms. Certainly the data available point toward differences in the intracellular changes that occur in response to the reduction of a_w in the cells' microenvironment. These, as we will see, vary with type of solute used and the microorganism involved.

Combinations of Factors

The preservation of foods through a_w reduction has been practiced for centuries. Jams, jellies, some types of sausages, honey, salted fish and salted meats have been the mainstays of many primitive diets throughout history. Most of these foods do not rely solely on a_w reduction for microbial control, however. Other preservative factors usually are present which act in association with reduced a_w to circumvent the growth of microorganisms.

The term 'hurdle' has been applied (Leistner *et al.*, 1981) to the exploitation of combination factors to circumvent the growth of microorganisms in foods. Unfortunately, this term, is not entirely satisfactory because it implies a series of inhibitory events occurring in stepwise fashion. In fact, this does not occur. The combinative inhibitory factors present in a medium or food product usually confront the organism simultaneously and, logically, enter the cell in the same manner or sequentially as determined by membrane transport factors which, as yet, are largely unknown.

The combination of pH and a_w reductions is especially common. Many types of cheese are examples, sauerkraut is another. In both of these examples, the primary preservative focus rests on pH, however neither product would be stable were it not for salt content (a_w reduction). The safety of many foods also is assured by combinative activity of pH and a_w inhibiting food-borne pathogens (Troller, 1986). Other combinations with a_w that are common in foods are controlled atmospheres, oxidation–reduction potential and chemical preservatives (nitrite, propionic acid, sorbic acid, etc.).

Thermal effects also work interactively with a_w. Storage or holding temperatures of food in which it is desired only to prevent microbial growth usually can be decreased as a_w is lowered, however, thermal destruction values are quite another matter. In these cases, heating conditions required to attain a specific level of thermal kill increase with decrease in a_w. Although this has been shown for several organisms, probably the most complete data relating to this phenomenon have been published by Corry (1974, 1978). These reports also contain data which indicate that the extent of this protective effect is related to cell plasmolysis and shrinkage which the cell undergoes at low a_w.

Mechanisms

Based on these and other data, it is obvious that our ability to exploit and gainfully utilize a_w-related concepts to create better and safer foods hinges significantly on the success that is obtained in understanding food/water/microorganism interactions at the molecular level. In other aspects of a_w that we have already discussed, this understanding has been aided by recent improvements in our ability to characterize water at the monomolecular level. In the case of microorganisms, however, information on the mechanisms involved in a_w-mediated inhibition of growth has been developed in a much more painstaking manner, partly because of the separation philosophically and geographically, of the two principal research groups in this area. To the reviewer of this intriguing work, it is somewhat comforting that the hypotheses and conclusions developed by the two groups assemble reasonably well to form a coherent hypothesis for the mechanisms by which microorganisms are inhibited by a_w reduction.

Generally, yeasts and molds survive and grow at lower a_w levels than nonhalophilic bacteria. Although reports are rare of the production of mycotoxins below $0{\cdot}81a_w$, most xerotolerant species of molds will grow at $0{\cdot}75a_w$ and there are reports that some yeasts have been found to grow at $0{\cdot}61a_w$.

The ability of yeasts to cope with water-limited extremes has been examined in detail by A. D. Brown and his associates in New Zealand who have dealt primarily with mechanisms. These workers (Brown and Simpson, 1972) noted that yeasts exposed to environments of low a_w accumulate intracellular polyols (by synthesis or uptake from the medium) such as arabitol with they described as 'compatible solutes' because they did not interfere with the vital metabolic functions of the yeast cell. The primary purpose of this accumulation is to create isoosmotic conditions across the cell membrane and in this way to maintain cell turgor. All fungi may not react to a_w reduction in an identical manner, however, Edgeley and Brown (1978) reported that arabitol content of yeasts does not change appreciably with changes in a_w, but that glycerol content does increase with a_w reductions. Both xerotolerant and nontolerant yeasts respond to a_w reductions by accumulating glycerol when exposed to low a_w. The key appears to be in the manner in which glycerol is accumulated. In xerotolerant yeasts, glycerol pools are created by permeation/transport whereas in nontolerant species glycerol builds up, but through the intervention of metabolic processes which synthesize the humectant. This latter process is expedient from the standpoint of tolerance to water stress, however it is extremely costly to the cell because precious energy must be diverted from cellular metabolism to the synthesis of glycerol. Reduced growth is a consequence.

For several years, it was assumed that molds responded to environments of low a_w in a similar fashion to yeasts (Corry, 1978), however, more recent data have brought this supposition into question. Luard and her coworkers (1982) showed that very little polyol of any type accumulated in molds and that proline and the humectant used to reduce a_w (sucrose or KCl) accumulated in hyphae exposed to low a_w levels. It is postulated that the accumulation of these, presumably, compatible materials would create isoosmotic conditions within mold cells.

The situation with bacteria is somewhat more developed. Halophilic and halotolerant organisms among the *Eubacteriales* have been known for some time. Important primarily as biological curiosities producing colored tints in salt lakes and salt pans, these organisms are not considered to be important food spoilage bacteria. They do however grow on some types of salt fish and meats and, by virtue of the pigments that they produce, make the product unappetizing. Many of these organisms are obligately halophilic and require NaCl for growth. Their a_w range for growth in NaCl-containing menstrua is between 0·75 and 0·88. Early research by Christian and Waltho (1962) had indicated that the halophiles and 'moderate' halophiles respond to low a_w levels by accumulating high intracellular concentrations of potassium. Subsequent research (Masui and Wada, 1973), however, has shown that in some halophilic bacteria the internal salt concentrations may be significantly lower than those occurring externally. Nevertheless, it cannot be denied that the metabolic enzymes of halophilic bacteria are elegantly adapted to function in high salt environments (Lanyi, 1974) and some of these enzymes, in fact, require substantial salt concentrations for activity. Other unique aspects of halophilic bacteria are the requirement of ribosomes for K^+ for stability and the fact that Na^+ is required to maintain cell wall stability. An excellent discussion of this interesting subject can be found in the review by Kushner (1978) which should be consulted by those interested.

Information on the mechanisms by which nonhalophilic bacteria adapt to environments of low a_w has come from many sources, some of which have been reviewed recently by Troller (1986). Although much has been learned about these processes, many key features of this adaptation continue to elude the efforts of scientists throughout the world.

To attempt to understand these mechanisms, one must realize that vital life functions in bacteria, as in yeasts and molds, are dependent upon the establishment of a positive turgor pressure within the cell. This pressure is maintained by the accumulation of any number of solutes (Gould and Measures, 1977) although K^+ is probably most common. When a cell is exposed to environments containing high concentrations of solutes, as when the a_w level is reduced, the cell undergoes plasmolysis which results in the loss of water and the effective concentration of K^+. In this conformation, the cell is unable to grow unless deplasmolyzed. More frequently, intracellular levels of K^+ (and internal osmolarity) are regulated by a series of physiological events that only now are beginning to be understood (Helmer *et al.*, 1982).

To a large extent, the a_w of the environment determines the amount of K^+ taken into the cell and the amount which is required to maintain cell turgor. The movement or flux of K^+ appears to be under the control of two distinct systems (Epstein, 1986); a constitutive system with low affinity for K^+ (Trk) and a repressible, high-affinity system (Kdp). Usually the Trk system is the sole source of intracellular K^+, however if K^+ is in short supply in the medium, both the Kdp system and a reserve uptake system become operative. Trk and Kdp stimulate influx without changes in efflux, however if the a_w of the medium is increased, K^+ is exported. A diagram showing the operation of these systems is shown in Fig. 4, however control of Tdk and Kdp expression is not shown because neither have been determined as yet. Epstein (1986) believes that expression is, in some way, inherent in the transport systems.

Under these circumstances, the accumulation of intracellular K^+ to high concentrations could result in excessively high internal ionic strength. Control of this ion, therefore, is necessary and is obtained

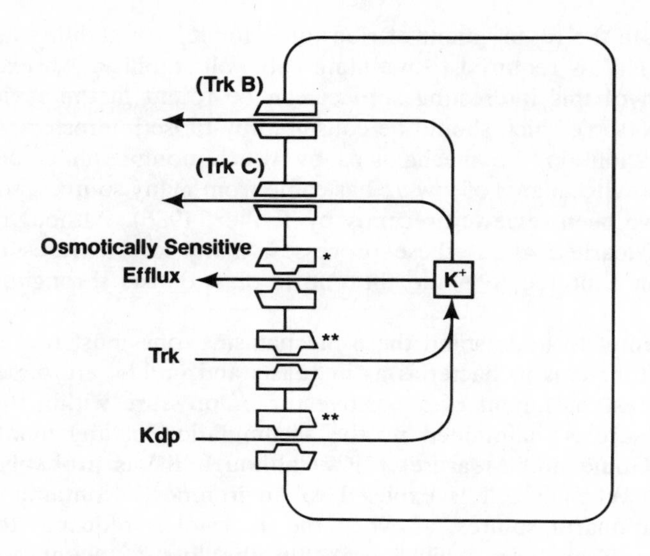

FIG. 4. Control systems for osmoregulation (Epstein, 1986).

through the synthesis of zwitterionic or neutral solutes such as trehalose and the increased transport of betaine. Both of these 'secondary' solutes reduce the need for K^+, however, the presence of cellular K^+ is necessary for their appearance.

Betaine and trehalose are not the only solutes that appear within cells that are adapting to limited moisture conditions. Gould and Measures (1977) have shown that any number of amino acids may be synthesized and create isoosmotic conditions across the membrane. Proline, at this time, appears to be one of the most important of these solutes in both bacteria and higher organisms, although γ-aminobutyric and glutamic acids may function in this way (Fig. 5). Proline can either be synthesized (under feedback control) if it is not present in the medium or it can be taken up by the cell via three active permeases if it is present in the medium, however, only one of these permeases is osmotically stimulated (Csonka, 1982).

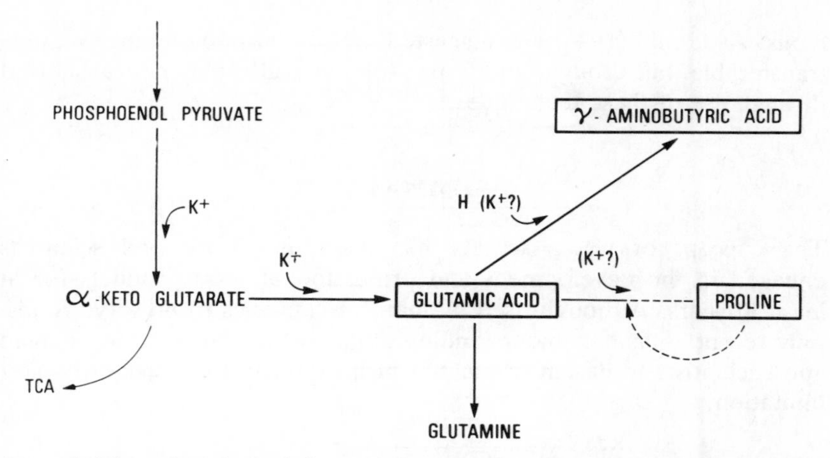

FIG. 5. The physiological interrelationships of the major osmoregulating amino acids (Gould and Measures, 1977).

A key characteristic of these osmotically active compounds is, as stated earlier, their noninterference in the physiological functioning of the cell despite relatively high concentrations. Gould (1985) has indicated that the physiological compatibility of these solutes is due to their ability to 'interfere minimally with biologically important macromolecules'. He further states that noninterference (probably by virtue of their exclusion from the hydration spheres of enzymes and other proteins), as opposed to interaction, is the source of their effectiveness. Schobert (1977) on the other hand disputes this indicating that proline does, in fact, interact with membranes and/or proteins to stabilize their configurations (and, presumably, reactivity) at low a_w levels.

A fascinating corollary to this work is the exploitation of mutants containing the third proline permease mentioned earlier (Csonka, 1982). LeRudelier and Valentine (1982) describe the circumvention of inhibitory a_w levels by the addition of such osmoregulatory solutes as proline and betaine. The genes which control and, thereby, confer osmotic tolerance in bacteria (probably the ProU-encoded betaine transport system described by Epstein, 1986) have been isolated, cloned and transmitted to an osmotically sensitive, nitrogen fixing bacterium which then became capable of nitrogen fixation under conditions of water stress. The implications of the transfer of genetic elements possessing tolerance to water stress, of course, are enormous. As Gould (1983) has suggested, not only osmotolerance may be transferable, but drought and freeze-tolerance also may be transmitted through use of these techniques.

SUMMARY

The concept of water activity has been useful to food scientists engaged in the development and protection of food products for at least 30 years. Although used by man for centuries to preserve foods, only recently has our understanding of a_w and its implications enabled good scientists to design products which optimize this aspect of water limitation.

REFERENCES

Anand, J. C. and Brown, A. D. (1968). Growth rate patterns of the so-called osmophilic and non-osmophilic yeasts in solutions of polyethylene glycol. *J. Gen. Microbiol.*, **52,** 205–12.

Bone, D. P. (1987). Practical applications of water activity and moisture relations in foods. In: *Water Activity: Theory and Applications to Food,* Marcel Dekker, Inc., NY, pp. 369–95.

Brockman, R. and Acker, L. (1977). Lipoxygenase activity and water activity in systems of low water content. *Ann. Tecnol. Agricole,* **26,** 167–74.

Brown, A. D. and Simpson, J. R. (1972). Water relations of sugar-tolerant yeasts: the role of intracellular polyols. *J. Gen. Microbiol.* **72,** 589–91.

Burvall, A., Asp, N., Bosson, A., San Jose, C. and Dahlquist, A. (1978). Storage of lactose-hydrolysed dried milk: effect of water activity on the protein nutritional value. *J. Dairy Res.*, **45,** 381–9.

Chirife, J., Scorza, O. C., Vigo, M. S., Bertoni, M. H. and Cattaneo, C. (1979). Preliminary studies on the storage stability of intermediate moisture beef formulated with various water binding agents. *J. Fd Technol.*, **14,** 421–8.

Chirife, J., Ferro Fontan, C. and Scorza, O. C. (1980). A study of the water activity lowering behavior of some amino acids. *J. Fd Technol.*, **15,** 383–7.

Chirife, J., Resnik, S. L. and Ferro Fontan, C. (1985). Application of Ross' equation for prediction of water activity in intermediate moisture food systems containing a non-solute solid. *J. Fd Technol.*, **20,** 773–9.

Christian, J. H. B. and Waltho, J. A. (1962). The water relations of staphylococci and micrococci. *J. Appl. Bacteriol.*, **25,** 369–77.

Chung, C. Y. and Toyomizu, M. (1976). Studies on the browning of dehydrated foods as a function of water activity. I. Effect of a_w on browning in amino acid-lipid systems. *Bull. Jap. Soc. Sci. Fisheries,* **42,** 697–702.

Corry, J. E. L. (1974). The effect of sugars and polyols on the heat resistance of salmonellae. *J. Appl. Bacteriol.*, **37,** 31–43.

Corry, J. E. L. (1978). Relationships of water activity to fungal growth. In: *Food and Beverage Mycology*, L. R. Beuchat (Ed.) AVI, Westport, CN, pp. 45–82.

Csonka, L. N. (1982). A third L-proline permease in *Salmonella typhimurium* which functions in media of elevated osmotic strength. *J. Bacteriol.*, **151,** 1433–43.

Dennison, D. B. and Kirk, J. R. (1982). Effect of trace mineral fortification on the storage stability of ascorbic acid in a dehydrated model food system. *J. Fd Sci.*, **47,** 1198–1200, 1217.

Drapron, R. (1972). Reactions enzymatiques en milieu peu hydrate. *Ann. Tecnol. Agricole,* **21,** 167–74.

Drapron, R. (1985). Enzyme activity as a function of water activity. In: *Properties of Water in Foods,* D. Simatos and J. L. Multon (Eds), Martinus Nijhoff Pub., Dordrecht, The Netherlands, pp. 171–90.

Duckworth, R. B. (1981). Solute mobility in relation to water content and water activity. In: *Water Activity: Influences on Food Quality,* L. B. Rockland and G. F. Stewart (Eds), Academic Press, NY, pp. 295–317.

Edgeley, M. and Brown, A. D. (1978). Response of xerotolerant and nontolerant yeasts to water stress. *J. Gen. Microbiol.*, **104,** 343–5.

Eichner, K. (1981). Autooxidative effect of Maillard reaction intermediates. In: *Maillard Reactions in Food,* C. Erickson (Ed.), Pergamon Press, Oxford, pp. 115–23.

Eichner, K., Laible, R. and Wolf, W. (1985). The influence of water content and temperature on the formation of Maillard reaction intermediates during drying of plant products. In: *Properties of Water in Foods,* D. Simatos and J. L. Moulton (Eds), Martinus Nijhoff Pub., Dordrecht, The Netherlands, pp. 191–210.

Epstein, W. (1986). Osmoregulation by potassium transport in *Escherichia coli. FEMS Microbiology Rev.*, **39,** 73–8.

Favetto, G. and Chirife, J. (1985). Simplified method for the prediction of water activity in binary aqueous solutions. *J. Fd Technol.*, **20,** 631–6.

Favetto, G., Resnik, S., Chirife, J. and Ferro Fontan, C. (1983). Statistical evaluation of water activity measurements obtained wth the Vaisala Humicap humidity meter. *J. Fd Sci.*, **48,** 534–8.

Ferro Fontan, C., Chirife, J. and Bouquet, R. (1981). Water activity in multicomponent non-electrolyte solutions. *J. Fd Technol.*, **18,** 553–9.

Funes, J. and Karel, M. (1981). Free radical polymerization and lipid binding of lysozyme reacted with peroxidizing linoleic acid. *Lipids,* **16,** 347–50.

Furuya, E. M., Wartheseu, J. J. and Labuza, T. P. (1984). Effects of water activity, light intensity and physical structure of food on the kinetics of riboflavin photodegradation. *J. Fd Sci.*, **49,** 525–8.

Goldblith, S. A. and Tannenbaum, S. R. (1966). The nutritional aspects of the freeze-drying of foods. In: *Proc. Intl. Cong. Nutr.*, J. Knehhau and H. D. Cremer (Eds), Pergamon, Oxford.

Gould, G. W. (1985). Present state of knowledge of a_w effects on microorganisms. In: *Properties of Water in Foods,* D. Simatos and J. L.

Multon (Eds), Martinus Nijhoff Pub., Dordrecht, The Netherlands, pp. 229–45.
Gould, G. W. and Measures, J. C. (1977). Water relations in single cells. *Phil. Trans. Royal Soc., Lond. B*, **278,** 151–66.
Greenspan, L. (1977). Humidity fixed points of binary saturated aqueous solutions. *J. Res. Nat. Bureau Stds, A. Phys. and Chem.*, **81A,** 89–96.
Hahn-Hagerdahl, B. (1986). Water activity: a possible external regulator in biotechnical processes. *Enzyme and Microbial Proc.*, **8,** 322–7.
Halim, D. H. and Montgomery, M. W. (1978). Polyphenol oxidase of d'Anjou pears (*Pyrus communis* L.). *J. Fd Sci.*, **43,** 603–5, 608.
Hardman, T. M. (1976). Measurement of water activity. Critical appraisal of methods. In: *Intermediate Moisture Foods,* R. Davis, G. G. Birch and K. J. Parker (Eds), Applied Sci. Pub., London.
Helmer, G. L., Laimius, L. A. and Epstein, W. (1982). Mechanisms of potassium transport in bacteria. In: *Membranes and Transport,* A. N. Martonosi (Ed.), Plenum Press, NY.
Hodge, J. E. and Osman, E. M. (1976). Carbohydrates. In: *Principles of Food Science. Part I. Food Chemistry,* O. R. Fennema (Ed.), Marcel Dekker Inc., NY.
Iglesias, H. and Chirife, J. (1982). *Handbook of Food Isotherms,* Academic Press, NY.
Jakobsen, M. and Trolle, G. (1979). The effect of water activity on growth of Clostridia. *Nord. Vet.-Med.*, **31,** 206–13.
Jensen, A. (1967). Tocopherol content of seaweed and seameal. 3. Influence of processing and storage on the content of tocopherol, carotenoids and ascorbic acid in seaweed meal. *J. Sci. Fd Agric.*, **20,** 622–33.
Kapsalis, J. G. (1975). The influence of textural parameters in foods at intermediate moisture levels. In: *Water Relations of Foods,* R. B. Duckworth (Ed.), Academic Press, NY.
Karel, M. (1975). Free radicals in low moisture systems. In: *Water Relations of Foods,* R. B. Duckworth (Ed.), Academic Press, NY.
Karel, M. (1985). Environmental effects on chemical changes in foods. In: *Chemical Changes in Food During Processing,* T. Richardson and J. W. Finley (Eds), AVI, Westport, CN, pp. 153–69.
Katz, E. E. and Labuza, T. P. (1981). Effect of water activity on the sensory crispness and mechanical deformation of snack food products. *J. Fd Sci.*, **46,** 403–9.
Kirk, J., Dennison, D., Kokoszka, P. and Heldman, D. (1977). Degradation of ascorbic acid in a dehydrated food system. *J. Fd Sci.*, **42,** 1274–9.
Koiwa, J. and Ohta, S. (1978). Lowering of water activity accompanying emulsion formation. *Proc. V Int. Cong. Food Science & Technol. Kyoto.*
Kreisman, L. N. and Labuza, T. P. (1978). Storage stability of intermediate moisture food process cheese food products. *J. Fd Sci.*, **43,** 341–4.
Kushner, D. J. (1978). Life in high salt and solute concentrations: halophilic bacteria. In: *Microbial Life in Extreme Environments,* D. J. Kushner (Ed.), Academic Press, London, pp. 318–68.
Labuza, T. P. (1985). Water binding of humectants. In: *Properties of Water in*

Foods, D. Simatos and J. L. Multon (Eds), Martinus Nijhoff Pub., Dordrecht, The Netherlands, pp. 421–45.

Labuza, T. P. and Saltmarch, M. (1981). Kinetics of browning and protein quality loss in whey powders under steady state and nonsteady state storage conditions. *J. Fd Sci.,* **47,** 92–6, 113.

Labuza, T. P. and Kamman, J. F. (1982). Comparison of stability of thiamin salts at high temperature and water activity. *J. Fd Sci.,* **47,** 525–8.

Labuza, T. P., McNally, L., Gallagher, D., Hawkes, J. and Hurtado, F. (1982). Stability of intermediate moisture foods. I. Lipid oxidation. *J. Fd Sci.,* **37,** 154–7.

Lanyi, J. K. (1974). Salt dependent properties of proteins from extremely halophilic bacteria. *Bacteriol. Rev.,* **38,** 272–90.

Leistner, L., Rodel, W. and Krispien, K. (1981). Microbiology of meat and meat products in high- and intermediate-moisture ranges. In: *Water Activity: Influences on Food Quality,* L. B. Rockland and G. F. Steward (Eds), Academic Press, NY, pp. 855–916.

LeRudelier, D. and Valentine, R. C. (1982). Genetic engineering in agriculture: osmoregulation. *Trends in Biochem. Sci.,* **7,** 431–3.

Luard, E. J. (1982). Accumulation of intracellular solutes by two filamentous fungi in response to growth at low steady state osmotic potential. *J. Gen. Microbiol.,* **128,** 2563–74.

Marshall, B. J., Ohye, D. F. and Christian, J. H. B. (1971). Tolerance of bacteria to high concentrations of NaCl and glycerol in the growth medium. *Appl. Microbiol.* **21,** 363–4.

Masui, M. and Wada, S. (1973). Intracellular concentrations of Na^+, K^+, and Cl^- of a moderately halophilic bacterium. *Can. J. Microbiol.,* **19,** 1181–6.

Mildenhall, J. P., Prior, B. A. and Trollope, L. A. (1981). Water relations of *Erwinia chrysanthemi*: growth and extracellular pectin acid lyase production. *J. Gen. Microbiol.,* **127,** 27–34.

Mishkin, M., Karel, M. and Seguy, I. (1982). Applications of optimization in food dehydration. *Fd Technol.,* **35,** 244–8.

Mossel, D. A. A. (1975). Water and microorganisms in foods—a review. In: *Water Relations of Foods,* R. B. Duckworth (Ed.), Academic Press, NY, pp. 347–61.

Mossel, D. A. A. and Westerdijk, J. (1949). The physiology of microbial spoilage in foods. *Leeuwenhoek ned. Tijdschr.,* **15,** 190–202.

Nashif, S. A. and Nelson, F. E. (1953). The lipase of *Pseudomonas fragi.* II. Factors affecting lipase production. *J. Dairy Sci.,* **36,** 471–80.

Petriella, C., Resnik, S. L., Lozano, R. D. and Chirife, J. (1985). Kinetics of deteriorative reactions in model food systems of high water activity: color changes due to nonenzymatic browning. *J. Fd Sci.,* **50,** 622–6.

Pollio, M. L., Kitic, D., Favetto, G. J. and Chirife, J. (1986). Effectiveness of available filters for an electric hygrometer for measurement of water activity in the food industry. *J. Fd Sci.,* **51,** 1358–9.

Potthast, K. (1978). Influence of water activity on enzymic activity in biological systems. In: *Dry Biological Systems,* Academic Press, NY, pp. 323–42.

Prabhakar, J. V. and Amla, B. C. (1978). Influence of water activity on the

formation of monocarbonyl compounds in oxidizing walnut oil. *J. Fd Sci.*, **43,** 1839–43.
Priestly, D. A., Werner, B. G. and Leopold, A. C. (1985). The susceptibility of soybean seed lipids to artificially-enhanced atmospheric oxidation. *J. Exptl Bot.*, **36,** 1653–9.
Prior, B. A. (1979). Measurement of water activity in foods: a review. *J. Fd Prot.*, **42,** 668–74.
Prior, B. A. and Kenyon, C. P. (1980). Water relations of glucose-catabolizing enzymes in *Pseudomonas fluorescens*. *J. Appl. Bacteriol.*, **48,** 211–22.
Prior, B. A., Kenyon, C. P., Vanderveen, M. and Mildenhall, J. P. (1987). Water relations of solute accumulation in *Pseudomonas fluorescens*. *J. Appl. Bacteriol.*, **62,** 119–28.
Quast, D. G. and Karel, M. (1972). Effects of environmental factors on the oxidation of potato chips. *J. Fd Sci.*, **37,** 584–8.
Reidy, G. A. and Heldman, D. R. (1972). Measurement of texture parameters of freeze-dried beef. *J. Texture Studies*, **3,** 218–26.
Resnik, S. L., Favetto, G., Chirife, J. and Ferro Fontan, C. (1984). A world survey of water activity of selected saturated salt solutions used as standards at 25°F. *J. Fd Sci.*, **49,** 510–13.
Rodel, W. and Leistner, L. (1971). Ein einfacher a_w-Wert-Messer fur die Praxis. *Fleischwirtsch.*, **51,** 1800–2.
Rodel, W., Krispien, K. and Leistner, L. (1979). Messung der Wasseraktivitat (a_w-Wert) von Fleisch und Fleisch-erzeugnissen. *Fleischwirtsch.*, **59,** 831–6.
Ross, K. D. (1975). Estimation of water activity in intermediate moisture foods. *Fd Technol.*, **29,** 26–30.
Schobert, B. (1977). Is there an osmotic regulatory mechanism in algae and higher plants? *J. Theor. Biol.*, **68,** 17–26.
Scott, W. J. (1953). Water relations of *Staphylococcus aureus* at 30°C. *Aust. J. Biol Sci.*, **6,** 549–64.
Scott, W. J. (1957). Water relations of food spoilage microorganisms. *Adv. Fd Res.*, **7,** 83–127.
Seow, C. C. and Teng, T. T. (1981). The prediction of water activity of some supersaturated non-electrolyte aqueous binary solutions from ternary data. *J. Fd Technol.*, **16,** 597–607.
Simatos, D., LeMeste, M., Petroff, D. and Halpen, B. (1981). Use of electron spin resonance for the study of solute mobility in relation to moisture content in model food systems. In: *Water Activity*: *Influences on Food Quality* L. B. Rockland and G. F. Stewart (Eds), Academic Press, NY, pp. 319–46.
Solomon, M. E. (1951). Control of humidity with potassium hydroxide, sulphuric acid or other solutions. *Bull. of Ent. Res.*, **42,** 543–54.
Stokes, R. H. and Robinson, R. A. (1949). Standard solutions for humidity control at 25°F. *Ind. Eng. Chem.*, **41,** 2013–16.
Taylor, J. A. (1961). Determination of moisture equilibria in hydrated foods. *Fd Technol.*, **15,** 536–7.
Toribio, J. L., Nunes, R. V. and Lozano, J. E. (1984). Influence of water activity on the nonenzymatic browning of apple juice concentrate during storage. *J. Fd Sci.*, **49,** 1630–1.

Troller, J. A. (1977). Statistical analysis of a_w measurements obtained with the Sina Scope. *J. Fd Sci.*, **42,** 86–90.

Troller, J. A. (1983*a*). Water activity measurements with a capacitance manometer. *J. Fd Sci.*, **48,** 739–41.

Troller, J. A. (1983*b*). Methods to measure water activity. *J. Fd Prot.*, **46,** 129–34.

Troller, J. A. (1986). Adaptation and growth of microorganisms in environments with reduced water activity. In: *Water Activity: Theory and Applications,* L. R. Beuchat and L. Rockland (Eds), Marcel Dekker Inc., NY, pp. 101–17.

Troller, J. A. and Christian, J. H. B. (1978). *Water Activity and Food,* Academic Press, NY.

Troller, J. A. and Stinson, J. V. (1978). Influence of water activity on the production of extracellular enzymes by *Staphylococcus aureus*. *Appl. Environ. Microbiol.*, **35,** 521–6.

Verma, M. M. and Prabhakar, J. V. (1982). Effect of water activity on the autoxidation of safflower seed oil. *Indian Food Packer,* **36,** 77–80.

Vertucci, C. W. and Leopold, A. C. (1984). Bound water in soybean seed and its relation to respiration and imbibitional damage. *Plant Physiol.*, **75,** 114–17.

Warmbier, H. C., Schnickels, R. A. and Labuza, T. P. (1976). Nonenzymatic browning kinetics in an intermediate moisture model system: effect of glucose to lysine ratio. *J. Fd Sci.*, **41,** 981–3.

Wolf, W., Speiss, W. E. L. and Jung, G. (1985). *Sorption Isotherms and Water Activity of Food Materials,* Elsevier, NY.

Woodcock, E. A., Warthesen, J. J. and Labuza, T. P. (1982). Riboflavin photochemical degradation in pasta measured by high performance liquid chromotography. *J. Fd Sci.*, **47,** 545–9, 555.

Zabik, M. E., Fierke, S. G. and Bristol, D. K. (1979). Humidity effects on textural characteristics of sugar snap cookies. *Cereal Chem.*, **56,** 29–33.

Chapter 2

DEHYDRATION OF FOODSTUFFS

J. G. Brennan
*Department of Food Science and Technology,
University of Reading, UK*

INTRODUCTION

Drying is the oldest method of food preservation practiced by man and is still widely used today. The low water activity attained by drying extends the shelf lives of dried foods without the need for refrigerated storage or transport. Usually significant reductions in weight and bulk occur during drying which can lead to savings in the costs of transport and storage. The rapid reconstitution characteristics and good organoleptic quality of many modern dehydrated products make them very acceptable as convenience foods. A quick look around a modern supermarket will reveal a very wide range of dried foods including: coffee, tea, milk, chocolate and malt based 'instant' drinks; baby foods containing dried cereals, fruits, vegetables and meats; dried cereals and cereal products such as rice, pasta and breakfast foods; dried vegetables such as potato flakes or granules, peas, beans, carrots and many more for use in home cooking; dried fruits for use as snacks or in desserts or baked products; soup mixes and instant meals containing dried vegetables, meat and fish ingredients. To provide such a comprehensive range of products the dehydration section of the food industry is large and extends to all countries of the globe. Many very high capacity, sophisticated drying facilities exist throughout the world.

On the other hand drying is a very suitable method of food preservation in hot climates where cold chains have not been established and relatively complex equipment for heat processing is scarce and expensive. In such situations solar drying may be employed or comparatively simple hot air driers constructed from local materials. On an industrial scale, in hot climates, dehydration is often preferred because dried products pose less problems during storage and distribution as compared to those resulting from alternative methods of preservation. Dried foods are also suitable for use in emergencies such as famines and floods, for supplying troops in the field and as rations for explorers, climbers and the like.

In this chapter some recent developments in industrial food dehydration are discussed and some of the trends in research and development into that method of food preservation are outlined.

HOT AIR DRYING

When a wet material is placed in a stream of heated air the sensible and latent heat of evaporation is transferred, mainly by convection, from the air stream to the wet solid. The air stream also acts as a carrier removing the water vapour from the vicinity of the drying surface thus permitting further evaporation. In a model system in which a wet material, consisting of inert solids wetted with pure water, is placed in contact with a stream of heated air of constant temperature, humidity and velocity the drying cycle may be represented by a graph as shown in Fig. 1. After an initial settling down period (A–B) a period in which the rate of drying remains constant occurs (B–C). During this period the rate of heat transfer to the solid and the rate of mass transfer to the air are in equilibrium. The temperature of the surface of the wet solid also remains constant during this period and corresponds to the wet-bulb temperature of the drying air. The factors which control the rate of drying during this period are the air temperature, humidity, velocity and the dimensions of the slab. Many food materials do not exhibit a constant rate period. In the case of foods that do, this period usually accounts for a relatively small proportion of the total drying time, and the surface temperature, although constant, will be higher than the wet-bulb temperature of the air.

Below a certain moisture content, known as the critical moisture content, the rate of drying begins to decrease. In this so-called falling rate period (C–D) the rate of drying is influenced mainly by the factors which control the movement of water within the solid and external factors become less important. Numerous attempts have been made to represent the falling rate period by means of mathematical

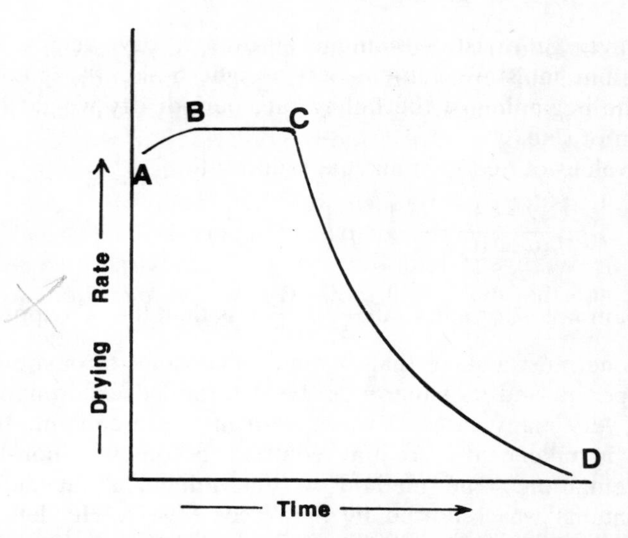

FIG. 1. Drying curve for a wet material drying in heated air at constant temperature, humidity and velocity.

models. There have been two main approaches to this. The first is to assume that a certain mechanism of moisture movement prevails within the solid and to develop expressions to represent that mechanism. The second approach is to construct drying curves from experimental data and to fit expressions to those curves. Several mechanisms of moisture movement within a wet solid during drying have been proposed by Van Arsdel *et al.* (1973), Brennan *et al.* (1976), Charm (1979) and Green (1984). The explanation which has received the widest acceptance is that moisture moves within the solid by diffusion as a result of concentration gradients. Such movement may be represented by Fick's second law of diffusion thus:

$$\frac{dW}{dt} = D\frac{d^2W}{dl^2} \tag{1}$$

where W = moisture content, dry weight basis; t = time; l = distance; and D = diffusivity.

One well known solution to this equation for a slab-shaped solid, drying from one large face only is as follows:

$$\frac{W - W_e}{W_c - W_e} = \frac{8}{\pi^2}\left[\exp\left\{-Dt\left(\frac{\pi}{2l}\right)^2\right\} + \frac{1}{9}\exp\left\{-9Dt\left(\frac{\pi}{2l}\right)^2\right\}\right] \tag{2}$$

where W = average moisture content at time t, dry weight basis; W_e = equilibrium moisture content, dry weight basis; W_c = moisture content at the beginning of the falling rate period, dry weight basis; and l = depth of slab.

For large values of t eqn (2) may be reduced to:

$$\frac{W - W_e}{W_c - W_e} = \frac{8}{\pi^2}\left[\exp\left\{-Dt\left(\frac{\pi}{2l}\right)^2\right\}\right] \tag{3}$$

This expression holds for values of $\frac{W - W_e}{W_c - W_e}$ less than 0·6. To apply eqn (3) directly one must assume that D remains constant throughout the falling rate period and its value must be determined experimentally. However in very many cases D varies with moisture content. If this relationship is taken into account eqn (2) becomes a non-linear differential equation. One method of determining D at different moisture contents was reported by Saravacos (1967). He defined a term known as the half equilibrium time, i.e. the time required to reach a moisture content half way between the moisture content at the start of the falling rate period and equilibrium moisture content. Thus eqn (2) may be written in a more general form as:

$$0{\cdot}5 = 1 - \frac{8}{\pi^2}\sum_{n=0}^{\infty}\frac{1}{(2n+1)^2}\exp\left\{-(2n+1)^2 Dt\left(\frac{\pi}{2l}\right)^2\right\} \tag{4}$$

If the first term only of this equation is used it reduces to:

$$D = \frac{0{\cdot}194 l^2}{t_{(0{\cdot}5)}} \tag{5}$$

where $t_{(0{\cdot}5)}$ is the half equilibrium time. Approximate values of D can be obtained by applying eqn (5) at various humidities or water contents in which the diffusivity is assumed to remain constant.

Diffusivity also varies with temperature. A relationship of the form of eqn (6) was found to apply in studies of drying of tapioca root (Chirife, 1971).

$$\bar{D}_e = D_0 \exp\left(-\frac{Q}{RT}\right) \tag{6}$$

where $\bar{D}_e$ = average effective diffusivity; D_0 = temperature independent constant; Q = energy of activation for diffusion; R = gas constant; and T = absolute temperature.

The effect of slab depth on drying rate has also been studied (Vaccarezza and Chirife, 1978). The value of the exponent of the depth in eqns (2), (3) and (4) was found to be lower than 2. This deviation was explained in terms of heat transfer effects. Numerous examples of the application of the diffusional model to describe drying in the falling rate period are to be found in the literature. In two recent examples it was applied to model gels and pasta respectively by Biquet and Guilbert (1986), and Andrien and Stamatopoulos (1986).

Many other analytical approaches to represent the drying of wet solids during the falling rate period have been reported. Detailed discussion of these is outside the scope of this chapter. Numerous papers on this topic can be found in reports of drying symposia held during the last few years (Ashworth, 1982; Mujumdar, 1980, 1982, 1984, 1985).

Empirical models have been used to represent falling rate drying from the early 1920s. The logarithmic model was used by Lewis (1921), i.e.

$$\frac{\mathrm{d}W}{dt} = -K_c(W - W_e) \tag{7}$$

or

$$\frac{W - W_e}{W_c - W_e} = \exp(-K_c t) \tag{8}$$

K_c came to be known as the mass transfer coefficient and was related to temperature by an expression of the form:

$$K_c = a \exp\left(\frac{-b}{T}\right) \tag{9}$$

where a and b are material constants.

Equation (8) was used to describe the drying of agricultural materials but it did not apply to the whole curve. To widen its applicability an empirical exponent n in the time term was proposed thus:

$$\frac{W - W_e}{W_c - W_e} = \exp(-K_c t^n) \tag{10}$$

This gave good results when applied to the drying of shelled corn and soybeans. Another empirical model was used to represent thin layer drying of shelled corn:

$$t = A_c \ln\left(\frac{W - W_e}{W_c - W_e}\right) + B_c\left\{\ln\left(\frac{W - W_e}{W_c - W_e}\right)\right\}^2 \tag{11}$$

where A_c and B_c are empirical coefficients which vary with temperature (Thompson *et al.*, 1968). An expression obtained from experimental data from drying peas was reported by Escardino (1979, 1980):

$$\frac{dW}{dt} = -1{\cdot}010 \times 10^{-4}(H_s - H_a)^{1{\cdot}69}G^{2{\cdot}02} + 0{\cdot}112(H_s - H_a)^{0{\cdot}925}(W - W_e) \tag{12}$$

where H_a = air humidity; H_s = humidity at the surface of the food; and G = the mass flowrate of air.

This equation was a good predictor of drying behaviour within a limited range of operating conditions. Many other empirical expressions have been reported, see symposia reports above and Sharaf-Elden *et al.* (1979). While such an approach can yield useful equations they are usually applicable only under conditions close to those used when obtaining the experimental data and they do not explain the behaviour of the materials under test.

A relatively new approach to this problem is to try to develop hybrid models incorporating both analytical and empirical elements. A recent example of this was reported by Alvarez and Legues (1986). They developed a model for the drying of seedless grapes based on the diffusion mechanism and accounting for variations of diffusivity with time. An effective diffusion coefficient, D_e, was proposed as described by:

$$D_e = D_0(1 - F_0)^b \tag{13}$$

where D_0 and b are constants and F_0 is a dimensionless number thus:

$$F_0 = \frac{D_0 t}{l^2} \tag{14}$$

The full expression was as follows:

$$\frac{W - W_e}{W_c - W_e} = \frac{6}{\pi^2}\sum_{n=1}^{\infty}\frac{1}{n^2}\exp\left\{-\frac{n^2\pi^2}{(1+b)}(1 + F_0)^{(1+b)} - 1\right\} \tag{15}$$

This model was simplified by taking only the first exponential term of eqn (15) to give:

$$\frac{W - W_e}{W_c - W_e} = \exp\left\{\frac{\pi^2}{(1+b)}(1+F_0)^{(1+b)} - 1\right\} \quad (16)$$

Drying curves constructed using eqns (15) or (16) closely matched those obtained from experimental data. This concept of a hybrid model is interesting and worthy of further investigation.

Hot Air Drying Systems

For drying piece-form materials such as sliced fruit, sliced or diced vegetables considerable use is still made of cabinet and tunnel driers. A *cabinet drier* consists of an insulated chamber into which tray-loads of prepared food are placed. A fan pushes or pulls air through a heater and then either horizontally between the trays or vertically through them. By means of a damper system part of the air may be recycled and part discharged to the atmosphere. Such driers vary in size from experimental models containing just one tray of food to very large units used singly or in groups with throughputs of fresh material up to 20 tonnes per day. A *tunnel drier* is an elongated cabinet which may be over 20 m long. Trays of wet food, stacked on trolleys are introduced at one of the tunnel, the 'wet' end, and when dry discharged from the other end, the 'dry' end. The drying air and food may move parallel to each other either concurrently or countercurrently. Two-stage driers are also used featuring a short concurrent stage followed by a longer countercurrent stage. In crossflow designs the air moves at right angles to the path of the trays of food. Each of the above designs results in its own characteristic drying pattern and product properties (Van Arsdel *et al.*, 1973; Brennan *et al.*, 1976; Williams-Gardner, 1976; Charm, 1979).

In *conveyor or belt driers* the food is carried through a tunnel on a perforated belt. Heated air is directed up or down through the bed of food, usually up in the early stages and down towards the 'dry' end of the tunnel. Some models consist of two or more belts in series. They make more economic use of the belt surface and lead to more uniform drying. Such driers are limited to food particles that form a porous bed. To reduce costs the food may be partly dried in a belt drier and finished off in a less expensive drier. If the product requires to be dried gently over a long period, e.g. pasta products, a multiple conveyor drier may be used. This consists of a number of conveyors located one above the other in an insulated chamber. The product is introduced onto the top belt and progresses downwards transferring from one conveyor to the next. Air circulation is usually a combination crossflow and throughflow (Van Arsdel *et al.*, 1973; Brennan *et al.*, 1976; Farrall, 1976; Williams-Gardner, 1976; Sturgeon, 1987).

In *fluidised-bed drying* a bed of food particles is fluidised with heated air. Heat and mass transfer are rapid and the drying process can be readily controlled. Particle diameters are usually in the range 20 μm to 10 mm. Such driers may be operated as batch units in which good mixing leads to uniform drying. When operated continuously mixing may result in some incompletely dried particles being discharged in the product. This problem can be reduced by using two or more beds in series. When handling fine powders or when fine particles are likely to be formed during drying, covered beds are used and cyclones or filters included to recover the fines. Heated panels may be introduced into the beds particularly when drying small particles, less than 300 μm. These can lead to a reduction in bed size and an improvement in the economy of the operation. Fluidised-bed driers are finding increasing use for food dehydration. Applications include: peas, beans, diced carrot, onions, potatoes, grains, sugar crystals and flours. If the particle size of the food covers a wide range uniform fluidisation in a stationary bed may be difficult. Using a shallow bed, mechanically vibrated, may overcome this problem. The frequency of vibration is usually in the range 5–25 Hz with a half amplitude of a few millimetres. Such beds, known as *vibro-fluidisers*, are frequently used in series with spray driers, see below (Van Arsdel *et al.*, 1973; Brennan *et al.*, 1976; Williams-Gardner, 1976, Charm, 1979; Hovmand, 1987).

A *spouted bed drier* is a modified form of a fluidised-bed drier. Instead of distributing the heated air uniformly across the bottom of the bed of particles it enters as a jet through the conical base of the chamber. This creates a spout of fast moving particles up through the centre of the bed. The particles return to the bottom of the chamber in the form of an annular slow moving bed surrounding the spout (Fig. 2). Good mixing of the heated air with the particles occurs leading to high rates of heat and mass transfer. Mild drying conditions may be achieved and less heat damage to the product may occur as compared with a conventional fluidised-bed drier. Spouted bed driers are particularly useful for drying relatively large particles, greater than 5 mm. Many papers analysing the mechanics of the spouted bed and its

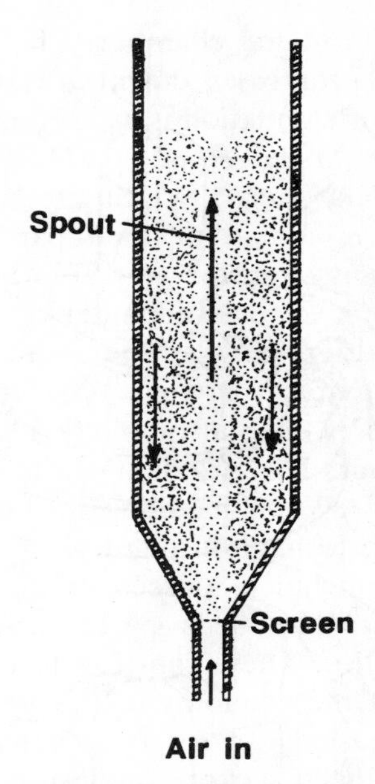

FIG. 2. Principle of the spouted bed drier.

applications are to be found in the symposia reports mentioned above and in Pallai *et al.* (1987). A draft tube may be located in the bed, surrounding the spout. This permits an increase in bed height without entrainment of particles in the air. A screw conveyor can be introduced into the spout zone to control the circulation of the particles independently of the air velocity.

Apart from straightforward drying of food particles fluidised-bed and spouted bed driers have been used for some novel duties. Baxeires *et al.* (1983) used a whirling fluidised-bed containing inert particles such as glass beads to dry relatively large food particles including diced potato and carrot and sticky products such as pre-cooked rice. Boeh-Ocansey (1986) carried out successfully low temperature drying of mushrooms, carrot, beef and shrimp in a fluidised-bed of activated alumina. Both types of bed have been used to dry paste and liquid products. Fane *et al.* (1980) dried cattle blood and other liquid food products in a spouting bed of inert particles by spraying the liquids onto the particles. Kutsakoua *et al.* (1984) dried protein hydrolysate in a fluidised-bed of metal balls. These techniques may prove to be viable

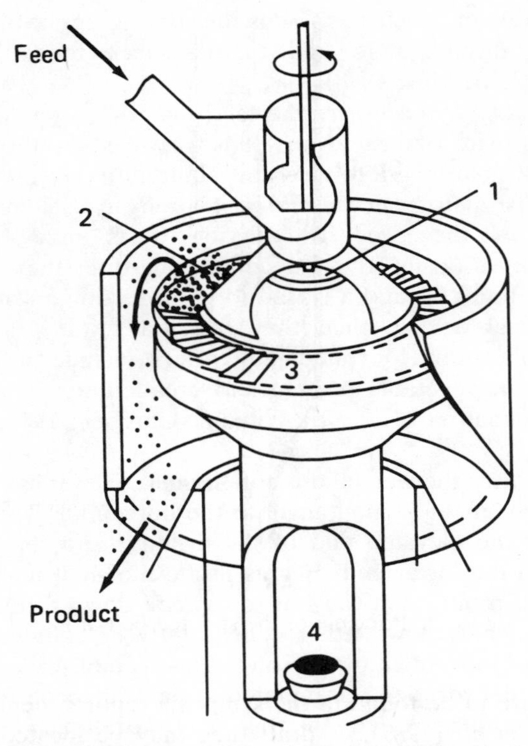

FIG. 3. The toroidal fluidised-bed drier. (1) Rotating disc distributor to deliver raw material evenly into processing chamber; (2) rotating bed of particles; (3) fixed blades with hot gas passing through at high velocity; (4) burner assembly. (By courtesy of Torftech Ltd.)

alternatives to spray drying, see below.

A novel type of fluidised-bed drier is being manufactured in the UK. Known as the *toroidal fluidised-bed* it can be used for a range of processes including cooking, expanding, roasting and drying. A high velocity stream of heated air enters the base of the process chamber through blades or louvres which impart a rotary motion to the air (Fig. 3). This creates a compact, rotating bed of particles which varies in depth from a few millimetres to in excess of 50 mm. High rates of heat and mass transfer are said to be attainable resulting in rapid drying. The drier can accommodate a wider range of particle sizes and shapes as compared with a conventional fluidised-bed and may be operated on a continuous or batch basis. Possible drying applications include: peas, beans, diced potato and carrot, snack foods and slurries (Dodson, 1987; Grikitis, 1988).

In a *pneumatic or flash drier* the feed material is introduced into a fast moving stream of heated air. It is conveyed by the air through ducting of sufficient length to give the required drying time. The dried product is separated from the exhaust air by a cyclone, filter or a combination of the two. The drying ducts may be arranged horizontally or vertically. In a pneumatic ring drier the food particles travel several times around a closed loop of ducting and are removed by a centrifugal device when they reach a suitably low moisture content. Applications for pneumatic driers include: grains, flours, starches, casein, vegetable protein and breadcrumbs (Van Arsdel *et al.,* 1973; Brennan *et al.,* 1976; Williams-Gardner, 1976; Kisakurek, 1987).

In a *rotary drier* the wet solid is rotated in a cylindrical shell usually mounted on rollers at a small angle to the horizontal. The wet feed is introduced at the elevated end of the cylinder and the dry product removed from the lower end. Flights inside the shell lift the material up as the shell rotates and cause it to cascade down through a stream of heated air which flows either concurrently or countercurrently to the direction of movement of the solid. This type of drier is not widely used for food dehydration. Granulated sugar and cocoa beans are the main food products dried in this way (Brennan *et al.,* 1976; Williams-Gardner, 1976). Mathematical models for this type of drier have been published by Kelly (1987).

Spray drying is by far the most common method used for drying food liquids and slurries. The principle of this method is shown in Fig. 4. The feed is introduced into a drying chamber in the form of a fine mist or spray where it contacts heated air. Very rapid drying takes place and the feed is converted into a powder. The bulk of the powder is usually removed directly from the chamber. Fines entrained in the exhaust air from the chamber are recovered in an air/powder separator. The advantages of this method for drying heat sensitive foods is the very short drying time and the evaporative cooling which occurs during drying. Provided the powder is removed quickly from the drying chamber low product temperatures can be maintained. The general principles of spray drying are well described in the literature (Van Arsdel *et al.,* 1973; Brennan *et al.,* 1976; Farrall, 1976; Williams-Gardner, 1976; Masters, 1985; Filkova and Mujumdar,

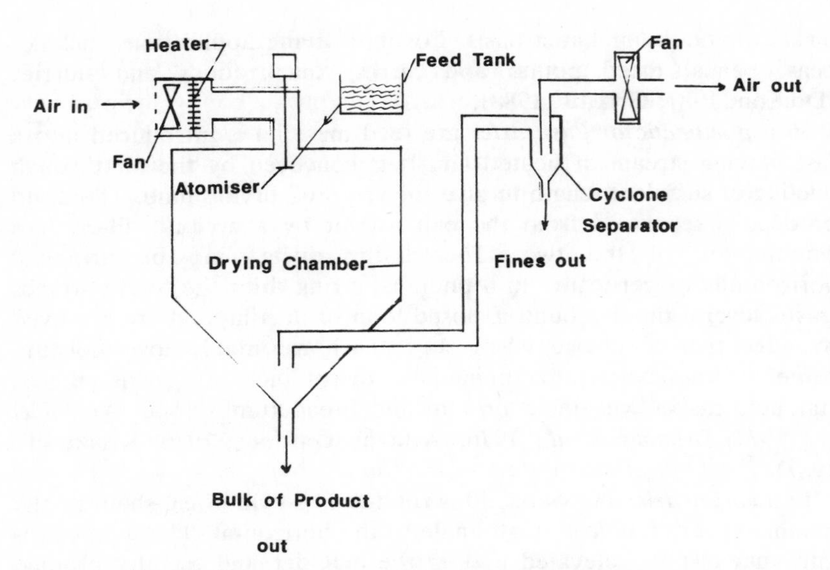

FIG. 4. The main components in a spray drying system.

1987). Mathematical treatment of the spray forming and drying stages have been published by Kerkhof and Schoeber (1974) and Masters (1985).

The important components of a spray drying system are the spray forming device, the drying chamber and the air/powder separator. Spray forming may be achieved by means of a pressure nozzle, a two-fluid nozzle or a spinning disc or wheel known as a centrifugal atomiser. The centrifugal device is the most widely used in food drying as it can handle more viscous feeds and feeds containing larger suspended solid particles than can the nozzles. Numerous designs of atomiser are available including abrasion and corrosion resistant types, designs to produce high bulk density powders and large particles. A steam-swept atomiser enables steam to be introduced to the wheel body together with the liquid feed. This reduces particle expansion.

Nozzle atomisers are still used for some food liquids such as coffee extract. Some highly viscous liquids are difficult to atomise by conventional means and the use of sonic devices has been investigated. A field of high frequency sound waves is created by a sonic resonance cup located in front of a nozzle. This brings about the break-up of the liquid. A few different types of sonic device have been studied. However such devices are not yet viable alternatives to conventional atomisers for use on an industrial scale.

The very many designs of drying chambers available may be classified into three groups featuring (1) concurrent, (2) countercurrent, and (3) mixed flow patterns of air/product movement. Those featuring mixed flow patterns are most widely used but downward concurrent designs are used for some sticky and thermoplastic products. Countercurrent patterns of flow are not commonly used for food applications because of the high risk of heat damaging the powder.

In most designs of spray driers the bulk of the powder is removed from the drying chamber via a rotary valve or vibrating feeder. With products that are thermoplastic and sticky when hot some additional assistance may be required. This may be in the form of pneumatically or electromagnetically operated hammers that strike the outside of the drier wall at predetermined intervals to loosen wall deposits. Rotating chains, brushes or air brooms within the chamber have also been used for this purpose. Cooling the chamber wall by removing some insulating material or passing cool air through a jacket surrounding the wall may also be effective. This prewarmed air may be introduced to the drying chamber as part of the drying air. The air leaving the chamber will contain some fine particles of powder which need to be recovered either to improve the economy of the operation and/or to avoid contaminating the environs of the factory. Dry cyclone separators, bag filters, wet scrubbers and electrostatic precipitators have been used for this purpose. Cyclones and filters are most commonly used, often in series. The use of scrubbers has an additional advantage of improving the thermal efficiency of the drier if the heat in the exhaust air is used to preheat or preconcentrate the feed. Electrostatic precipitation is not common in food applications.

In order to improve the thermal efficiency of spray driers up to 50% of the total drying air may be recycled through the heater. This can result in up to 20% saving in fuel costs. A semi-closed system can be used if a scrubber-condensor is introduced to remove water vapour from the air before it is recycled. Only a small amount of air is exhausted to the atmosphere. This is useful for handling materials that have odour problems. A totally enclosed system may be used to remove non-aqueous solvents or where a gas other than air is to be used e.g. nitrogen for drying products highly susceptible to oxidation or pyrophoric materials such as some pre-gelatinised starches. Aseptic spray drying is also feasible using presterilised air and discharging the product into a 'clean air' room (Kjaergaard, 1974; Brennan, 1978); Masters, 1985; Filkova and Mujumdar, 1987).

It is now quite common to dry liquid foods in a two or three stage operation. This is said to give better control over product quality and to improve the thermal efficiency of the operation. Product with a higher than normal moisture content is discharged from the drying chamber into a vibro-fluidiser, see above, where the drying is completed using a fresh supply of warm air. Some agglomeration of the powder particles will occur in the vibro-fluidiser. If, in addition, the fines from the cyclones are returned to the wet zone of the drier the overall size of the powder particles will be increased and its reconstitution characteristics improved. A second vibro-fluidiser may be used to cool the product. Spray driers with built in fluid-beds are now available, see Fig. 5. In this design the drying chamber features concurrent air/product flow with rotary atomisation. There is a stationary fluid-bed in the form of a ring at the bottom of the chamber surrounding the outlet duct. In addition to the air supply from the top

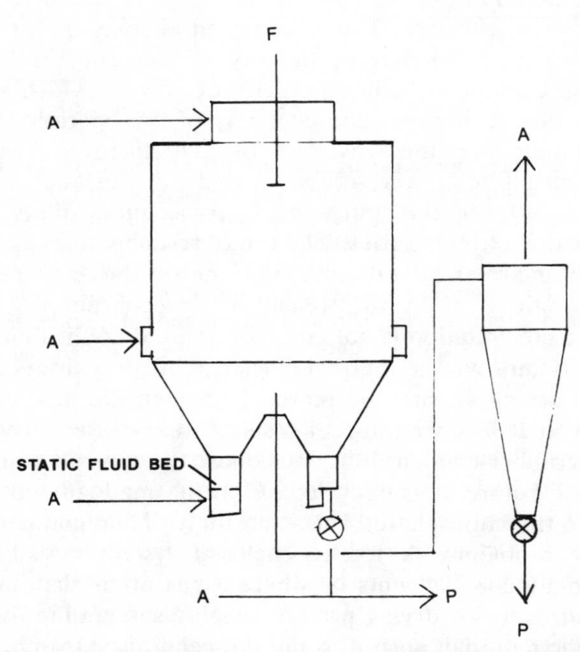

FIG. 5. Two-stage spray drier layout with integrated static fluid bed in drying chamber. A = airflow, F = feed, P = dried product. (From Masters and Pisecky (1983), with permission.)

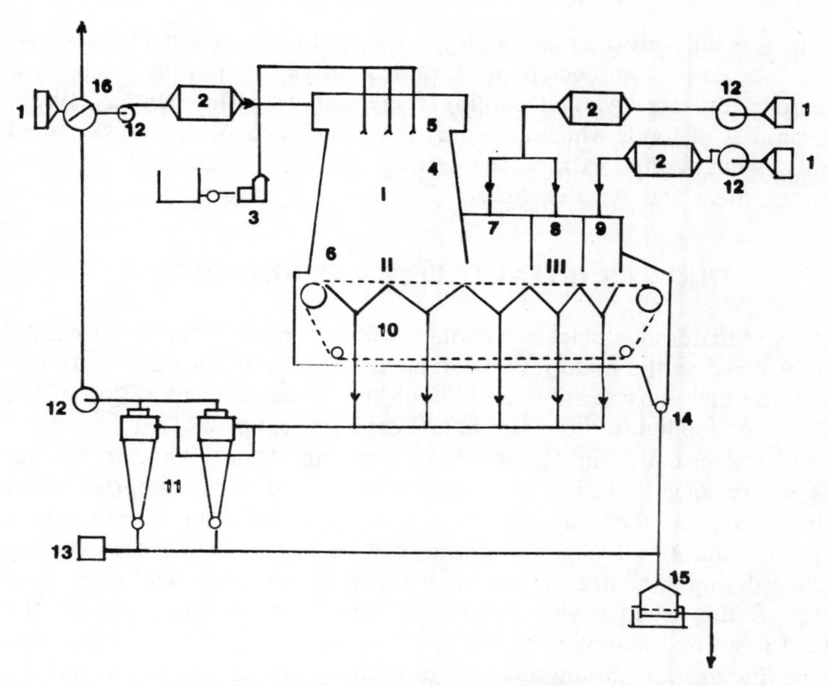

FIG. 6. A three-stage spray drying system (FILTERMAT): (1) air filter; (2) heater-cooler; (3) high-pressure pump; (4) nozzle system; (5) air distributor; (6) primary drying chamber; (7) retention chamber; (8) final drying chamber; (9) cooling chamber; (10) FILTERMAT belt assembly; (11) cyclones; (12) fan; (13) fines recovery system; (14) FILTERMAT powder discharge; (15) sifting system; (16) heat recovery system; I first drying stage; II second drying stage; III third drying stage. (From Filkova and Mujumdar (1987), with permission.)

and bottom air is introduced tangentially to sweep the vertical wall of the chamber. Because of the strong rotary movement of the air powder separation is good and the fraction going to the cyclone low. The product is partly dried in the main body of the chamber, then in the fluid bed. Drying may be completed in these two stages or alternatively the product may be discharged into a vibro-fluidiser for final drying and cooling (Masters and Pisecky, 1983). Another multistage drier design, known as the Filtermat, is shown in Fig. 6. In the main chamber the product is reduced to between 10 and 20% moisture content and deposited onto a moving belt. Secondary drying occurs as air is directed through the belt. After a short holding period the powder is subjected to a third stage of drying using low temperature air. A final cooling stage follows (Rheinlander, 1982). Examples of foods which are spray dried include: whole and skimmed milk, whey, coffee extract, tea extract, whiteners, fruit and vegetable juices, meat and yeast extracts.

DIRECT CONTACT (CONDUCTION) DRYING

If a wet material is placed in contact with a heated surface heat will be transferred to the material by conduction from the hot surface. In this way the necessary sensible and latent heat to cause evaporation of the water is supplied to the material and drying takes place. The pattern of drying is usually similar to that occurring during hot air drying. However since there is little evaporative cooling of the material, when drying by conduction at atmospheric pressure, the temperatures reached during all stages of drying will be higher than those attained when drying in heated air at similar rates. To minimise heat damage to the product when drying by conduction at atmospheric pressure the feed is normally applied to the hot surface in the form of a thin film resulting in short drying times. Alternatively drying may be carried out under reduced pressure conditions i.e. vacuum drying. Assuming that drying takes place from one face only and neglecting shrinkage effects the average drying rate for a complete cycle may be represented by the expression:

$$\frac{dw}{dt} = \frac{(W_0 - W_f)M}{t} = \frac{K_0 A(\theta_W - \theta_E)}{L} \quad (17)$$

where $\frac{dw}{dt}$ = rate of change of weight; W_0 = initial moisture content, dry weight basis; W_f = final moisture content, dry weight basis; M = mass of dry solids; t = total drying time; A = drying surface area; θ_E = drying surface temperature; θ_W = temperature of heating surface; L = latent heat of evaporation; and K_0 = overall heat transfer coefficient for the cycle.

Drum (film, roller) drying is a long established method of drying liquids and slurries. In this method one or two hollow metal cylinders rotate on horizontal axes and are heated internally, usually by steam. The feed material is applied to each drum surface in the form of a thin

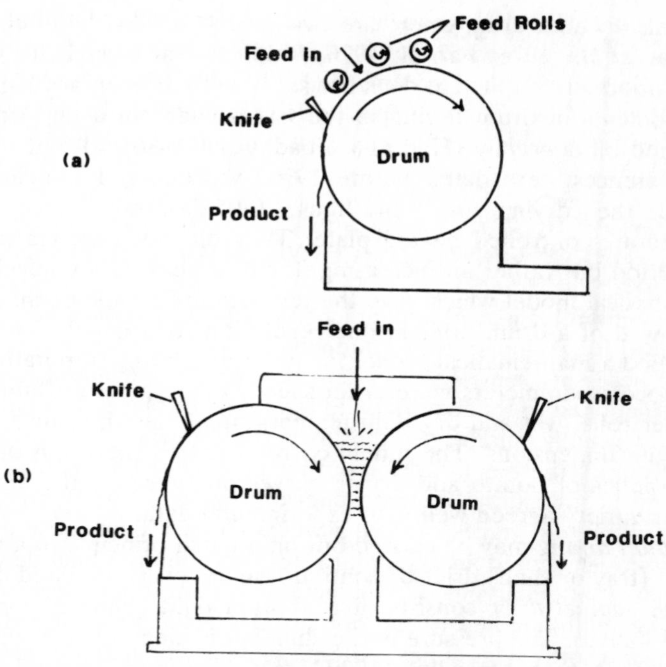

FIG. 7. Drum driers. (a) Single drum; (b) double drum.

film. Drying takes place as the drum rotates and a knife removes the dry product from the drum surface at a point 2/3–3/4 of a revolution from the point of application of the feed. Single and double drum driers are commonly used for food dehydration (Fig. 7). In the case of the single drum drier the feed is applied by dipping the drum into the liquid contained in a trough or tray, by spraying or splashing it onto the drum surface or by the use of applicator rolls as in Fig. 7(a). In the case of a double drum drier the feed is introduced into the trough between the drums (Fig. 7(b)). Factors which affect the drying rate for a particular feed material are drum surface temperature, speed of rotation of the drum and the thickness and uniformity of the film on the drum surface. The method of application and the surface tension, density and viscosity of the feed influence the thickness and uniformity of the film. Products commonly produced by drum drying include; potato flakes, cereal flakes for inclusion in breakfast and baby foods, vegetable and fruit flakes and milk powders. For very heat sensitive materials vacuum drum driers are available (Van Arsdel *et al.*, 1973; Brennan *et al.*, 1976; Farrall, 1976; Williams-Gardner, 1976; Charm, 1979; Moore, 1987). There does not seem to have been any significant developments in drum drying in the last decade. In a fairly recently published bibliography (Hall and Upadhyaya, 1986), all but a few of the references were dated before 1970. Vasseur and Loncin (1984) studied the drying of thin films, 0·05–0·20 mm thick, on a temperature-controlled heated plate. They did not detect a constant rate period but rather an increasing flux to a peak. They developed a mathematical model which gave the temperature profile of the surface of the wall of a drum drier in one revolution. Kozempel *et al.* (1986) developed a mathematical model for the drum drying of potato flakes. The process parameters were drum speed, steam pressure, number of spreader rolls, wet and dry bulb temperatures, mass moisture content and drum dimensions. The model correlated well the drum drying of two varieties of potato and drying curves predicted by the model for another variety agreed well with experimental data.

Vacuum drying may be carried out on a batch principle in a vacuum cabinet (tray or shelf) drier or continuously in a vacuum band drier. A *vacuum cabinet drier* consists of a vacuum tight chamber containing heated shelves. The pressure in the chamber is maintained in the range 0·133–9310 N m^{-2}, typically *c.* 1000 N m^{-2}. The food is spread on trays which are placed between the shelves. Heat is transferred to the food by conduction through the bottom of each tray and by radiation from the underside of the shelf above. Drying can be carried out at temperatures in the range 40–90°C. High initial rates of drying are attainable but as drying proceeds thermal contact between the tray and the food may diminish and rates of heat transfer decline. In one type of *vacuum band drier* a continuous metal band moves within a vacuum chamber. The food in the form of a liquid or paste is applied to the belt in a thin layer. Heating is usually by a combination of conduction and radiation. Dried product is scraped off the band by a knife and removed from the vacuum chamber via a rotary valve. Kumazawa *et al.* (1980) developed a mathematical model for a vacuum band drier based on the 'uniformly retreating ice front' principle applied in freeze drying, see below. Using the model they were able to optimise the heating temperature to give good product quality for meat paste and cheese. Vacuum drying is an expensive method of removing moisture and is only used for very heat sensitive foods. Examples of its use include: fruit pieces, fruit juices, fish and meat pastes and extracts, egg yolk and white, coffee extract and malt beverages (Van Arsdel *et al.*, 1973; Brennan *et al.*, 1976; Williams-Gardner, 1976; Charm, 1979).

RADIATIVE DRYING

In hot air and direct contact driers some of the energy is supplied to the food by radiation from the hot internal surfaces of the drier, drier shelves and trays. However the use of radiant energy as the major source of heat in dehydration of foods is rare. *Infrared driers* are used for removing small amounts of water from granular or powder materials such as breadcrumbs, starches and spices. These foods are conveyed beneath infrared radiators on belts or vibrating decks. Radiant energy is also used in vacuum shelf and band driers, see above and in freeze driers, see below. Ginsberg (1969) is the most frequently quoted author in discussions of the use of infrared radiation in food processing. He outlines the complex relationships between the physical, thermal and optical properties of foods and their influence on absorption of infrared radiation. Dagerskog and Osterstrom (1979) reported on studies of the optical properties of potatoes, pork and bread in the near infrared part of the spectrum. They found that the penetration capacity of short wave radiation (less than 1·25 μm) was about ten times that of longer wave (greater than 1·25 μm) radiation. However, body reflection was greater at the shorter wavelengths and the resulting heat effect was almost the same for the two wavelength ranges at equal energy flux. Karuri (1984) carried out a study of the application of radiant energy to the dehydration of foods. He used black body and infrared radiators as well as solar and simulated solar radiation. He found that the transmissivity and absorptivity of potatoes and carrots decreased during drying in the wavelength range 1·10–2·50 μm. He developed empirical models to represent the falling rate period of drying under different conditions. Sandu (1986) published a process analysis of infrared radiative drying of foods. He concluded that the wavelength region of interest in respect of foodstuffs was 0·75–15·00 μm. Each of the major chemical components of food, proteins, fats and carbohydrates exhibits its own characteristic absorption pattern. Superimposed on these in foods is the absorption characteristics of water in the liquid, vapour or frozen state. Foodstuffs generally are expected to exhibit two spectral regions where transmissivity to infrared radiation is high, less than 2·20 μm and between 4·00 and 5·50 μm. Sandu discussed the potential advantages of radiative drying and in particular the technique of intermittent radiation. In this method the drying surface is exposed to radiation at different time intervals with relaxation periods in between. This results in lower product temperatures and savings in energy as compared to continuous radiation.

SOLAR DRYING

For centuries man has used sun drying as a means of food preservation. Fruit and vegetables have been spread on the ground on leaves or mats and strips of meat and fish have been hung on racks and allowed to dry slowly under the heat from the sun. The obvious disadvantages of this method of drying include; the susceptibility of the food to contamination by insects, birds and animals, the probability of it being exposed to rain, the lack of control over the drying conditions and the possibility of chemical, enzymic and microbiological spoilage because of the long drying times. Consequently the practice of covering the food with a transparent material was introduced. In addition to reducing the risks of contamination and spoilage this practice can also lead to the attainment of higher temperatures in such enclosures as compared with outside temperatures. Transparent materials such as glass and clear plastics will transmit most solar radiation incident on them. However, radiation produced by heated surfaces within the enclosure will, in most cases, be of longer wavelength and will not pass out through the transparent covering. This is known as the 'greenhouse effect'. Glass is usually a better heat trap than plastics. The simplest type of solar drier is a plastic tent erected over the food material which is spread on a perforated tray raised some few centimetres off the ground. The panels facing the sun are transparent while the others are made of black pigmented plastic. A simple solar box as shown in Fig. 8 is widely used. Its limitations are that it has a relatively small capacity, the air flowrate is low and hence so is the drying rate. To increase capacity food may be placed on perforated shelves located one above the other in the drying chamber (Fig. 9). Inlet air is heated by a solar collector and percolates up through the layers of food on the shelves. Additional heat may enter by making the panels facing the sun of transparent material. To induce higher air flowrate a chimney may be provided. The taller the chimney

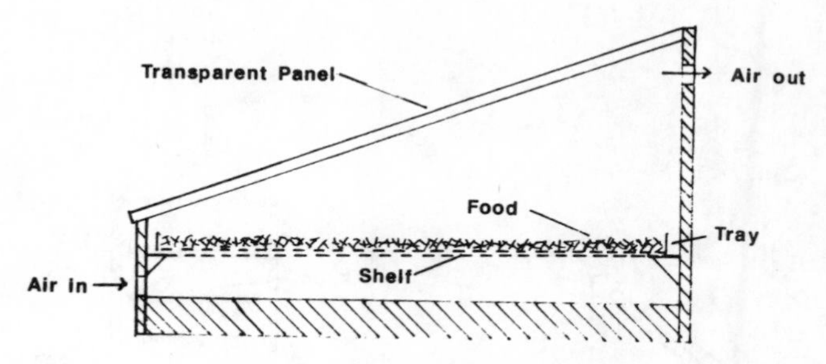

FIG. 8. A simple solar 'box' drier.

the higher the flowrate. A fan may be included in the system to increase the air flowrate further. This design has the disadvantage that a power supply is required which limits its use in rural areas. More sophisticated designs of flat plate solar collectors are now available. Imrie (1987) described such collectors and their efficiency. Inflatable solar collectors are also used as preheaters for driers. Bolin and Salunkhe (1982) describe such collectors. One consisted of a black polythene collector tube inside a larger diameter clear tube which acted as an insulator. A vane blower kept the tubes inflated and provided the necessary flow of air. Another consisted of a long polythene dome with a transparent top and black base. This collector was situated between rows of grape vines and was connected to a cabinet drier. A fan inflated the tube and directed the air through the

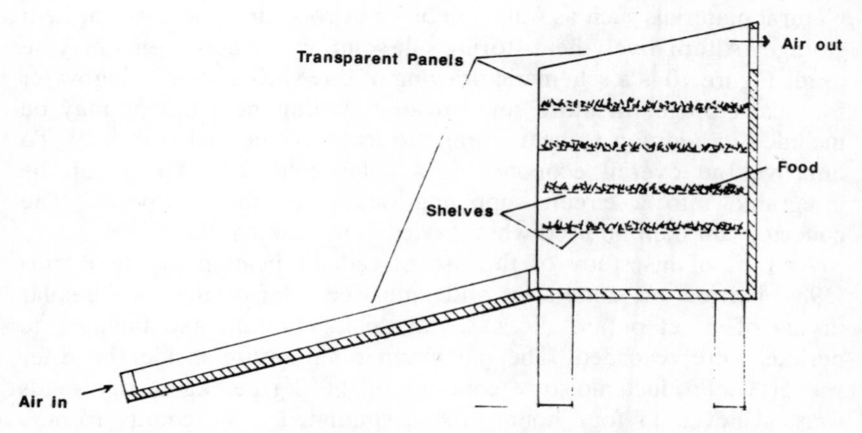

FIG. 9. A solar drier featuring a solar collector for preheating the air.

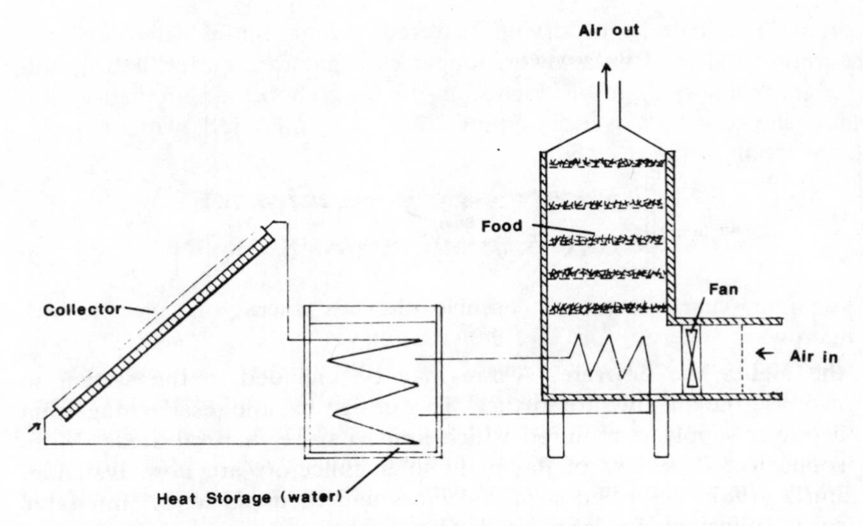

FIG. 10. A solar drier featuring heat storage in water.

cabinet containing the grapes. In order to achieve higher drying rates and/or to continue drying in the absence of sunlight an alternative source of heat such as a gas or oil burner may be incorporated into a solar drier. Some form of heat storage may also be included. This could extend the drying period, make use of surplus energy collected during the day and facilitate control over the drying temperature. Natural materials such as water, pebbles or rocks may be used for heat storage, Alternatively heat-storing salt solutions or adsorbents may be used. Figure 10 is a schematic drawing of one such system using water for heat storage. In more sophisticated systems heat pumps may be included in addition to heat storage to improve thermal efficiency. To improve the overall economy of a solar collector/drier it can be integrated into a circuit supplying heat for other purposes. The collector can then be used when drying is not taking place.

As part of his study of the use of radiant heat in drying Karuri (1984) carried out some solar and simulated solar drying. In the solar drying of sweet potato slices, 2-mm thick, constant and falling rate periods were recorded. The maximum temperature inside the drier was 50°C. Product moisture contents of 0·05 kg per kg of dry solids were achieved in four hours. Using simulated solar energy from a Compact Source Iodide lamp small quantities of sliced potato were dried. The pattern of drying featured a high initial rate of short duration followed by a lower, longer constant rate period and a still longer falling rate period. Inclusion of a fan to boost air circulation did not shorten the overall drying time as compared with natural convection.

DIELECTRIC AND MICROWAVE DRYING

For many years there has been an interest in the use of dielectric and microwave energy for the dehydration of foods. It is generally accepted that dielectric heating is done in the frequency range 1–100 MHz and microwave between 300 MHz and 300 GHz. Frequency allocations for industrial, scientific and medical uses of dielectric and microwave are agreed internationally (Schiffmann, 1987). The mechanism of drying with dielectric and microwave energy differs from that which occurs when drying in heated air or in contact with a hot surface. With rapid internal heat generation moisture is vaporised quickly within the material. This creates a total pressure gradient which promotes rapid movement of moisture vapour and perhaps liquid water to the surface. This 'pumping action' leads to relatively short drying times and lower product temperatures as compared with conventional methods of drying. Other advantages may include uniform heating, reduction in movement of solutes within the food and energy savings. The general principles of dielectric and microwave heating of foods have been discussed by Brennan *et al.* (1976) and Schiffmann (1987). To date the application of these methods of heating to drying of foods has been limited to removing relatively small amounts of water. Dielectric heating has been used to remove the final traces of moisture from starch-reduced bread rolls, biscuits and cereals. Microwave heating has been used to dry pasta products, potato crisps and onions. Microwave energy may also be used to dry heat sensitive foods under vacuum such as in a vacuum band drier, see above. Fruit juice concentrates, tea extract and malt based drinks have been dried in this way. Final drying of cereals such as malt flakes can also be achieved under vacuum using microwaves (Anon., 1985). Pilot scale drying of grain by this method has also been reported (Forwalter, 1978). Microwave heating has also been considered for use in freeze drying, see below. Potentially it suits this application well. Loss factors for ice and liquid water are considerably higher than for dry tissue and so in a frozen food, ice should absorb energy preferentially. However difficulties arise in controlling the heat input as the loss factor increases with increase in temperature and so as the ice core warms up it absorbs more energy. Due to the heterogeneous nature of foods this may lead to local melting of ice. The water thus formed will absorb energy very quickly. This can lead to sudden vaporisation of the water and cause explosive damage to the food. At the low pressures used in freeze drying ionisation of the rarefied gases can occur which can cause plasma discharge and may overheat the food. This problem can be reduced by using high frequency (2450 MHz) microwaves. The technology is available to carry out microwave freeze drying on an industrial scale provided it can be done economically. Sunderland (1982) published a study of the economics of this process. He concludes that it compares favourably with conventional freeze drying. That author refers to one industrial application for microwave freeze drying for coffee extract.

FREEZE DRYING

Freeze drying has been in commercial use for about thirty years. There are three stages in the freeze drying process, (i) the food is frozen and most of the water converted to ice, (ii) the bulk of the ice is sublimed and the vapour removed from the vicinity of the drying surface, (iii) final traces of bound moisture are removed in the freeze drying chamber by vacuum drying or the final drying is carried out in another type of drier. The advantages of freeze drying over other methods of drying lie in the quality of the product. Little or no shrinkage occurs. Movement of soluble solids within the food is minimised. The porous structure of the dried product facilitates rapid rehydration. Retention of volatile flavour compounds is high. However, not all foods can be successfully freeze dried. In some structural damage occurs during the freezing stage and this results in poor texture in the dried product. If too high a temperature is reached during drying the structure of the food may collapse. The main limitation of freeze drying is that it is a very expensive process. Hence its commercial application has been confined to expensive food products including coffee extract, meat and fish for inclusion in instant meals and soups and exotic fruits. Protein hydrolysates, enzymes and microoorganisms have also been freeze dried.

The principles of freeze drying, the procedures and equipment used have been well described in the literature (King, 1971; Van Arsdel *et al.*, 1973; Karel, 1974; Goldblith *et al.*, 1975; Brennan *et al.*, 1976; Mellor, 1978; Charm, 1979; Liapis, 1987). The food is frozen by the method most suited to that product. Blast freezing is frequently used. Plate freezing, immersion freezing and liquid gas freezing have also been used. There is some limited evidence that low rates of freezing lead to shorter drying times because of the open structure which is formed when the large ice crystals sublime. In order to promote sublimation of the ice the partial water vapour pressure in the atmosphere surrounding the frozen food must be reduced to less than the vapour pressure of ice at the same temperature. The vapour pressure of ice at −20°C is about 1 torr (135 N m^{-2}). Hence very low partial water vapour pressures must be created and maintained. There are many reports in the literature of laboratory trials in which refrigerated, dehumidified air was used to freeze dry food pieces. The air was dried by passing through a molecular sieve, a desiccant or over refrigerated coils. Mellor (1978) discussed some of these reports. Drying rates comparable with vacuum freeze drying could be obtained only with small particles. Boeh-Ocansey (1984, 1986) described work in which atmospheric freeze drying was carried out in a fluidised bed. In his early work the quality of the foods dried in this way was not as good as those dried by vacuum freeze drying. However in the later paper the quality of the two groups of foods was reported to be comparable. No information on the application of atmospheric freeze drying on a commercial scale could be found by the author. Commercial freeze drying is normally carried out under vacuum. Cabinets or tunnels are evacuated down to pressures in the range 0·1–2·0 torr (13·5–270·0 N m^{-2}). At these very low pressures the partial water vapour pressure in the atmosphere inside the chamber will be less than that of ice at the same temperature and sublimation will occur. Most commercial freeze drying is carried out on a batch principle. High performance *vacuum cabinets* are fitted with heated shelves on which trays of frozen food are placed. The cabinets are closed and the pressure reduced quickly to the desired level. The most common system of creating the low pressure required is to have a refrigerated plate or coil located in the drying cabinet (−10 to −40°C) or in a chamber directly connected to the cabinet. The water vapour produced by the subliming food condenses on the refrigerated surface as ice. A mechanical pumping system removes the non-condensible gases (Fig. 11). These pumps are usually oil sealed and two of them

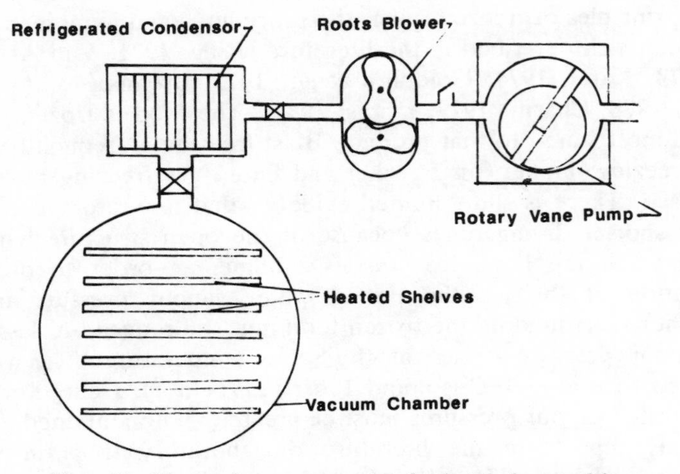

FIG. 11. The main components of a batch freeze drier.

are used in series, the second being gas ballasted to reduce the danger of water vapour condensing and making the oil seals ineffective. Roots blowers may be used in the first stage of the pumping system. Multistage steam ejectors have also been used to evacuate freeze drying chambers. Heating usually occurs by conduction from the hot plates through the trays and by radiation from the underside of the trays above. Shelves and trays are carefully designed to promote rapid heat transfer. Other forms of heating, including microwaves, have been studied, see above. A typical temperature–time pattern for a cycle in a batch freeze drier is shown in Fig. 12.

Unlike hot air or conduction drying there should be no significant constant rate period during a freeze drying cycle. If it is assumed when drying a slab-shaped solid from one large face only, that the ice front retreats uniformly into the body of the slab, that the dry layer is at its equilibrium moisture content and that heat is supplied only through the dry layer then the rate of loss of weight of the slab, $\mathrm{d}w/\mathrm{d}t$, at any time t can be represented by the expression:

$$\frac{\mathrm{d}w}{\mathrm{d}t} = \frac{Ak_{\mathrm{D}}(\theta_{\mathrm{D}} - \theta_{\mathrm{i}})}{L_{\mathrm{S}}l} = \frac{Ag(p_{\mathrm{i}} - p_{\mathrm{D}})}{l}$$

$$= A\rho_{\mathrm{m}}(W_0 - W_{\mathrm{e}})\frac{\mathrm{d}l}{\mathrm{d}t} \qquad (18)$$

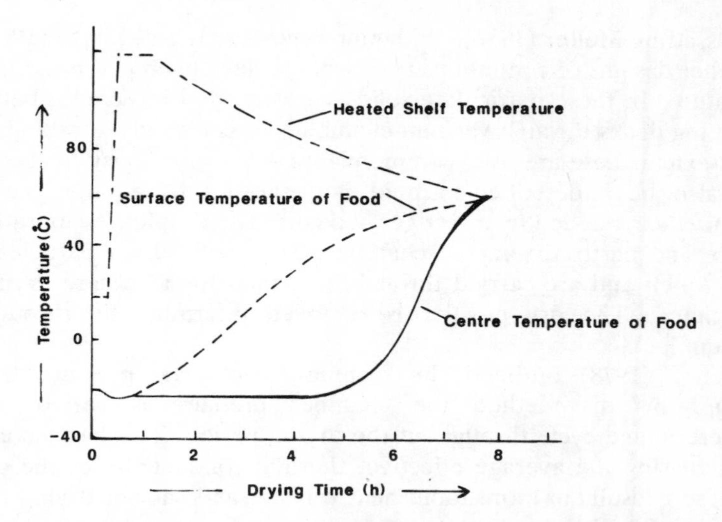

FIG. 12. A typical time–temperature pattern for a batch freeze drying cycle.

where A = the drying surface area; k_D = the thermal conductivity of the dry layer; L_S = the latent heat of sublimation; l = the thickness of the dry layer; θ_D and θ_i = the temperature at the dry surface and ice front respectively; g = the permeability of the dry layer; p_D and p_i = the partial water vapour pressure at the dry surface and ice front respectively; ρ_m = the density of the dry layer; and W_0 and W_e = the initial and equilibrium moisture contents of the material respectively, dry weight basis.

Note that θ_i and p_i are related thermodynamically and if the dry surface temperature and chamber pressure are fixed so is θ_i. If the heat is supplied through the ice layer the relationship becomes more complex as p_i is then a function of l. However such relationships can be solved with the aid of computers.

In a *tunnel freeze drier* the food is placed on specially designed trays which are assembled on trolleys. The trolleys enter the tunnel through a vacuum lock. In the main tunnel body the trays move between stationary heated plates and exit through another lock at the dry end. The drying conditions are carefully controlled along the length of the tunnel to optimise product quality and to make efficient use of the vacuum equipment and the heating system. Methods of controlling conditions during batch and continuous freeze drying have been discussed by Mellor (1978), Willemer *et al.* (1983) and Liapis (1987).

Other designs of continuous freeze driers have been described in the literature. In these particulate solids are conveyed by screws, belts or vibrating decks through vacuum chambers. It is difficult to establish to what extent these are used commercially. A *vacuum spray freeze drier* has also been described. Liquid is sprayed into a large vacuum chamber surrounded by a refrigerated coil. The droplets evaporatively freeze and partly dry in the chamber. The partly dried particles fall onto a belt and are carried through a vacuum tunnel where drying is completed. The dry powder is removed intermittently through a vacuum lock.

Mellor (1978) outlined the technique of cyclic pressure freeze drying. In this method the chamber pressure is varied on a predetermined cycle throughout the drying period. This has the effect of increasing the average effective, thermal conductivity of the dried layer and results in more rapid heat transfer and shorter drying times as compared with steady-state pressure operation.

OSMOTIC DRYING

Osmotic drying of fruits and vegetables has been the subject of investigation for over thirty years (Ponting *et al.*, 1966), and many papers on this topic have been published in recent times. Examples of these are discussed below. This method of drying, or more accurately concentrating, involves immersing pieces of fresh food in a solution with a higher osmotic pressure and hence lower water activity than the food. Sugar solutions, with or without small quantities of added salt, are usually used for fruit while salt solutions are used for vegetables. Under the influence of osmotic pressure water will pass from the food pieces into the solution. The natural cell walls of the food act as semipermeable membranes. However, these membranes are not completely selective and some solute will also pass into the food from the solution and vice versa. Osmotic dehydration is a simultaneous water and solute diffusion process. Up to a 50% reduction in the fresh weight of the food can be achieved by osmosis. The potential advantages of this method of drying as compared with hot air or direct contact techniques are: less heat damage and enzymic browning, better retention of flavour compounds and energy savings. On the other hand products cannot be dried to completion by this method and some means of stabilising them is required to extend their shelf lives.

In osmotic drying usually the rate of diffusion is rapid in the first one or two hours and it slows up markedly thereafter. The main variables of the process are solution concentration and temperature. An increase in either of these factors will increase the initial rate of diffusion and the maximum water loss from the food. These effects are illustrated by unpublished data on apples (Lu and Brennan, 1987) (Fig. 13). In this work the minimum water activity attained by the apples was about 0·875. Conway *et al.* (1983) studied the osmotic drying of apples using sugar solutions in a circulating system. They found that the solids gain by the fruit reached a maximum after 30 min drying time and remained constant for the remainder of the cycle, up to three hours. They proposed a diffusional model for the process and data obtained experimentally and theoretically correlated well. An apparent diffusion coefficient was evaluated and was found to depend on temperature and sugar concentration. Lerici *et al.* (1985), again using apple, studied the effects of using different sugar solutions, with and without added salt, on osmotic drying. Their conclusions were that the final water activity of the apple depended not only on the water activity of the osmotic solution but also on the gain of solids. In different solutions the products showed different relationships between water loss, solids gain and weight reduction. The addition of salt increased the driving force for the drying process and as the percentage of salt added increased, in the range 0·5–2·0, the final water activity of the product was reduced.

Lenart and Flink (1984) studied osmotic concentration of potato slabs in sucrose solution, in salt solution and in solutions containing both sucrose and salt. The variables investigated were solute concentration, time and temperature, with and without agitation. They followed the process mainly by measuring the depth of penetration of the osmosis effect (the depth to which solids concentration was higher than that in the original tissue) and the surface moisture content relative to its initial value. They found that all the variables studied influenced the spatial distribution of solids and moisture in the potato flesh. With a few exceptions increasing the solute concentration in the solution and the time of immersion increased the penetration depth and decreased the relative moisture content for all locations in the sample. An increase in temperature also gave similar results but it was found that in the case of sucrose solutions this effect was related

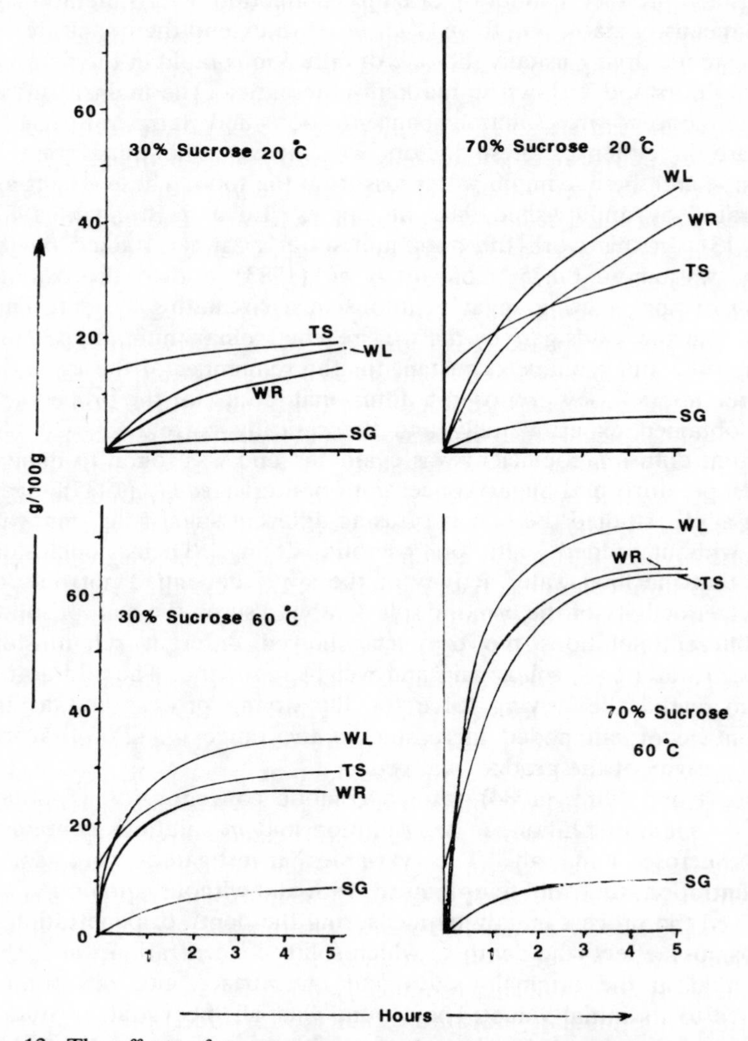

FIG. 13. The effects of sucrose concentration and temperature on the changes with time of total solids (TS), water loss (WL), weight reduction (WR) and solids gain (SG) of apples during osmotic drying. (From Lu and Brennan (1987).)

mainly to the rate of water loss while in the case of salt solutions it related both to water loss and salt penetration. They discussed the different behaviours of sucrose and salt and proposed models to account for these. Using these models they explained why combinations of sucrose and salt are so effective in osmotic concentration. Lewicki *et al.* (1984*a*) dehydrated potato halves by dipping them in 15% salt solution for different times at different temperatures. Maximum penetration of the salt into the potato was 18 mm. The most effective drying conditions were at 30°C for one hour. A value for the diffusivity coefficient of salt in potato was calculated.

Some pretreatments of the fruit or vegetable to speed up osmosis have also been proposed. Riva and Peri (1983) dipped grapes in an ethyloleate alkaline solution. This caused solubilisation of the waxy coating of the grapes and modified the permeability of the skin without damaging its mechanical structure. This treatment reduced drying times five to ten fold. Lewicki *et al.* (1984*b*) coated apple cubes, previously treated with sulphite, with 2·5% pectin solution and 1% starch solution. The precoated cubes were air dried for 5 min prior to osmotic drying. The starch coating promoted greater weight loss and lower moisture content in the final product than did the pectin.

Some osmotically dried products can be stabilised by a mild heat treatment. Maltini *et al.* (1983) described a process which involved blanching, osmotic drying, vacuum packaging in pouches or sealing into glass jars topped up with isotonic syrup and pasteurisation of the packaged project. Excellent results were obtained with peaches and apricots which had shelf lives of 12 and 6 months respectively at room temperature. Results with other fruits, pears, apples and cherries were not so good. Mastrocola *et al.* (1987) osmosed carrot cubes in hydrolysed corn starch syrups at 25°C. Then the cubes in syrup were heated to between 70 and 90°C for 1–20 min. The rate of osmosis increased markedly at the high temperature and the heat treatment inactivated the enzymes in the product.

Other ways of stabilising osmotically dried foods are: to formulate them into intermediate moisture products with added preservative; to complete the drying in hot air, under vacuum or by freeze drying; to freeze them or to can and heat process them.

While there are numerous reports in the literature of experimental work on osmotic drying there is no real indication of the extent to which it has been applied commercially, apart from candied fruits, nor could any information on the economics of the process be found.

ACOUSTIC DRYING

There are some reports in the literature on the use of sound waves to accelerate drying. Muralidhara and Ensminger (1986) applied acoustic energy at frequencies of 12–19 kHz and intensities of between 128 and 132 dB to the drying of green rice in a fluidised bed at 20 and 40°C. The acoustic energy was more effective at the lower temperature. Shaw *et al.* (1986) applied for a patent for a method and apparatus for removing volatiles from or dehydrating liquid products. A high pulse combustor would be used to generate high temperature and acoustic resonance in a drying chamber. High fructose syrup, or other liquid foods, would be sprayed into the chamber and the droplets would dry to a powder. This could lead to less heat damage as compared with spray drying.

Muralidhara *et al.* (1985) published a state of the art review of acoustic dewatering and drying. Among the food materials that are listed as having been dried with the aid of acoustic waves are yeast cake, granulated sugar, wheat and rice. While this is an interesting area of research it is unlikely that any industrial scale applications for this technique will be found in the food industry in the foreseeable future.

EXPLOSION PUFFING

In this process diced vegetable pieces are partly air dried to between 15 and 30% moisture content. The rate of drying is usually still reasonably high at that point in the cycle. These partly dried pieces are put into a puffing gun. The gun is sealed and pressurised by heating internally with superheated steam or externally with a gas flame. As the pressure increases the water in the food heats up to above 100°C. The chamber pressure is suddenly released, the superheated water converts rapidly to steam and flashes off creating a porous structure in the food pieces. The pieces are returned to the air drier where drying is completed. The advantages of this technique include shorter drying times which may be reduced by a factor of two or three as compared with conventional air drying, good reconstitution characteristics because of the porous structure and good organoleptic properties.

Work done on puffing in the 1960s and early 1970s was carried out in batch guns. A continuous explosion puffing system was designed and manufactured in the mid-1970s (Heiland *et al.*, 1977). A paper outlining the development of explosion puffing was published by Sullivan and Craig (1984). Products which have been successfully puff dried include potatoes, onions, beet, turnips and pears.

CONCLUSION

Dehydration continues to be an important method of food preservation and attracts considerable interest from researchers world wide. Much work has been reported on fluidised-bed, including spouted bed, drying and its potential for drying liquid foods seems promising. The introduction of multistage spray drying, in particular the integral fluid-bed design, has led to savings in energy and better control over product characteristics. Direct contact drying and freeze drying continue to fulfill certain important roles in food dehydration but no significant developments were identified. Solar drying methods have become very sophisticated and should find increasing use in countries with adequate amounts of sunshine. There is considerable interest by researchers in osmotic drying. However its commercial exploitation seems to be very limited and is likely to remain so for some years. The industrial scale use of dielectric and microwave heating for food dehydration, particularly under vacuum, has become more feasible in the last decade or so and an increase in its use is anticipated. A great deal of investigative work has been directed towards energy conservation in drying. Drying is an energy intensive operation. A loss in excess of 30% of the energy supplied to an hot air drier is not unusual. While a detailed discussion of this topic is not included in this review it is worth noting that remedies which have been under study in recent years include: the use of direct methods of heating the air supply to driers, in particular using natural gas; the recovery of heat in the exhaust air from driers by the use of wet scrubbers or dry heat exchangers; better control over drying conditions so as to improve the thermal efficiency of driers and the use of combinations of drying methods to reduce energy requirements while maintaining product quality.

REFERENCES

Alvarez, P. I. and Legues, P. (1986). A semi-theoretical model for the drying of Thompson seedless grapes. *Drying Technol.*, **4,** 1–17.

Andrien, J. and Stamatopoulos, A. (1986). Durum wheat pasta drying kinetics. *Lebensm.-Wiss. u-Technol.*, **19,** 448–56.

Anon. (1985). Advances in microwave technology. *Food Engn. Int.*, **10,** 78, 81.

Ashworth, J. C. (1982). *Proceedings of the Third International Drying Symposium, Vols 1 and 2,* Drying Research Limited, Wolverhampton, England.

Baxeires, J. L. Yow, Y. S. and Gilbert, H. (1983). Study of the fluidised bed drying of various food products. *Lebensm.-Wiss. u-Technol.*, **16,** 27–31.

Biquet, B. and Guilbert, S. (1986). Relative diffusivities of water in model intermediate moisture food. *Lebensm.-Wiss. u-Technol.*, **19,** 208–14.

Boeh-Ocansey, O. (1984). Effects of vacuum and atmospheric freeze drying on the quality of shrimp, turkey flesh and carrot samples. *J. Fd. Sci.*, **49,** 1457–61.

Boeh-Ocansey, O. (1986). Low temperature fluidised-bed drying of mushrooms, beef and shrimp samples. *Acta Alimentara,* **15,** 79–82.

Bolin, H. R. and Salunkhe, D. K. (1982). Food dehydration by solar energy. *CRC Critical Reviews in Food Science and Technology,* **16,** 327–54.

Brennan, J. G. (1978). Developments in food dehydration. *Process Chem. Engn.*, **31,** 11–12, 14, 16–17, 20, 24.

Brennan, J. G., Butters, J. R., Cowell, N. D. and Lilly, A. E. V. (1976). *Food Engineering Operations,* 2nd Edn, Applied Science Publishers, London, pp. 313–59.

Charm, S. E. (1979). *Fundamentals of Food Engineering,* 3rd Edn, Avi Publishing Company Inc., Westport, Connecticut, pp. 298–433.

Chirife, J. (1971). Diffusional process in the drying of tapioca root. *J. Fd Sci.*, **36** 327–30.

Conway, J., Castaigne, F., Picard, G. and Vovan, X. (1983). Mass transfer considerations in osmotic dehydration of apples. *Can. Inst. Fd Sci. Technol. J.*, **16,** 25–9.

Dagerskog, M. and Osterstrom, L. (1979). Infra-red radiation for food processing 1. A study of the fundamental properties of infra-red radiation. *Lebensm.-Wiss. u-Technol.*, **12,** 237–42.

Dodson, C. (1987). Application of the Torbed technology in the snacks industry, including the flash expansion of pellets without using fats or oils. Torftech Ltd, Reading, UK. Private communication.

Escardino, A., Monton, J. and Front, R. (1979). Desecacion de granos de leguiminosas II Estudio cinetico del secado de guisantes. *Rev. Agroquim Technol. Aliment.*, **19,** 61–74.

Escardino, A., Monton, J. and Front, R. (1980). Desecacion de granos de leguiminosas III Estudio cinetico del secado de granos de habas. *Rev. Agroquim Technol. Aliment.*, **20,** 112–24.

Fane, A. G., Stevenson, T. R., Lloyd, C. J. and Dunn, M. (1980). The spouted bed drier—an alternative to spray drying. Paper presented at the eighth Australian Chemical Engineering Conference, Melbourne, Australia.

Farrall, A. W. (1976). *Food Engineering Systems Vol. I,* Avi Publishing Company Inc., Westport, Connecticut, pp. 181–215.

Filkova, P. J. A. and Mujumdar, A. S. (1987). Industrial spray drying systems. In: *Handbook of Industrial Drying,* A. S. Mujumdar (Ed.), Marcel Dekker Inc., New York, pp. 243–93.

Forwalter, J. (1978). Microwave/vacuum dryer cuts drying time by 1/2. *Food Processing,* **39,** 176–7.

Ginsburg, A. S. (1969). *Applications of Infra-red Radiation In Food Processing,* Leonard Hill Books, London.

Goldblith, S. A., Rey, L. and Rothmayor, W. W. (1975). *Freeze Drying and Advanced Food Technology,* Academic Press, London.

Green, D. W. (Ed.) (1984). *Perry's Chemical Engineers' Handbook,* 6th Edn, McGraw-Hill, London.

Grikitis, K. (1988). Crispy fillings. *Food Processing,* Sept., 12–13.

Hall, C. W. and Upadhyaya, R. L. (1986). Roller drum dryers—a bibliography. *Drying Technol.,* **4,** 477–89.

Heiland, W. K., Sullivan, J. F., Konstance, R. P., Craig, J. C. Jr, Conding, J. Jr, and Aceto, N. C. (1977). A continuous explosion puffing system. *Food Technol.,* **31,** 32.

Hovmand, S. (1987). Fluidized-bed drying. In: *Handbook of Industrial Drying,* A. S. Mujumdar (Ed.), Marcel Dekker Inc., New York, pp. 165–225.

Imrie, L. L. (1987). Solar drying. In: *Handbook of Industrial Drying,* A. S. Mujumdar (Ed.), Marcel Dekker Inc., New York, pp. 357–417.

Karel, M. (1974). Fundamentals of dehydration processes. In: *Advances in Preconcentration and Dehydration of Foods,* A. Spicer (Ed.), Applied Science Publishers, London, pp. 45–90.

Karuri, E. G. (1984). The application of radiant energy to the dehydration of foods. PhD Thesis, University of Reading, UK.

Kelly, J. J. (1987). Rotary drying. In: *Handbook of Industrial Drying,* A. S. Mujumdar (Ed.), Marcel Dekker Inc., New York, pp. 133–54.

Kerkhof, P. J. A. and Schoeber, W. J. A. (1974). Theoretical modelling of the drying behaviour of droplets in spray dryers. In: *Advances in the Preconcentration and Dehydration of Foods,* A. Spicer (Ed.), Applied Science Publishers Ltd, London, pp. 491–7.

King, C. J. (1971). *Freeze-drying of Foods,* CRC Monoscience Series, Butterworths, London.

Kisakurek, B. (1987). Flash drying. In: *Handbook of Industrial Drying,* A. S. Mujumdar (Ed.), Marcel Dekker Inc., New York, pp. 475–99.

Kjaergaard, O. G. (1974). Effects of latest developments on design and practice of spray drying. In: *Advances in the Preconcentration and Dehydration of Foods,* A. Spicer (Ed.), Applied Science Publishers Ltd, London, pp. 321–48.

Kozempel, M. F., Sullivan, J. F., Craig, J. C. Jr and Heiland, W. W. (1986). Drum drying of potato flakes—a predictive model. *Lebensm.-Wiss. u-Technol.,* **19,** 193–7.

Kumazawa, E., Saiko, Y., Ishioka, Y., Taneya, S. and Hayashi, H. (1980). Continuous vacuum band drying for high viscous and heat sensitive foods. In: *Drying '80, Vol. 2,* A. S. Mujumdar (Ed.), Hemisphere Publishing Corporation, London, pp. 244–50.

Kutsakoua, V. E., Utkin, Yu. V. and Kupanov, V. Yu. (1984). Drying of protein hydrolysate solutions in a fluidised-bed of metal balls. *Izvestiya Vysshikh Uchebnykh Zavedenii Pishchevaya Tekhnologiya No. 1,* 94–6.

Lenart, A. and Flink, J. M. (1984). Osmotic concentration of potato. II Spatial distribution of the osmotic effect. *J. Fd Technol.*, **19,** 65–9.

Lerici, C. R., Pinnavaia, G., Dalla Rosa, M. and Bartolucci, L. (1985). Osmotic dehydration of fruit: Influence of osmotic agents on drying behaviour and product quality. *J. Fd Sci.*, **50,** 1217–19.

Lewicki, P. P., Lenart, A. and Pakula, W. (1984*a*) Influence of artificial semi-permeable membranes on the process of osmotic drying of apples. *Annals of Warsaw Agricultural University SGGW-AR Food Technology and Nutrition,* **No. 1,** 17–24.

Lewicki, P. P., Lenart, A. and Turska, D. (1984*b*) Diffusive mass transfer in potato tissue during osmotic drying. *Annals of Warsaw Agricultural University SGGW-AR Food Technology and Nutrition,* **No. 1,** 25–32.

Lewis, W. K. (1921). The rate of drying of solid materials. *J. Ind. and Engn. Chem.*, **13,** 427–32.

Liapis. L. I. (1987). Freeze drying. In: *Handbook of Industrial Drying,* A. S. Mujumdar (Ed.), Marcel Dekker Inc., New York, pp. 295–326.

Lu, Q. and Brennan, J. G. (1987). Department of Food Science and Technology, University of Reading, UK, Unpublished data on osmotic drying, paper in preparation.

Maltini, E., Torreggiani, D., Bertolo, G. and Stecchini, M. (1983). Recent developments in the production of shelf stable fruit by osmosis. In: *Proceedings of the Sixth International Congress in Food Science and Technology, Dublin, Vol. 1,* J. V. McLoughlin and B. M. McKenna (Eds), Boole Press, Dublin, pp. 177–8.

Masters, K. (1985). *Spray Drying Handbook,* 4th Edn, George Godwin, London.

Masters, K. and Pisecky, J. (1983). New spray dryer concept for the food industry. In: *Proceedings of the Sixth International Congress in Food Science and Technology, Dublin, Vol. 5,* J. V. McLoughlin and B. M. McKenna (Eds), Boole Press, Dublin, pp. 379–86.

Mastrocola, D., Severini, C., Lerici, C. R. and Sensidoni, A. (1987). Osmotic drying of carrots. *Industrie Alimentari,* **26,** 133–8.

Mellor, J. D. (1978). *Fundamentals of Freeze-Drying,* Academic Press, London.

Moore, J. D. (1987). Drum dryers. In: *Handbook of Industrial Drying,* A. S. Mujumdar (Ed.), Marcel Dekker Inc., New York, pp. 227–42.

Mujumdar, A. S. (Ed.) (1980). *Drying '80, Vol. 1 and 2,* Hemisphere Publishing Corporation, New York.

Mujumdar, A. S. (Ed.) (1982). *Drying '82,* Hemisphere Publishing Corporation, New York.

Mujumdar, A. S. (Ed.) (1984). *Drying '84,* Hemisphere Publishing Corporation, New York.

Mujumdar, A. S. (1985). *Selection of papers from the Fourth International Drying Symposium, Kyoto,* Hemisphere Publishing Corporation, New York.

Muralidhara, H. S. and Ensminger, D. (1986). Acoustic drying of green rice. *Drying Technol.,* **4,** 137–43.

Muralidhara, H. S., Ensminger, D. and Putnam, A. (1985). Acoustic dewatering and drying (low and high frequency): state of the art review. *Drying Technol.,* **3,** 529–66.

Pallai, E., Nemeth, J. and Mujumdar, A. S. (1987). Spouted bed drying. In: *Handbook of Industrial Drying,* Marcel Dekker Inc., New York, pp. 419–60.

Ponting, J. D., Watters, G. G., Forret, R. R. Jackson, R. and Stanley, W. L. (1966). Osmotic dehydration of fruits. *Fd Technol.,* **29,** 125.

Rheinlander, P. M. (1982). Filtermat—the 3-stage dryer from DEC. In: *Proceedings of the Third International Drying Symposium, Birmingham, Vol. 1,* J. C. Ashworth (Ed.), Drying Research Limited, Wolverhampton, pp. 528–34.

Riva, M. and Peri, C. (1983). Osmotic dehydration of grapes, In: *Proceedings of the Sixth International Congress of Food Science and Technology, Dublin, Vol. 1,* J. V. McLoughlin and B. M. McKenna (Eds), Boole Press, Dublin, pp. 179–80.

Sandu, C. (1986). Infrared radiative drying in food engineering: a process analysis. *Biotechnology Progress,* **2,** 109–29.

Saravacos, G. D. (1967). Effect of drying method on the water sorption of dehydrated apple and potato, *J. Fd Sci.,* **32,** 81–4.

Schiffmann, R. F. (1987). Microwave and dielectric drying. In: *Handbook of Industrial Drying,* A. S. Mujumdar (Ed.), Marcel Dekker Inc., New York, pp. 327–56.

Sharaf-Elden, I., Hamdy, M. T. and Blaisdell, J. L. (1979). Falling rate drying of fully exposed biological materials: a review of mathematical models. American Society of Agricultural Engineers, USA, Paper No. 79–6522.

Shaw, A. J., Marks, J. S., Gahagam, H. E. and Bowles, A. J. G. (1986). Method and apparatus for removing volatiles from or dehydrating liquid products. PCT International Patent Application, WO 86/06746A1.

Sturgeon, L. F. (1987). Conveyor dryers. In: *Handbook of Industrial Drying,* A. S. Mujumdar (Ed.), Marcel Dekker Inc., New York, pp. 501–13.

Sullivan, J. F. and Craig, J. C. Jr. (1984). The development of explosion puffing. *Fd Technol.,* **38,** 52–5, 131.

Sunderland, J. E. (1982). An economic study of microwave freeze-drying. *Fd Technol.,* **36,** 50–6.

Thompson, T. L., Peart, R. M. and Foster, G. M. (1968). Mathematical simulation of corn drying. A new model. *Trans. ASAE,* **11,** 582–6.

Vaccarezza, L. M. and Chirife, J. (1978). On the application of Fick's for the kinetic analysis of air drying of foods. *J. Fd Sci.,* **43,** 238–40.

Van Arsdel, W. B., Copley, M. J. and Morgan, A. I. Jr (1973). *Food Dehydration, Vols. 1 and 2,* 2nd Edn, Avi Publishing Company Inc., Westport, Connecticut.

Vasseur, J. and Loncin, M. (1984). High heat transfer coefficient in thin film drying: application to drum drying. In: *Engineering and Food Vol. 1,* B. M McKenna (Ed.), Elsevier Applied Science Publishers, London, pp. 217–25.

Willemer, H., Lentges, G. and Honrath, M. (1983). Developments in freeze-drying. *Proceedings Institute of Refrigeration*, **79**, 11–15.
Williams-Gardner, A. (1976). *Industrial Drying*, George Godwin Ltd, London.

Chapter 3

INTERPRETING THE BEHAVIOR OF LOW-MOISTURE FOODS

HARRY LEVINE* and LOUISE SLADE*

General Foods Corp., Central Research, Tarrytown, New York, USA

SYMBOLS AND ABBREVIATIONS

Ap	Amylopectin
C	Concentration
$C_{g'}$	Concentration of solute in the glass formed at $T_{g'}$
C_1, C_2	Universal constants in the WLF equation
DE	Dextrose equivalent
DP	Degree of polymerization
$\overline{DP}_n$	Number-average DP
$\overline{DF}_w$	Weight-average DP
DSC	Differential scanning calorimetry (or calorimeter)
IMF	Intermediate moisture food
$\bar{M}_n$	Number-average MW
$\bar{M}_w$	Weight-average MW
MW	Molecular weight
MWD	Molecular weight distribution, expressed as $\bar{M}_w/\bar{M}_n$
n-mer	Oligomer of $DP = n$
PEG	Poly(ethylene glycol)
PVP	Poly(vinyl pyrrolidone)
Q_{10}	Rate law of Arrhenius kinetics
r	Linear correlation (regression) coefficient
RH	Relative humidity
SHP	Starch hydrolysis product
T	Temperature
T_{am}	Antemelting temperature
T_c	Collapse temperature

* Present Address: Nabisco Brands, Inc., Corporate Technology Group, PO 1943, East Hanover, New Jersey 07936-1943, USA.

T_d	Devitrification temperature
T_e	Eutectic melting temperature
T_f	Freezer storage temperature
T_g	Glass-to-rubber transition temperature
$T_{g'}$	Subzero T_g of a maximally-freeze-concentrated aqueous solution
T_h	Homogeneous nucleation temperature
T_{im}	Incipient melting temperature
T_m	Crystalline melting temperature
T_r	Recrystallization temperature
T_{sp}	Sticky point temperature
$T_{storage}$	Storage temperature
T_v	Vaporization temperature
ΔT	Difference in temperature between T and T_g (e.g. $T_f - T_{g'}$)
UFW	Unfrozen water
$W_{g'}$	Concentration of 'unfreezable' water in the glass formed at $T_{g'}$
WLF	Williams–Landel–Ferry
η	Viscosity
η_g	Viscosity at T_g
ρ	Density
ρ_g	Density at T_g

INTRODUCTION

The key to interpreting the behavior of low-moisture foods lies in recognizing the widely-reported and amply-demonstrated fact that water is an effective plasticizer of hydrophilic food materials, both polymeric and monomeric (Sears and Darby, 1982; Karel, 1985). It has become well established that plasticization by water affects the glass-to-rubber transition temperature (T_g) of many natural and synthetic amorphous polymers (particularly at low moisture contents), and that the resulting T_g depression can be advantageous or disadvantageous to material properties, processing, and stability (Rowland, 1980; Eisenberg, 1984). The physico-chemical effect of water, acting as a plasticizer, on the T_g of starch (van den Berg, 1981, 1986; van den Berg and Bruin, 1981; Ablett *et al.*, 1986; Biliaderis *et al.*, 1986*a,b,c*; Blanshard 1986; Yost and Hoseney, 1986; Zeleznak and Hoseney, 1987) and other partially-crystalline or completely-amorphous polymeric food materials such as gelatin (Jolley, 1970; Yannas, 1972; Marshall and Petrie, 1980; Tomka, 1986), gluten (Hoseney *et al.*, 1986), lysozyme, and other enzymes and proteins (Bone and Pethig, 1982, 1985; Poole and Finney, 1983*a,b*; Finney and Poole, 1984; Morozov and Gevorkian, 1985) has been increasingly discussed in recent years, especially since 1980.

The critical role of water as a plasticizer of amorphous materials (both water-soluble and water-sensitive ones, collectively referred to as water-compatible) or amorphous regions of partially-crystalline materials has been a focal point of our research since 1980, and has developed into a central theme of our program in food polymer science. As reviewed elsewhere (Levine and Slade, 1987*a*), studies in our labs have been based on thermal and thermomechanical analysis methods used to illustrate and characterize the polymer physico-chemical properties of various food ingredients and products. For example, starch and rice (Slade, 1984; Slade and Levine, 1984*a,b,* 1987*a,b*; Biliaderis *et al.*, 1985; Maurice *et al.*, 1983), gelatin (Slade and Levine, 1984*a,b,* 1987*c*), gluten (Slade, 1984; Slade and Levine, 1987*a*), frozen aqueous solutions of small sugars, sugar derivatives, polyols, and starch hydrolysis products (SHPs) (Schenz *et al.*, 1984; Levine and Slade, 1986, 1987*b,* 1988; Slade and Levine, 1987*b,d*), and 'intermediate moisture food' (IMF) carbohydrate systems (Slade, 1982; Slade and Levine, 1987*b,d*) have all been described as systems of completely-amorphous or partially-crystalline polymers, oligomers, and/or monomers, soluble in and/or plasticized by water.

A new understanding of structure/property relationships of low-moisture foods and food ingredients has been approached from a new perspective based on food polymer science, a research discipline which emphasizes the fundamental similarities between synthetic polymers and food molecules. From a theoretical basis of established structural principles from the field of polymer science, functional properties of food materials during processing and finished-product storage have been explained and often predicted (Levine and Slade, 1987*a*). The discipline of food polymer science has developed to unify structural aspects of foods (conceptualized as completely-amorphous or partially-crystalline polymer systems, the latter typically based on the classical 'fringed micelle' structural model (Flory, 1953; Wunderlich, 1973; Billmeyer, 1984)) with functional aspects described in terms of 'water dynamics' and 'glass dynamics' (Slade and Levine, 1987*b,d*). These integrated concepts focus on the non-equilibrium nature of all 'real world' food products and processes, and stress the importance to ultimate product quality and stability of maintenance of food systems in kinetically-metastable 'states' (as opposed to equilibrium thermodynamic phases), which are always subject to potentially-detrimental plasticization by water. Through this unification, the kinetically-controlled behavior of low-moisture food systems has been described by a 'map' (which is derived from a solute–solvent 'state' diagram (Franks *et al.*, 1977; MacKenzie, 1977)), in terms of the critical variables of moisture content, temperature, and time (Levine and Slade, 1987*a*). Glass dynamics deals with the temperature dependence of relationships among composition, structure, thermomechanical properties, and functional behavior, and has been used to describe a unifying concept for interpreting 'collapse' phenomena, which govern, e.g. caking during storage of low-moisture food ingredients and products (Levine and Slade, 1986, 1987*b*).

The extensive literature on caking and other collapse-related processes in completely-amorphous or partially-crystalline food powders (reviewed by Flink (1983), Karel and Flink (1983), and Karel (1985)) supports our conclusion that collapse phenomena are diffusion-controlled consequences of a material-specific structural relaxation process. We have postulated that these consequences represent microscopic and macroscopic manifestations of an underlying molecular 'state' transformation from kinetically-metastable amorphous solid to unstable amorphous liquid, which occurs at T_g (Levine and Slade, 1986, 1987*b*). The critical effect of plasticization by water (leading to increased free volume and mobility in the dynamically-constrained amorphous solid) on T_g is a central element of our concept and the mechanism derived from it. A general physico-chemical mechanism for collapse has been described (Levine and Slade, 1986), based on the occurrence of a material-specific structural transition at T_g, followed by viscous flow in the rubbery liquid state. The mechanism was derived from Williams–Landel–Ferry (WLF) free volume theory for amorphous polymers (Ferry, 1980), and led to a conclusion of the fundamental identity of T_g with the transition temperatures observed for structural collapse (T_c) and recrystallization (T_r). The non-Arrhenius kinetics of collapse and/or recrystallization in the high-viscosity (η) rubbery state, which are governed by the mobility of the water-plasticized polymer matrix, depend on the magnitude of ΔT above T_g, as defined by an exponential relationship derived from WLF theory.

The key to our new perspective on water-plasticized food polymer systems at low moisture relates to recognition of the fundamental importance of the above-mentioned dynamic map. Our research has demonstrated that the critical feature of the map is identification of the glass transition as the reference surface which serves as a basis for description of non-equilibrium behavior of polymeric materials, in response to changes in moisture content, temperature, and time (Levine and Slade, 1987*a*). The kinetics of all diffusion-controlled relaxation processes, which are governed by the mobility and viscosity of the water-plasticized polymer matrix, vary (from Arrhenius to WLF-type) between distinct temperature/structural domains, which are divided by this glass transition. The viscoelastic, rubbery fluid state, for which WLF kinetics apply (Ferry, 1980), represents the most significant domain for study of water dynamics (Slade and Levine, 1987*b*). One particular location on the reference surface results from the behavior of water as a crystallizing plasticizer and corresponds to an invariant point on a state diagram for any particular solute. This location represents the practical glass with maximum moisture content as a kinetically-metastable, dynamically-constrained solid which is pivotal to characterization of structure and function of amorphous and partially-crystalline food materials, both polymeric and monomeric.

To illustrate our concept of collapse phenomena in low-moisture foods, the behavior of 80 commercial SHPs (of dextrose equivalent (DE) values 0·3–100) and 76 other polyhydroxy compounds (sugars, glycosides, and polyhydric alcohols), studied by a low-temperature Differential Scanning Calorimetry (DSC) technique, is described in this chapter. The method, based on analog derivative thermograms (for experimental details, see Levine and Slade (1986, 1987*b*)), has been used to measure two thermomechanical properties characteristic of, and physico-chemically invariant for, each non-crystallizing solute: $T_{g'}$, the subzero T_g of a maximally-freeze-concentrated aqueous solution; and $W_{g'}$, the amount of water rendered 'unfreezable' (expressed as g unfrozen water/g solute) by immobilization with the solute in a kinetically-metastable amorphous solid formed on slow cooling to $T < T_{g'}$ (Franks, 1982, 1985*a*; Levine and Slade, 1987*a*). Results of this study have demonstrated the classical behavior of SHPs as a homologous family of amorphous glucose polymers, and revealed an 'entanglement coupling' (Ferry, 1980; Graessley, 1984) capability for SHPs of $\leq$6 DE and $T_{g'} \geq -8$°C, which is not evidenced by polyhydroxy compounds of molecular weight (MW) < 1200 (Levine and Slade, 1986, 1987*b*). The relationship between intermolecular chain entanglement (leading to network formation (Flory, 1953, 1974; Mitchell, 1980)) and SHP functionality, for these common food ingredients, in applications involving gelation, encapsulation, frozen-storage stabilization, thermomechanical stabilization of low-moisture food glasses, and facilitation of drying processes, is reviewed here. The utility of low-DE SHPs (i.e. dextrins, maltodextrins) for inhibiting various collapse phenomena, which affect the processing, storage stability, and quality of many low-moisture foods, is discussed and explained, as is the contrasting role typically played by sugars and polyols in promoting these phenomena. An example of the inhibition of enzymatic activity in a model food system, stabilized with low-DE maltodextrin, is also described.

In this chapter, the materials (SHPs, sugars, and polyols, representing widely-used ingredients in low-moisture foods) and method (low-temperature DSC) described exemplify our model-system approach to the study and interpretation of the behavior of low-moisture, glass-forming foods. The validity of this approach, which is based on the use of concentrated aqueous solutions of water-soluble carbohydrates as model low-moisture systems, has been justified (Levine and Slade, 1987*b*) by the fact that, upon freezing a 20 weight percentage (w%) sugar solution, dehydration due to freeze-concentration produces a solute concentration $\gg$50 w% in most typical cases (Franks, 1985*b*). We have also noted previously (Levine and Slade, 1987*a*) that a 38 w% fructose solution at 20°C represents another example of a low-moisture situation, in the context of this review. This insight arose from results of a seminal study by Soesanto and Williams (1981) on viscosities of concentrated aqueous sugar solutions. They demonstrated that, for such glass-forming liquids, in their rubbery state ($\eta > 10$ Pa s) at $20 < T < 80$°C, the WLF equation (Ferry, 1980) characterizes $\eta(T)$ extremely well. Arrhenius kinetics would not be applicable to describe the behavior of such 'low-moisture' food systems (Slade and Levine, 1987*d*).

DSC CHARACTERIZATION OF STRUCTURE/THERMAL PROPERTY RELATIONSHIPS IN LOW-MOISTURE FOODS

Aside from crystalline hydrates, most common low-moisture polymeric food materials which are solids at room temperature exist in one of two possible structural forms, completely amorphous or partially crystalline. (The latter term, rather than semicrystalline, is preferred to describe polymers of relatively low percentage crystallinity, such as starch (~15–42% crystallinity) and gelatin (~20% crystallinity) (Levine and Slade, 1987*a*). The term semicrystalline is the preferred usage to describe polymers of ≥50% crystallinity (Flory, 1953; Wunderlich, 1973).) Completely-amorphous, homogeneous polymers manifest a single, 'quasi-2nd-order' transition from metastable glass to unstable amorphous liquid at a characteristic T_g. Partially-crystalline polymers show two characteristic transitions; a T_g for the amorphous component, and, at a crystalline melting temperature, T_m, which is always at a higher temperature than T_g for homopolymers, a 1st-order transition from crystalline solid to amorphous liquid (Wunderlich, 1981). The same is true for partially-crystalline oligomers and monomers (Soesanto and Williams, 1981; Levine and Slade, 1987*a*). Of course, some common, low-MW food materials, e.g. table sugar (sucrose) and salt (NaCl), can exist under anhydrous conditions as completely-crystalline solids.

Both T_g and T_m are measurable as thermal or thermomechanical transitions by various instrumental methods, including DSC, Differential Thermal Analysis, Thermomechanical Analysis, and Dynamic Mechanical Analysis (Fuzek, 1980). Since an extensive discussion of DSC theory and practice is beyond the scope of this chapter, the interested reader is referred to the book '*Thermal Characterization of Polymeric Materials*' edited by Turi (1981), a general review of DSC methods by Richardson (1978), and reviews of DSC in food research by Biliaderis (1983), Lund (1983), Wright (1984), and Donovan (1985). These references provide excellent descriptions of instrumentation and methods, plus illustrative DSC results and interpretations.

The following serves as a brief introduction to the DSC results reviewed in this chapter. For both completely-amorphous and partially-crystalline food systems at low moisture, T_g is manifested as a diagnostic, discontinuous change in heat capacity (signifying a 2nd-order transition), which is represented as an endothermic step-change (i.e. baseline shift) in the DSC heat flow curve (Wunderlich, 1981). For partially-crystalline systems, continued heating beyond T_g results in the appearance of an endothermic peak (signifying a 1st-order transition) in the heat flow curve, corresponding to T_m. Melting of a crystalline solid involves dissociation of an ordered molecular structure, generally concomitant with an increase in volume, as the lattice components move apart (Wunderlich, 1980). 'Softening' of a glass at T_g likewise involves an increase in volume (i.e. free volume) and mobility, as the system transforms to a liquid state characterized by greater degrees of translational and rotational freedom (Ferry, 1980). Hence, both T_g and T_m are endothermic events, because heat input is required to raise the energy level of a solid to that of the higher-energy, generally higher-entropy liquid state (Wunderlich, 1981). Once melted, crystalline or partially-crystalline materials can usually be immobilized in a completely-amorphous solid form by sufficiently rapid cooling of the melt from $T > T_m$ to $T < T_g$, which 'freezes in' the disordered molecular structure of the liquid state. This technique allows analysis of T_g for metastable glasses. Subsequent rewarming of such a completely-amorphous but crystallizable material, to the so-called devitrification temperature (T_d, which is $>T_g$ but $<T_m$ (Luyet, 1960)), allows crystallization to occur, as manifested by an exothermic peak in the DSC heat flow curve. In contrast to the rapid cooling technique, slower cooling would have allowed crystallization to occur during cooling from the undercooled melt (Wunderlich, 1976). Crystallization from the liquid state, whether it occurs during cooling of a melt or heating of a glass, is an exothermic event, opposite in a thermodynamic sense to the process of crystalline melting, since it involves a 1st-order transition from a disordered liquid to an ordered solid state (Wunderlich, 1973). The DSC thermograms in Fig. 1 (described on page 81) illustrate many of these introductory remarks.

DSC Characterization of Low-Moisture Carbohydrate Model Systems

The physico-chemical properties of commercial SHPs (i.e. dextrins, maltodextrins, corn syrup solids, corn syrups) represent an important, but sparsely-researched, subject within the food industry (Murray and Luft, 1973; Dziedzic and Kearsley, 1984). To and Flink (1978) reported a correlation between increasing T_c and increasing number-average degree of polymerization, $\overline{DP}_n$, for a series of SHP powder samples of $2 \leq \overline{DP}_n \leq 16$ (calculated DE = 52·6–6·9). (By convention, DE is defined as $100/(\bar{M}_n/180 \cdot 16)$, where $\bar{M}_n$ is number-average MW, since reducing sugar content (in terms of number of reducing end groups) of a known weight of sample is compared to an equal weight of glucose of DE 100 and $\bar{M}_n$ 180·16.) In contrast, low-MW sugars and polyols have been studied extensively, and limited compilations of

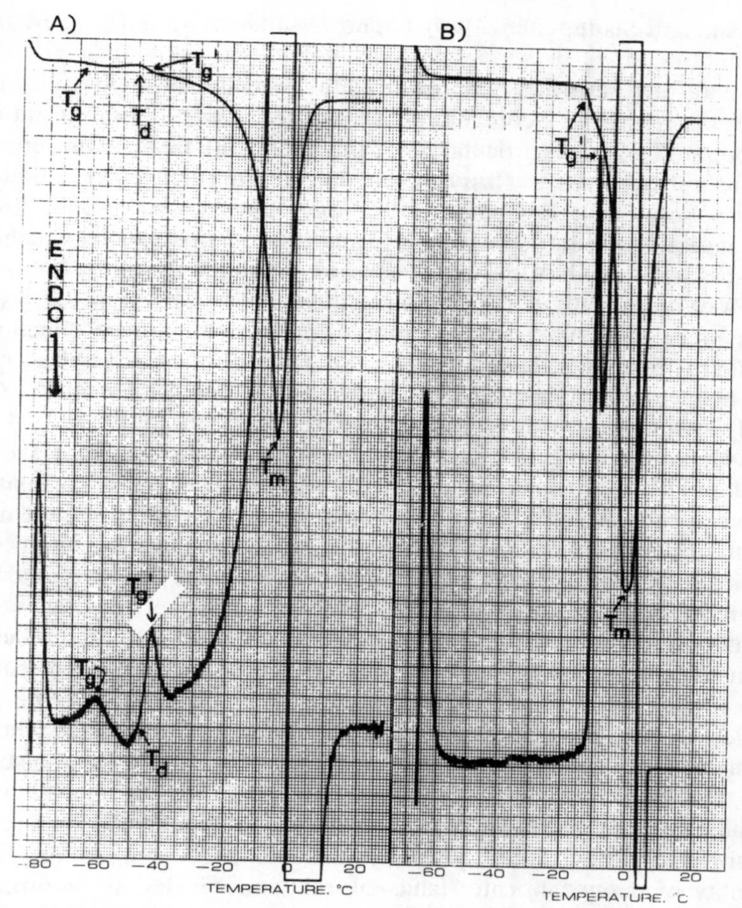

FIG. 1. Typical DSC thermograms for 20 w% solutions of (A) glucose, and (B) Star Dri 10 10-DE maltodextrin (Staley, 1984). In each, the heat flow curve begins at the top (endothermic down), and the derivative trace (zeroed to the temperature axis) at the bottom. Reproduced from Levine and Slade (1986).

their characteristic T_c values are available (Franks, 1982, 1985*b*). Our research has demonstrated that much can be learned about functional attributes of SHPs, sugars, and polyols, many of which are common ingredients in both fabricated and natural foods, from a polymer physico-chemical approach to systematic studies of their thermo-mechanical properties and effects thereon of plasticization by water

(Levine and Slade, 1986, 1987*b*). These studies have led to predictive capabilities.

We review here our DSC results for $T_{g'}$ values of 80 SHPs and 76 other polyhydroxy compounds. (As defined by Franks (1982, 1985*a,b*), $T_{g'}$ is the particular T_g of the maximally freeze-concentrated solute/unfrozen water glassy matrix surrounding the ice crystals in a frozen solution.) For the SHPs, our analysis yielded a linear correlation between decreasing DE and increasing $T_{g'}$, from which we constructed a calibration curve used to predict DE values for other SHPs of unknown DE (Levine and Slade, 1986). The same DE vs $T_{g'}$ data were also used to construct a predictive map of functional attributes for SHPs, based on a demonstration of their classical T_g vs $\bar{M}_n$ behavior as a homologous series of amorphous polymers. Our study, which covered a more extensive range of SHP DE values, provided a theoretical basis for interpreting previous results of To and Flink (1978). For the 76 polyhydroxy compounds, a linear correlation between increasing $T_{g'}$ and decreasing value of 1/MW was demonstrated (Levine and Slade, 1987*b*). As expected, this correlation was not quite as good as the one for SHPs, because these 76 compounds do not represent a single homologous family of monomer and oligomers.

The demonstration that SHP functional behavior can be predicted from the correlation between DE (or $\overline{DP}_n$ or $\bar{M}_n$) and T_g, and that of sugars and polyols from their T_g vs 1/MW relationship (Levine and Slade, 1987*b*), has important implications for a better understanding of the collapse mechanism in polymeric food systems at low moisture. For the food industry, such predictive capabilities are valuable, because various non-equilibrium collapse phenomena (which are all sensitive to plasticization by water) affect the processing, quality, and stability of many fabricated and natural foods, including amorphous powders, candy glasses, and frozen products. We review here the potential (and frequently demonstrated) utility of SHPs in preventing structural collapse, within a context of collapse processes which are often promoted by formulation with large amounts of low-MW saccharides. We also relate these insights to our interpretation of collapse phenomena, within the context of glass dynamics.

Low-Temperature Thermal Properties of Polymeric vs Monomeric Carbohydrates

Figure 1 (Levine and Slade, 1986) shows two typical low-temperature DSC thermograms for 20 w% solutions: (A) glucose, and (B) Star Dri

10 (1984) 10-DE maltodextrin. In each, the heat flow curve begins at the top (endothermic down), and the analog derivative trace (zeroed to the temperature axis) at the bottom (endothermic up). For both thermograms, instrumental amplification and sensitivity settings were identical, and sample weights comparable. As illustrated by Fig. 1, the direct analog derivative feature of the DuPont 990 DSC greatly facilitates identification of sequential thermal transitions, assignment of precise transition temperatures (to ±0·5°C for duplicate samples), and thus overall interpretation of thermal behavior, especially for such frozen aqueous solutions exemplified by Fig. 1(A). Surprisingly, we had found no other reported use of derivative thermograms, in the many previous DSC studies of such systems (see Franks (1982) for extensive bibliography), to sort out the small endothermic and exothermic changes in heat flow that typically occur below 0°C.

Despite the handicap of such instrumental shortcomings in the past, the theoretical basis for the thermal properties manifested by aqueous solutions at subzero temperatures has come to be well understood (Luyet, 1960; Rasmussen and Luyet, 1969; MacKenzie and Rasmussen, 1972; Mackenzie, 1977, 1981; Franks *et al.*, 1977; Franks, 1982, 1985*a,b*). As shown in Fig. 1(A), after rapid cooling (about 50°C/min) of the glucose solution to <-80°C, slow heating (5°C/min) revealed a minor T_g at −61·5°C, followed by an exothermic devitrification (a crystallization of some of the previously-unfrozen water) at −47·5°C, followed by another (major) T_g, namely $T_{g'}$ at −43°C, and then finally the melting of ice at T_m. In Fig. 1(B), the maltodextrin solution thermogram shows only the obvious $T_{g'}$ at −10°C, in addition to T_m. These assignments of transitions and temperatures have been reconciled definitively (Levine and Slade, 1986) with state diagrams previously reported for such materials (MacKenzie and Rasmussen, 1972; Franks, 1982). In such diagrams (e.g. Fig. 2 for poly(vinyl pyrrolidone) (PVP), from Levine and Slade, 1986), the different cooling/heating paths that can be followed by solutions of monomeric (glucose) vs polymeric (maltodextrin) solutes are revealed. However, as demonstrated by the thermograms in Fig. 1, in either general case, rewarming forces the system through a glass transition at $T_{g'}$. In many earlier DSC studies (e.g. MacKenzie, 1977, 1981; Maltini, 1977), performed without benefit of derivative thermograms, a pair of transition temperatures (each independent of initial concentration), called *T* antemelting and *T* incipient melting, were reported in place of a single $T_{g'}$. In fact, for the many cases that we

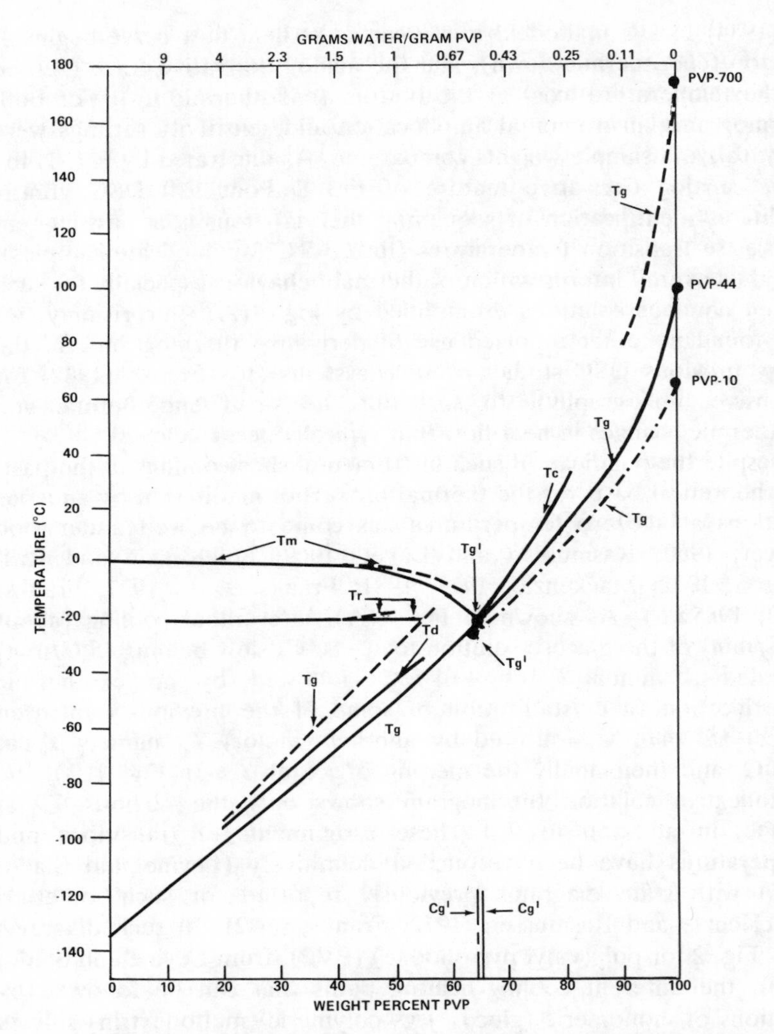

FIG. 2. Solid–liquid state diagram for water–PVP, showing the following transitions: T_m, T_g, $T_{g'}$, T_d, T_r, T_c. (——— = PVP-44, ----- = PVP-700, –·–·– = PVP-10 (MacKenzie and Rasmussen, 1972; Franks *et al.*, 1977); •----• = data from Olson and Webb (1978); other • refer to data from Levine and Slade (1986).) Reproduced from Levine and Slade (1987*b*).

have studied, the reported values of T_{am} (Virtis, 1983) and T_{im} bracket that of $T_{g'}$ (as we measure it), which led us to suggest that T_{am} and T_{im} actually represent the temperatures of onset and completion of the single thermal event (a glass transition) that must occur at $T_{g'}$, as defined by the state diagram (Levine and Slade, 1986). Even today, many workers in this field still do not recognize the subzero $T_g < T_d < T_{g'}$ sequence characteristic of frozen solutions of low-MW polyhydroxy compounds. Instead, they refer to 'anomalous double glass transitions' manifested by aqueous solutions of propylene glycol and glycerol (Boutron and Kaufmann, 1979; MacFarlane, 1985; Vassoille *et al.*, 1986). Far from anomalously, for each solute, the higher T_g of the doublet coincides with our measured $T_{g'}$ value (Levine and Slade, 1987*b*). Similarly, T_r, manifested as the onset temperature for opacity during warming of vitrified aqueous solutions and known to be independent of initial concentration, is still a topic of current interest and discussion as to its origin (Forsyth and MacFarlane, 1986; Thom and Matthes, 1986), but is not yet widely recognized to coincide with $T_{g'}$ for water-soluble carbohydrates, including several much-studied polyols.

Our previous studies (Levine and Slade, 1986, 1987*a,b*) have helped explain why $T_{g'}$ is the most noteworthy feature of the thermograms in Fig. 1. We have demonstrated the significance of T_g in general, and $T_{g'}$ in particular, to interpreting the behavior of low-moisture foods, and established that $T_{g'}$ is of fundamental technological importance to the quality and stability of frozen foods. The matrix surrounding the ice crystals in a maximally-frozen solution is a supersaturated solution (hence, in our perspective, a low-moisture system) of all the solute in the fraction of water remaining unfrozen. This matrix exists as a kinetically-metastable amorphous solid (a glass of constant composition) at any $T < T_{g'}$, but as a viscoelastic liquid (a rubbery fluid) at $T_{g'} < T <$ ice T_m. Again with regard to a state diagram for a typical solute that does not readily undergo eutectic crystallization (e.g. Franks, 1982, or our Fig. 2), $T_{g'}$ corresponds to the intersection of an extension of the thermodynamically-defined equilibrium liquidus curve and the kinetically-determined supersaturated glass curve. As such, Franks (1982) described $T_{g'}$ as representing a quasi-invariant point in the state diagram, invariant in both its characteristic temperature ($T_{g'}$) and composition (i.e. $C_{g'}$, expressed as w% solute, or $W_{g'}$ expressed as g unfrozen water/g solute) for any particular solute. This glass, which forms, e.g. on slow cooling to $T_{g'}$, serves as a kinetic barrier (of high

TABLE 1
$T_{g'}$ values for commercial SHPs

SHP	*Manufacturer*	*Starch source*	*DE*	$T_{g'}$ (°C)	*Gelling*
AB 7436	Anheuser Busch	Waxy maize	0·5	−4	
Paselli SA-2	AVEBE (1984)	Potato (Ap)	2	−4·5	Yes
Stadex 9	Staley	Dent corn	3·4	−4·5	Yes
78NN128	Staley	Potato	0·6	−5	Yes
78NN122	Staley	Potato	2	−5	Yes
V-O Starch	National	Waxy maize	?	−5·5	Yes
N-Oil	National	Tapioca	?	−5·5	Yes
ARD 2326	Amaizo	Dent corn	0·4	−5·5	Yes
Paselli SA-2	AVEBE (1986)	Potato (Ap)	2	−5·5	Yes
ARD 2308	Amaizo	Dent corn	0·3	−6	Yes
AB 7435	Anheuser Busch	Waxy/dent blend	0·5	−6	
Star Dri 1	Staley (1984)	Dent corn	1	−6	Yes
Crystal Gum	National	Tapioca	5	−6	Yes
Maltrin M050	GPC	Dent corn	6	−6	Yes
Star Dri 1	Staley (1986)	Waxy maize	1	−6·5	Yes
Paselli MD-6	AVEBE	Potato	6	−6·5	Yes
Dextrin 11	Staley	Tapioca	1	−7·5	Yes
MD-6-12	V-Labs		2·8	−7·5	
Stadex 27	Staley	Dent corn	10	−7·5	No
MD-6-40	V-Labs		0·7	−8	
Star Dri 5	Staley (1984)	Dent corn	5	−8	No
Star Dri 5	Staley (1986)	Waxy maize	5·5	−8	No
Paselli MD-10	AVEBE	Potato	10	−8	No
Paselli SA-6	AVEBE	Potato (Ap)	6	−8·5	No
α-Cyclodextrin	Pfanstiehl			−9	
Capsul	National	Waxy maize	5	−9	
Lodex Light V	Amaizo	Waxy maize	7	−9	
Paselli SA-10	AVEBE	Potato (Ap)	10	−9·5	No
Morrex 1910	CPC	Dent corn	10	−9·5	
Star Dri 10	Staley (1984)	Dent corn	10	−10	No
Maltrin M040	GPC	Dent corn	5	−10·5	
Frodex 5	Amaizo	Waxy maize	5	−11	
Star Dri 10	Staley (1986)	Waxy maize	10·5	−11	No
Lodex 10	Amaizo (1986)	Waxy maize	11	−11·5	No
Lodex Light X	Amaizo	Waxy maize	12	−11·5	
Morrex 1918	CPC	Waxy maize	10	−11·5	
Mira-Cap	Staley	Waxy maize	?	−11·5	
Maltrin M100	GPC	Dent corn	10	−11·5	No
Lodex 5	Amaizo	Waxy maize	7	−12	No
Maltrin M500	GPC	Dent corn	10	−12·5	
Lodex 10	Amaizo (1982)	Waxy maize	12	−12·5	No
Star Dri 15	Staley (1986)	Waxy maize	15·5	−12·5	No
MD-6	V-Labs		?	−12·5	

TABLE 1—*contd.*

SHP	*Manufacturer*	*Starch source*	*DE*	$T_{g'}$ (°C)	*Gelling*
Maltrin M150	GPC	Dent corn	15	−13·5	No
Maltoheptaose	Sigma		15·6	−13·5	
MD-6-1	V-Labs		20·5	−13·5	
Star Dri 20	Staley (1986)	Waxy maize	21·5	−13·5	No
Maltodextrin syrup	GPC	Dent corn	17·5	−14	No
Frodex 15	Amaizo	Waxy maize	18	−14	
Maltohexaose	Sigma		18·2	−14·5	
Frodex 10	Amaizo	Waxy maize	10	−15·5	
Lodex 15	Amaizo	Waxy maize	18	−15·5	No
Maltohexaose	V-Labs		18·2	−15·5	
Maltrin M200	GPC	Dent corn	20	−15·5	
Maltopentaose	Sigma		21·7	−16·5	
Maltrin M250	GPC	Dent corn	25	−17·5	
N-Lok	National	Blend	?	−17·5	
Staley 200	Staley	Corn	26	−19·5	
Maltotetraose	Sigma		27	−19·5	
Frodex 24	Amaizo	Waxy maize	28	−20·5	
Frodex 36	Amaizo	Waxy maize	36	−21·5	
DriSweet 36	Hubinger	Corn	36	−22	
Maltrin M365	GPC	Dent corn	36	−22·5	
Staley 300	Staley	Corn	35	−23·5	
Globe 1052	CPC	Corn	37	−23·5	
Maltotriose	V-Labs		35·7	−23·5	
Frodex 42	Amaizo	Waxy maize	42	−25·5	
Neto 7300	Staley	Corn	42	−26·5	
Globe 1132	CPC	Corn	43	−27·5	
Staley 1300	Staley	Corn	43	−27·5	
Neto 7350	Staley	Corn	50	−27·5	
Maltose	Sigma		52·6	−29·5	
Globe 1232	CPC	Corn	54·5	−30·5	
Staley 2300	Staley	Corn	54	−31	
Sweetose 4400	Staley	Corn	64	−33·5	
Sweetose 4300	Staley	Corn	64	−34	
Globe 1642	CPC	Corn	63	−35	
Globe 1632	CPC	Corn	64	−35	
Royal 2626	CPC	Corn	95	−42	
Glucose	Sigma	Corn	100	−43	

Updated from Levine and Slade (1986).

TABLE 2
$T_{g'}$ values for sugars, glycosides, and polyhydric alcohols

Sugar or polyol	*MW*	$T_{g'}$ (°C)	$W_{g'}$ (g UFW/g)	Dry T_g (°C)	Dry T_m (°C)	T_m/T_g (°K)
Ethylene glycol	62·1	−85	1·90			
Propylene glycol	76·1	−67·5	1·28			
1,3-Butanediol	90·1	−63·5	1·41			
Glycerol	92·1	−65	0·85	−93	18	1·62
Erythrose	120·1	−50	1·39			
Threose	120·1	−45·5				
Erythritol	122·1	−53·5	(Eutectic)			
Thyminose (deoxyribose)	134·1	−52	1·32			
Ribulose	150·1	−50				
Xylose	150·1	−48	0·45	9·5	153	1·51
Arabinose	150·1	−47·5	1·23			
Lyxose	150·1	−47·5				
Ribose	150·1	−47	0·49	−10	87	1·37
Arabitol	152·1	−47	0·89			
Ribitol	152·1	−47	0·82			
Xylitol	152·1	−46·5	0·75	−18·5	94	1·44
Methyl riboside	164·2	−53	0·96			
Methyl xyloside	164·2	−49	1·01			
Quinovose (deoxyglucose)	164·2	−43·5	1·11			
Fucose (deoxygalactose)	164·2	−43	1·11			
Rhamnose (deoxymannose)	164·2	−43	0·90			
Talose	180·2	−44				
Idose	180·2	−44				
Psicose	180·2	−44				
Altrose	180·2	−43·5				
Glucose	180·2	−43	0·41	31	158	1·42
Gulose	180·2	−42·5				
Fructose	180·2	−42	0·96	100	124	1·06
Galactose	180·2	−41·5	0·77	110	170	1·16
Allose	180·2	−41·5	0·56			
Sorbose	180·2	−41	0·45			
Mannose	180·2	−41	0·35	30	139·5	1·36
Tagatose	180·2	−40·5	1·33			
Inositol	180·2	−35·5	0·30			
Mannitol	182·2	−40	(Eutectic)			
Galactitol	182·2	−39	(Eutectic)			
Sorbitol	182·2	−43·5	0·23	−2	111	1·42
2-*o*-Methyl fructoside	194·2	−51·5	1·61			
β-1-*o*-methyl glucoside	194·2	−47	1·29			
3-*o*-Methyl glucoside	194·2	−45·5	1·34			
6-*o*-Methyl galactoside	194·2	−45·5	0·98			
α-1-*o*-methyl glucoside	194·2	−44·5	1·32			

TABLE 2—*contd.*

Sugar or polyol	*MW*	$T_{g'}$ (°C)	$W_{g'}$ (g UFW/g)	*Dry* T_g (°C)	*Dry* T_m (°C)	T_m/T_g (°K)
1-*o*-Methyl galactoside	194·2	−44·5	0·86			
1-*o*-Methyl mannoside	194·2	−43·5	1·43			
1-*o*-Ethyl glucoside	208·2	−46·5	1·35			
2-*o*-Ethyl fructoside	208·2	−46·5	1·15			
1-*o*-Ethyl galactoside	208·2	−45	1·26			
1-*o*-Ethyl mannoside	208·2	−43·5	1·21			
Glucoheptose	210·2	−37·5				
Mannoheptulose	210·2	−36·5				
Glucoheptulose	210·2	−36·5	0·77			
Perseitol (mannoheptitol)	212·2	−32·5	(Eutectic)			
1-*o*-Propyl glucoside	222·2	−43	1·22			
1-*o*-Propyl galactoside	222·2	−42	1·05			
1-*o*-Propyl mannoside	222·2	−40·5	0·95			
2,3,4,6-*o*-Methyl glucoside	236·2	−45·5	1·41			
Isomaltulose (palatinose)	342·3	−35·5				
Nigerose	342·3	−35·5				
Cellobiulose	342·3	−32·5				
Isomaltose	342·3	−32·5	0·70			
Sucrose	342·3	−32	0·56	52	192	1·43
Gentiobiose	342·3	−31·5	0·26			
Laminaribiose	342·3	−31·5				
Turanose	342·3	−31	0·64			
Mannobiose	342·3	−30·5	0·91	90	205	1·32
Melibiose	342·3	−30·5				
Lactulose	342·3	−30	0·72			
Maltose	342·3	−29·5	0·25	43	129	1·27
Maltulose	342·3	−29·5				
Trehalose	342·3	−29·5	0·20			
Cellobiose	342·3	−29		77	249	1·49
Lactose	342·3	−28	0·69			
Maltitol	344·3	−34·5	0·59			
Isomaltotriose	504·5	−30·5	0·50			
Panose	504·5	−28	0·59			
Raffinose	504·5	−26·5	0·70			
Maltotriose	504·5	−23·5	0·45	76	133·5	1·16
Nystose	666·6	−26·5				
Stachyose	666·6	−23·5	1·12			
Maltotetraose	666·6	−19·5	0·55			
maltopentaose	828·9	−16·5	0·47			
α-Cyclodextrin	972·9	−9				
Maltohexaose	990·9	−14·5	0·50			
Maltoheptaose	1153·0	−13·5	0·27			

Updated from Levine and Slade (1987*b*).

activation energy) to further ice formation (within the experimental time frame), despite the continued presence of unfrozen water at all temperatures $<T_{g'}$, as well as to any other diffusion-controlled process (Franks, 1985*b*). Recognizing this, one begins to appreciate why the temperature of this glass transition is so important to different aspects of frozen food technology, e.g. freezer storage stability, freeze concentration, and freeze drying (Franks, 1982, 1985*b*), all of which can be subject to various recrystallization and collapse phenomena.

Measured $T_{g'}$ vaues for the 80 SHPs and 76 polyhydroxy compounds are listed in Tables 1 and 2 (updated from Levine and Slade, 1986 and 1987*b*), respectively. We had noted in these reports that our $T_{g'}$ values for various solutes fall between earlier literature values for T_{am} and T_{im} (Rasmussen and Luyet, 1969), and within a few degrees of values reported for T_c and T_r (as shown in Table 5).

Figure 3 (updated from Levine and Slade, 1986) shows $T_{g'}$ plotted vs DE for all SHPs with manufacturer-specified DE values. There is an excellent linear correlation between increasing $T_{g'}$ and decreasing DE (coefficient $r = -0{\cdot}98$). Since DE is inversely proportional to $\overline{DP}_n$ and $\bar{M}_n$ for this series of SHPs (Dziedzic and Kearsley, 1984), the results in

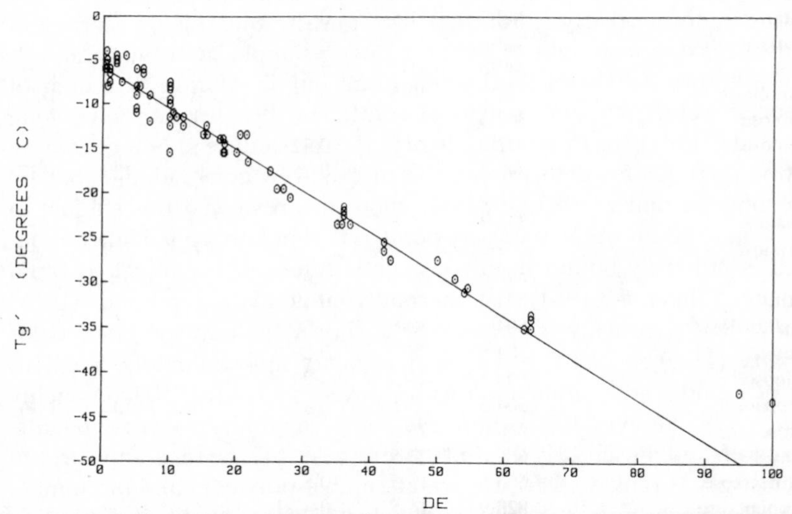

FIG. 3. Variation of the glass transition temperature, $T_{g'}$ for maximally-frozen 20 w% solutions against DE value for the commercial SHPs in Table 1. Reproduced from Levine and Slade (1986).

Fig. 3 demonstrated that $T_{g'}$ increases with increasing $\bar{M}_n$. Such a correlation between T_g and $\bar{M}_n$ is the general rule for any homologous family of glass-forming monomer, oligomers, and polymers (Billmeyer, 1984). The equation of the regression line is DE = −2·2 ($T_{g'}$, °C) − 12·8. We have shown (Levine and Slade, 1986) that Fig. 3 can be used as a calibration curve for interpolating DE values of new or 'unknown' SHPs, in preference to the time-consuming classical methods for DE determination (Murray and Luft, 1973).

For the 76 polyhydroxy compounds, the corresponding plot of $T_{g'}$ vs 1/MW is shown in the inset of Fig. 4 (updated from Levine and Slade, 1987*b*). Here, the regression line has an r value of −0·94, which is slightly lower than that for the homologous series of SHPs. The major cause of scatter in this plot is the series of non-homologous, chemically-different glycosides. In contrast, Fig. 5(B) (adapted from Levine and Slade, 1987*b*) shows the smooth curve of $T_{g'}$ vs MW (from Table 2) for the homologous malto-oligosaccharides from glucose to maltoheptaose, with $r = -0{\cdot}99$ for the corresponding plot of $T_{g'}$ vs 1/MW.

$W_{g'}$, the composition of the glass at $T_{g'}$, is calculated from a thermogram, from measurement of the area (enthalpy) under the ice melting endotherm. By calibration with pure water, this measurement yields a maximum weight of ice in a frozen sample, and by difference from a known weight of total water in an initial solution, a weight of unfrozen water, per unit weight of solute, in the glass at $T_{g'}$ (Levine and Slade, 1986). In the food industry, this procedure is one of several routine methods for determining the so-called 'water binding capacity' of a solute. Franks (1982, 1985*a,b,* 1986) has reviewed this subject in detail, and taken great pains to point out that this so-called 'bound' water is not truly bound in any energetic sense. It is subject to rapid exchange (Jin *et al.*, 1984), has thermally-labile hydrogen bonds (Biros *et al.*, 1979; Pouchly *et al.*, 1979), shows cooperative molecular mobility (Hoeve, 1980), has a heat capacity approximately equal to that of liquid water rather than ice (Hoeve and Hoeve, 1978; Pouchly *et al.*, 1979; Hoeve, 1980), and has some capability to dissolve salts (Burghoff and Pusch, 1980). Furthermore, it has been demonstrated conclusively (Franks, 1986), for water-soluble polymers and monomers alike, that such 'unfreezability' is not due to tight equilibrium binding by solute (Starkweather, 1980), but to purely kinetic retardation of diffusion of water and solute molecules at the low temperatures

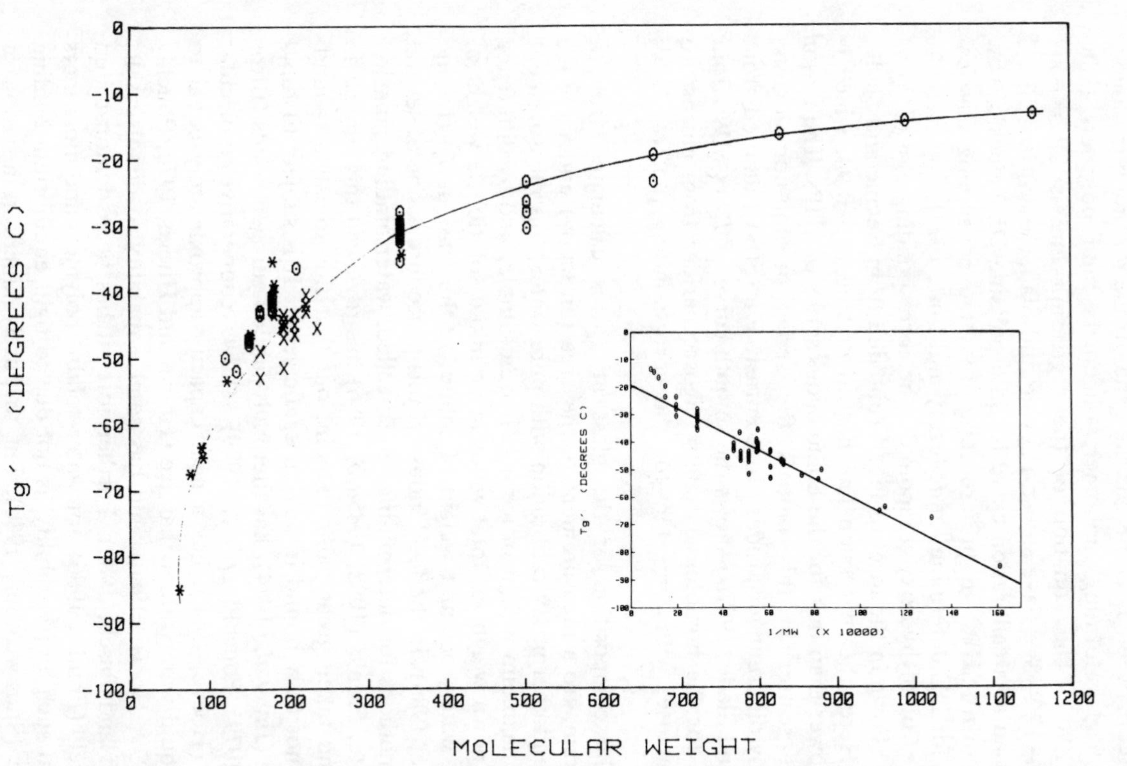

FIG. 4. Variation of the glass transition temperature, $T_{g'}$ for maximally-frozen 20 w% solutions against MW for the sugars (o), glycosides (x), and polyols (*) in Table 2. (Inset: a plot of $T_{g'}$ vs $1/MW \times 10^4$, illustrating the theoretically-predicted linear dependence.) Reproduced from Levine and Slade (1987*b*).

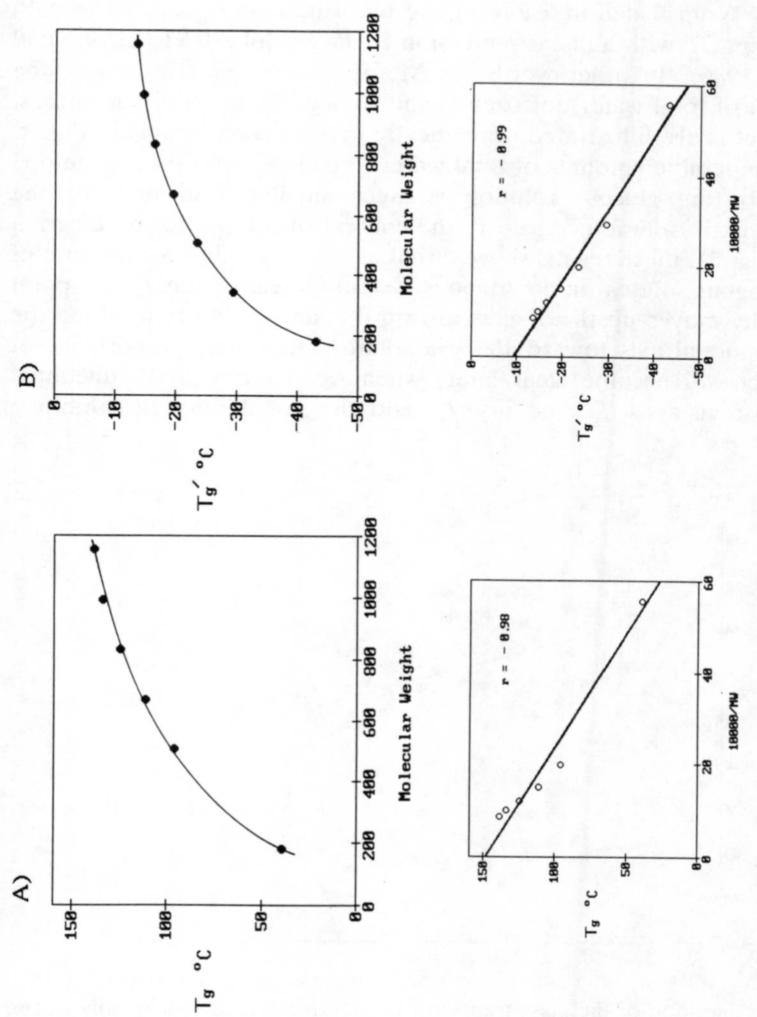

FIG. 5. Top: variation of the glass transition temperature (A) T_g, of dry powders, and (B) $T_{g'}$ for maximally-frozen 20 w% solutions, against MW for a homologous series of malto-oligosaccharides from glucose through maltoheptaose. Bottom: Corresponding plots of (A) T_g and $T_{g'}$ vs 10 000/MW. Adapted from Levine and Slade (1987*b*).

approaching the vitrification T_g of a solute/unfrozen water mixture (Biros *et al.*, 1979; Pouchly *et al.*, 1979).

For a homologous series of corn syrup solids solutions (i.e. from the 13 corn syrups listed in Table 1), we reported that $W_{g'}$ decreases with increasing $T_{g'}$ with a linear regression coefficient of $-0{\cdot}91$ (Levine and Slade, 1986). In other words, as $\bar{M}_n$ of the solute(s) increases, the fraction of total water unfrozen in the glass at $T_{g'}$ generally decreases. This fact is also illustrated dramatically by the thermograms in Fig. 1. For comparable amounts of total water, the area under the ice melting peak for the glucose solution is much smaller than that for the maltodextrin solution. Again in the context of a typical state diagram (e.g. Fig. 2), these results showed that as $\bar{M}_n$ of a solute (or mixture of homologous solutes) in an aqueous system increases, the $T_{g'}/C_{g'}$ point generally moves up the T axes toward 0°C and to the right along the compositional axis toward 100 w% solute. The critical importance of this fact will become clear later, when we describe SHP functional behavior vis-a-vis $T_{g'}$ and dry T_g, and the possibilities of inhibiting

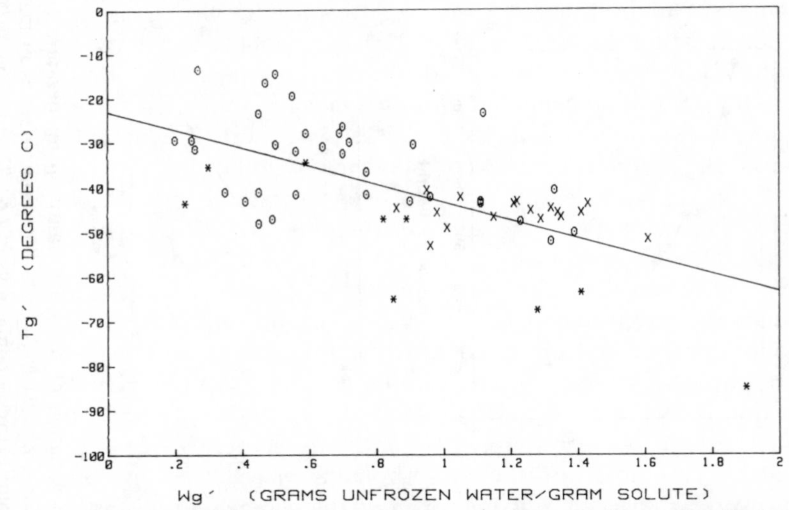

FIG. 6. Variation of the glass transition temperature, $T_{g'}$ for maximally-frozen 20 w% solutions against $W_{g'}$, the composition of the glass at $T_{g'}$, in g unfrozen water/g solute, for the sugars (o), glycosides (x), and polyhydric alcohols (∗) in Table 2. Reproduced from Levine and Slade (1987*b*).

collapse in low-moisture foods by formulating a fabricated product with the intent of elevating $T_{g'}$ and dry T_g.

In apparent contrast to the above results for a homologous series of mixed glucose monomer and oligomers, the results in Fig. 6 (Levine and Slade, 1987*b*), of $T_{g'}$ vs $W_{g'}$ for the diverse polyhydroxy compounds listed in Table 2, yielded a regression coefficient of only $-0{\cdot}64$. Thus, when Franks (1985*b*) noted that, among the (non-homologous) sugars and polyols most widely used as 'water binders' in fabricated foods, 'the amount of unfreezable water does not show a simple dependence on MW of the solute', he was sounding a necessary caution. (In fact, when the $W_{g'}$ data in Table 2 were plotted against 1/MW, $r = 0{\cdot}47$.) We concluded that the plot in Fig. 6, as shown, obviously could not be used for predictive purposes, so the safest approach would be to rely on measured $W_{g'}$ values for each potential 'water binding' candidate. However, we demonstrated that the situation is not quite as nebulous as represented by Fig. 6. When some of the same data were plotted (actually $T_{g'}$ vs $C_{g'}$, shown in Fig. 7 (Levine and Slade, 1987*a*)), but compounds were grouped by chemical classification into specific homologous series (e.g. polyols, glucose-only polymers, and fructose- or galactose-containing saccharides), obviously-better linear correlations became evident. The plots in Fig. 7 illustrated the same linear dependence of $T_{g'}$ on the composition of the glass at $T_{g'}$ (i.e. as the amount of unfrozen water in the glass decreases, $T_{g'}$ increases) as did the data for the series of corn syrup solids. Still, Franks' suggestion (personal communication, 1986) that investigations of T_m and η as functions of solute concentration, and the liquidus curve as a function of solute structure, would be particularly worthwhile, is a good one.

We have noted with interest the $W_{g'}$ results for the series of monomeric glycosides listed in Table 2, in terms of a possible relationship between glycoside structure (e.g. position and size of the hydrophobic aglycone, which is absent in the parent sugar) and the function reflected by $W_{g'}$ (Levine and Slade, 1987*b*). Clearly, the $W_{g'}$ values for all the methyl, ethyl, and propyl derivatives are much greater than those for the corresponding parent monosaccharides. However, as illustrated in Fig. 8, $W_{g'}$ values appeared consistently to be maximized for the methyl or ethyl derivatives, but somewhat decreased for the propyl derivatives. We suggested that these results could indicate that increasing hydrophobicity (of the aglycone) leads to

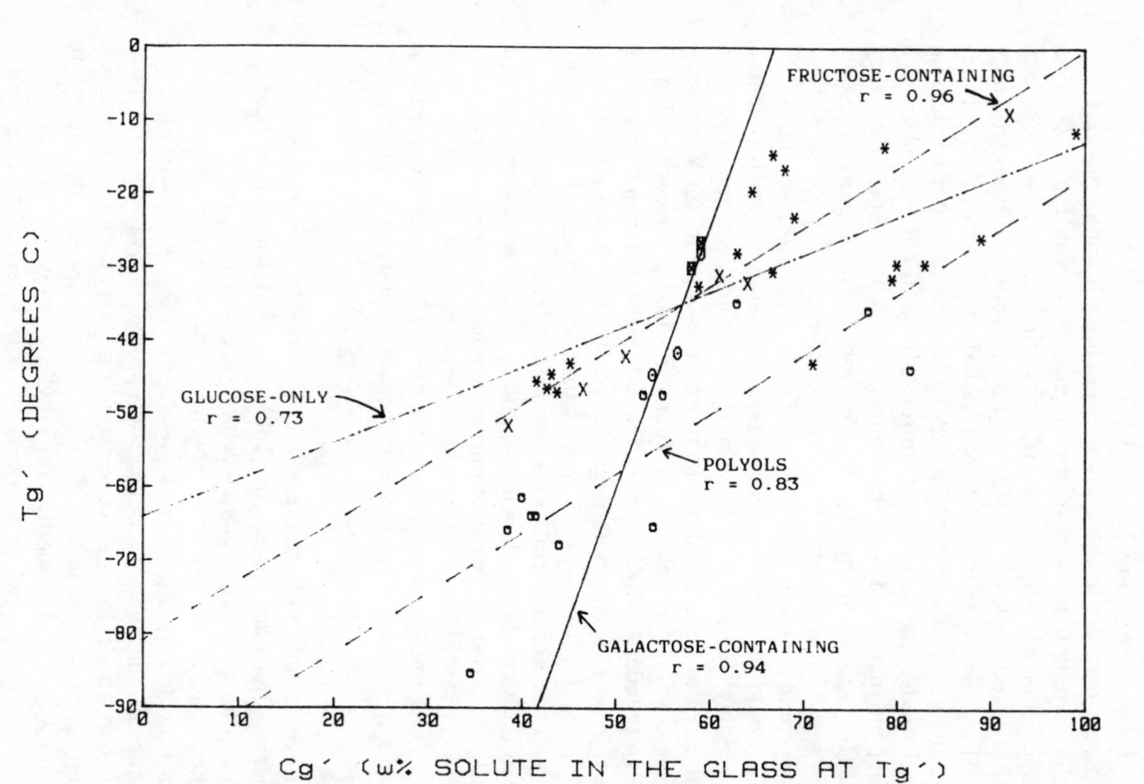

FIG. 7. Variation of the glass transition temperature, $T_{g'}$, for maximally-frozen 20 w% solutions against $C_{g'}$, composition of the glass at $T_{g'}$ in weight % solute, for homologous series of polyhydric alcohols (o), glucose-only polymers (*), fructose- (x), and galactose-containing saccharides (O) in Table 2. Reproduced from Levine and Slade (1987*a*).

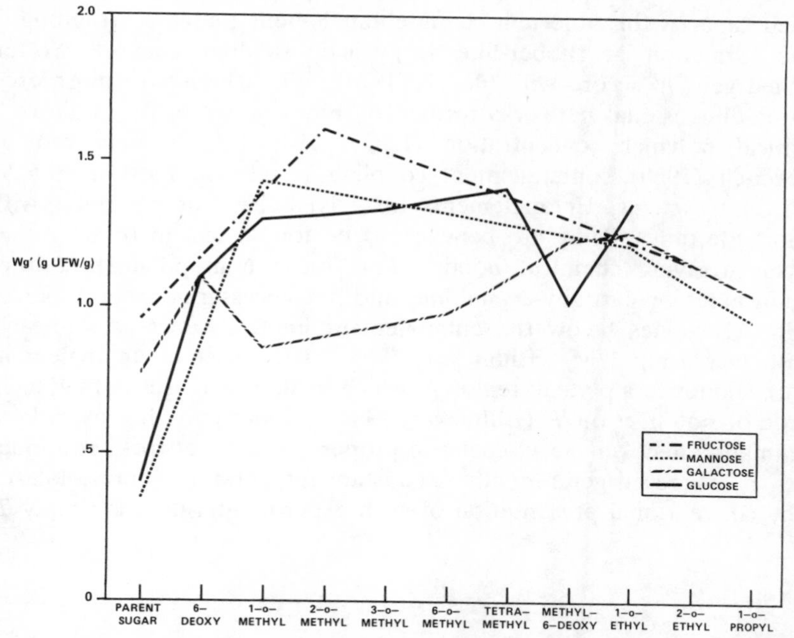

FIG. 8. Variation of $W_{g'}$, the composition of the glass at $T_{g'}$, in g unfrozen water/g solute, with glycoside structure for a series of monomeric glycosides produced from common monosaccharides.

both decreasing $W_{g'}$ and the demonstrated tendency toward increasing insolubility of propyl and larger glycosides in water.

STRUCTURE/PROPERTY RELATIONSHIPS FOR SHPs AND POLYHYDROXY COMPOUNDS

The straightforward presentation of DE vs $T_{g'}$ data in Fig. 3 is not the most rigorous theoretical treatment. Yet, the linear correlation of DE with $T_{g'}$ and the convenience for practical application in estimation of DE to characterize SHP samples recommend its use. The rigorous theoretical dependence of DE on $T_{g'}$ stems from the respective dependence of each of these parameters on linear DP and MW within a series of monodisperse (i.e. MW = $\bar{M}_n$ = weight-average MW, $\bar{M}_w$) homopolymers. High polymers can be distinguished from oligomers by their capacity for molecular chain 'entanglement coupling', resulting in the formation of rubber-like viscoelastic random networks (often called gels, in accord with Flory's (1974) nomenclature for disordered three-dimensional networks formed by physical aggregation) above a critical polymer concentration (Ferry, 1980). As summarized by Mitchell (1980), 'entanglement coupling is seen in most high MW polymer systems. Entanglements (in gels) behave as crosslinks with short life-times. They are believed to be topological in origin rather than involving chemical bonds'. For linear homopolymers (either amorphous or partially-crystalline, and not necessarily monodisperse) with $\bar{M}_n$ values below the entanglement limit, T_g decreases linearly with increasing $1/\bar{M}_n$ (Billmeyer, 1984). The onset of entanglement corresponds to a plateau region in which further increases in MW have little or no effect on T_g (Billmeyer, 1984). (There may, however, be a dramatic effect on the viscoelastic properties of a network, resulting, e.g. in increased gel strength at constant temperature (Ferry, 1980).) The conventional presentation of such experimental data is simply T_g

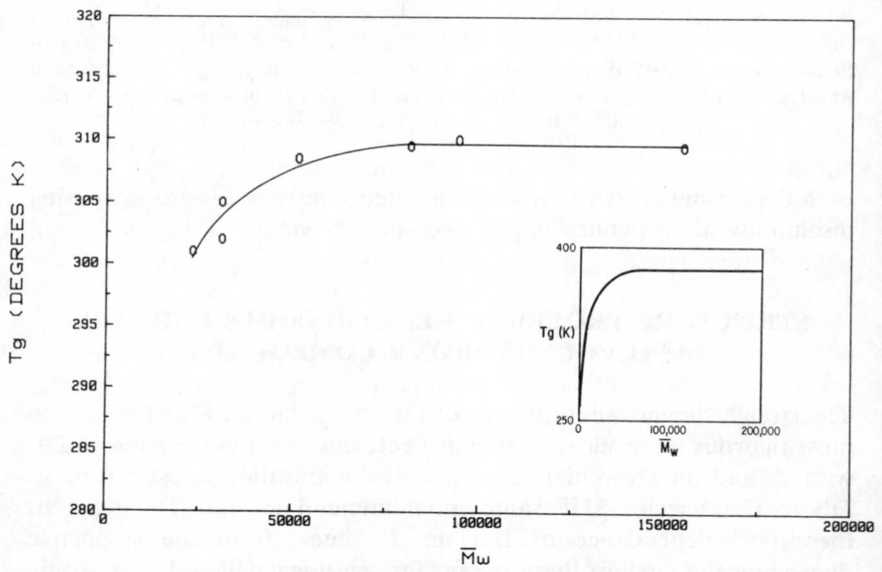

FIG. 9. Variation of the glass transition temperature, T_g, against $\bar{M}_w$ for a series of commercial poly(vinyl acetate) polymers. (Inset: an idealized plot of T_g vs $\bar{M}_w$ (Brennan, 1973).) Reproduced from Levine and Slade (1986).

vs $\bar{M}_w$ (Billmeyer, 1984), which conveniently displays the plateau region. An example is shown in Fig. 9 (Levine and Slade, 1986) for a homologous series of amorphous linear poly(vinyl acetates). The inset of Fig. 9 (Brennan, 1973) illustrates idealized behavior of T_g increasing with increasing $\bar{M}_w$, up to the plateau limit for entanglement and onset of viscoelastic rheological properties, then leveling off with further increases in $\bar{M}_w$ (Ferry, 1980; Billmeyer, 1984).

To a first approximation, DE has the simple inverse dependence on $\bar{M}_n$ defined earlier. Expressing that equation as $\bar{M}_n = 18016/\mathrm{DE}$, and using the conventional presentation to explore the behavior of $T_{g'}$ with MW, we have shown (Levine and Slade, 1986), in the main plot of Fig. 10, the $T_{g'}$ results for the SHPs in Table 1. After a steeply-rising portion, a plateau region was reached for SHPs with DE ≤ 6 and $T_{g'} \geq -8$°C. (The inset of Fig. 10 shows the linear relationship between decreasing $T_{g'}$ and increasing $1/\bar{M}_n$, for SHPs with $\bar{M}_n$ values below the entanglement limit. To and Flink's (1978) T_c/MW data showed the same correlation. In fact, the theoretical treatment in the inset is simply a modified version of the straightforward presentation in Fig. 3, with the same correlation coefficient of −0·98.) We suggested (Levine and Slade, 1986) that the most likely explanation for this plateau behavior is that such SHPs experience molecular entanglement in the 'low-moisture' freeze-concentrated glass that exists at $T_{g'}$ and $C_{g'}$. Consequently, we predicted that SHPs with DE ≤ 6 and $T_{g'} \geq -8$°C would be capable of forming gel networks (via entanglement), above a critical polymer concentration (which would be related to $C_{g'}$). Braudo *et al.* (1984), in their reports on the viscoelastic properties of thermoreversible maltodextrin gels (at $T > 0$°C), also implicated entanglement coupling above a critical polymer concentration. They concluded that the non-cooperative gelation behavior shown by maltodextrins is characteristic of semi-rigid chain polymers. This is consistent with Ferry's (1980) observation that 'molecules which are relatively stiff and extended (in concentrated solution) exhibit the effects of entanglement coupling even more prominently than do highly flexible polymers'. Additional information about thermoreversible maltodextrin (5–8 DE) gels came from Bulpin *et al.* (1984), who reported that such SHP gels are apparently composed of a network of high-MW (>10 000) branched molecules derived from amylopectin. These branched molecules represent the structural elements, which are aggregated with, and further stabilized by interactions with, short linear chains (MW < 10 000) derived from amylose.

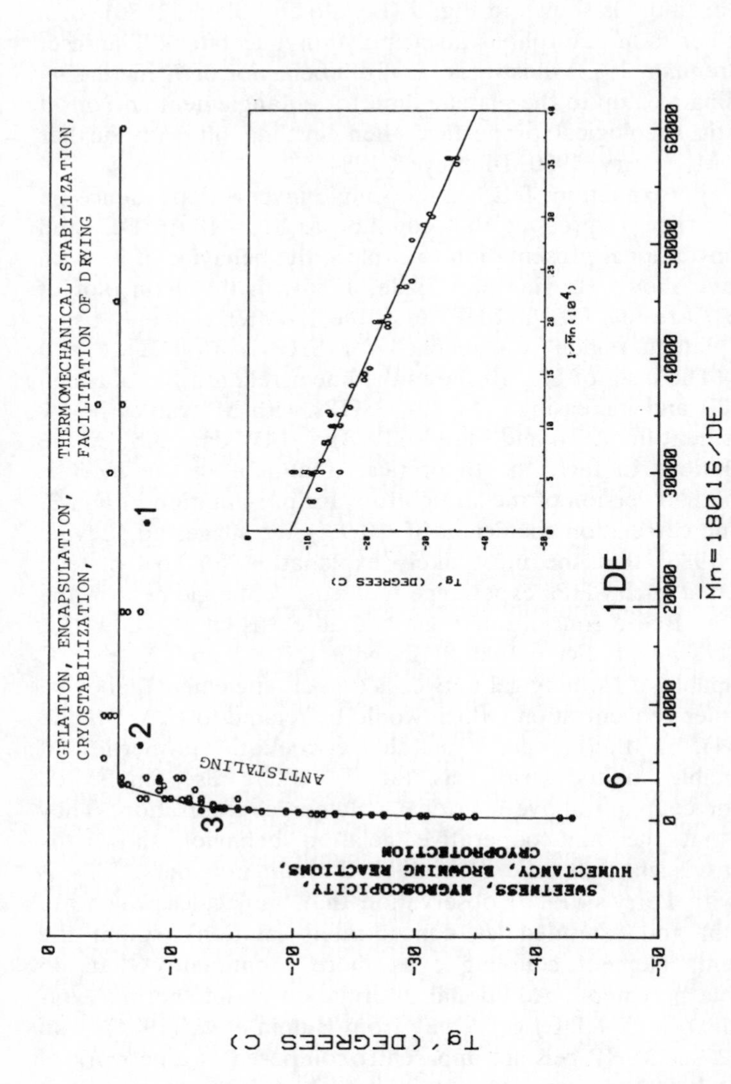

FIG. 10. Variation of the glass transition temperature, $T_{g'}$, against $\bar{M}_n$ (expressed as a function of DE) for commercial SHPs in Table 1. DE values are indicated by numbers marked above x-axis. Data points for maltodextrin MW standards are numbered 1, 2, and 3 to provide MW markers. Areas of specific functional attributes, corresponding to three regions of the diagram, are labeled. (Inset: plot of T_g vs $1/\bar{M}_n \times 10^4$ for SHPs with $\bar{M}_n$ values below entanglement limit, illustrating the theoretically-predicted linear dependence.) Updated from Levine and Slade (1986).

The implications of our finding and the conclusions we drew from it have allowed us to explain previously-observed but poorly-understood aspects of SHP functional behavior in various food applications (Levine and Slade, 1986, 1987*b*). The SHPs which fall on the plateau region in Fig. 10 have DEs from 6–0·3. These DEs correspond to $\overline{DP}_n$ values in the range 18–370, respectively, and $\bar{M}_n$ values between 3000 and 60 000. (Data points for three maltodextrin MW standards are numbered in Fig. 10 to provide MW markers. The points for #1) MD-6-40 ($\bar{M}_n = 27\,200$; $\bar{M}_w = 39\,300$) and #2) MD-6-12 ($\bar{M}_n = 6500$; $\bar{M}_w$ 13 000) fall on the plateau, while #3) MD-6-1 ($\bar{M}_n = 880$; $\bar{M}_w = 1030$) is below the entanglement limit.) Within this series of SHPs, the minimum linear chain length apparently required for intermolecular entanglement corresponds to $\overline{DP}_n \simeq 18$ and $\bar{M}_n \simeq 3000$. This fact explained why there is no plateau region in To and Flink's (1978) plot of T_c vs $\overline{DP}_n$ for SHPs of $\overline{DP}_n \leq 16$ and $DE \geq 6{\cdot}9$. Their Fig. 3 and the portion of our Fig. 10 for $DE \geq 7$ are similar in appearance; both show a steeply-rising portion for $DE \geq 20$, followed by a less steeply-rising portion for $20 \geq DE \geq 7$. Importantly, the entanglement capability evidenced by SHPs of $DE \leq 6$ (materials analyzed by our polymer characterization method (Levine and Slade, 1986), but not previously analyzed by To and Flink (1978)) underlies various aspects of their functional behavior. For example, as described by Slade (1984), sufficiently long linear chain lengths ($\overline{DP}_n \gtrsim 18$) of SHPs have been correlated with intermolecular network formation and thermoreversible gelation, and with SHP and starch (re)crystallization by a chain-folding mechanism in dilute solution. We suggested (Levine and Slade, 1986) that, in a partially-crystalline SHP gel network, the existence of random interchain entanglements in amorphous regions and 'fringed micelle' or chain-folded microcrystalline junction zones (Reuther *et al.*, 1984) each represents a manifestation of sufficiently long chain length. This suggestion was supported by recent work (Ellis and Ring, 1985; Miles *et al.*, 1985) which showed that amylose gels, found to be partially crystalline, were formed by cooling solutions of entangled chains. Miles *et al.* (1985) stated that amylose gelation requires network formation, and this network formation requires entanglement, and they concluded that 'polymer entanglement is important in understanding the gelation of amylose'.

The excellent fit of the experimental data in Fig. 10 to a conventional presentation of behavior expected for such a homologous family of oligomers and high polymers was gratifying, especially considering the numerous caveats that one must mention about commercial SHPs. For example, in Fig. 10, we used $\bar{M}_n$ (and implicitly $\overline{DP}_n$) values, calculated from DE, while in the conventional form (Fig. 9), $\bar{M}_w$ is used as a basis for specifying a typical MW range for the entanglement limit. Furthermore, for highly-polydisperse solutes such as commercial SHPs (for which MW distribution, MWD $= \bar{M}_w/\bar{M}_n$, is frequently a variable (Dziedzic and Kearsley, 1984)), the observed $T_{g'}$ is actually a weight-average $T_{g'}$ of the mixture of solutes (Franks, 1985*b*; Levine and Slade, 1987*a*). Despite these facts, the entanglement limit of $\bar{M}_n \simeq 3000$ for the SHPs in Fig. 10 is within the characteristic range of 1250–19 000 for the minimum entanglement MWs of many typical synthetic linear high polymers (Graessley, 1984). This result for the SHPs in Fig. 10 was corroborated by the behavior of the polyhydroxy compounds in Table 2, as illustrated by the $T_{g'}$ vs MW data in the main plot in Fig. 4. It was clear, from the shape of the curve in Fig. 4 (and the linearity of the $T_{g'}$ vs 1/MW plot in the inset), that these monodisperse sugars, glycosides, and polyols did not show evidence of entanglement coupling (Levine and Slade, 1987*b*). For these saccharide oligomers, none larger than a heptamer of MW 1153, the entanglement plateau has not been reached, a result in agreement with the MW range of entanglement limits cited above.

The variable polydispersity of commercial SHPs was mentioned above. Other largely-uncontrollable potential variables within the series of SHPs include (a) significant lot-to-lot variability of solids composition (i.e. saccharides distribution) for a single SHP, which would affect the reproducibility of $T_{g'}$; and (b) 'as is' moisture contents (in the generally-specified range of 5–10 w%) for different solid SHPs, which would not affect measured $T_{g'}$ (since, e.g. 15, 20, and 25 w% solutions would all freeze-concentrate to the same invariant $T_{g'}$ point on the state diagram), but would affect the calculated $W_{g'}$. Another source of variability among different SHPs of nominally-comparable DE concerns method of production, i.e. hydrolysis by acid, enzyme, or acid plus enzyme (Dziedzic and Kearsley, 1984; Medcalf, 1985). Especially with regard to enzymatic hydrolyzates, each particular enzyme produces a different set of characteristic breakdown products with a unique MWD (Krusi and Neukom, 1984).

Still another major variable among SHPs (and even for a single SHP) from different vegetable starch sources involves the original amylose/amylopectin ratio of a starch, and the consequent ratio of linear to branched polymeric chains in an SHP (Dziedzic and Kearsley, 1984; Medcalf, 1985). The influence of this variable can be particularly pronounced among a set of low-DE maltodextrins (which would contain higher-DP fractions), some from amylose-containing normal starches (e.g. dent corn) and others from all-amylopectin waxy starches (e.g. waxy maize). We demonstrated that the consequent range of $T_{g'}$ values can be quite broad , since, as a generally-observed rule, linear chains give rise to higher $T_{g'}$ than branched chains (with multiple chain ends (Blanshard, 1986)) of equal $\bar{M}_w$ (Levine and Slade, 1986). This behavior was shown by several pairs of SHPs in Table 1. For each pair, of the same DE and manufacturer (e.g. Star Dri 1 and Star Dri 10 (1984) vs (1986); Paselli MD-6 vs SA-6 and MD-10 vs SA-10; Morrex 1910 vs 1918), the hydrolyzate from amylose-free starch had a lower $T_{g'}$ than the corresponding one from a starch containing amylose (Levine and Slade, 1987*b*). This behavior was also exemplified by the $T_{g'}$ data for 13 10-DE maltodextrins in Table 1, for which $T_{g'}$ ranged from $-7{\cdot}5$°C for Stadex 27 from dent corn to $-15{\cdot}5$°C for Frodex 10 from waxy maize, a ΔT of 8°C. It is noteworthy that such a ΔT is comparable to the difference in $T_{g'}$ between oligomers of $\mathrm{DP} = n$ and $n + 1$. Further illustration was provided by the $T_{g'}$ data for some of the glucose oligomers in Table 2. Those results (see Table 3) demonstrated that, within such a homologous series, $T_{g'}$ appears to depend most rigorously on linear $\overline{\mathrm{DP}}_w$ of the solute. From comparisons of the significant $T_{g'}$ differences among maltose (1→4-linked glucose dimer), gentiobiose (1→6-linked), and isomaltose (1→6-linked), and among maltotriose (1→4-linked trimer), panose (1→4, 1→6-linked), and isomaltotriose (1→6, 1→6-linked), we suggested that 1→4-linked (linear amylose-like) glucose oligomers manifest greater 'effective' linear chain lengths in aqueous solution (and, consequently, greater hydrodynamic volumes) than oligomers of the same MW which contain 1→6 (branched amylopectin-like) links (Levine and Slade, 1987*b*). Table 3 (Levine and Slade, 1988) illustrates the intrinsic sensitivity of the $T_{g'}$ parameter to molecular configuration, in terms of linear chain length, as influenced by the nature of the glycosidic linkages in various non-homologous saccharide oligomers (not limited to glucose units) and the resultant effect on solution conformation. Another interesting comparison was between the $T_{g'}$ values for the linear and cyclic α-(1→4)-linked glucose hexamers, maltohexaose ($-14{\cdot}5$°C) and α-cyclodextrin (-9°C). In this case, the higher $T_{g'}$ of the cyclic

TABLE 3
Dependence of $T_{g'}$ on linear $\overline{DP}_w$ for sugars

Sugar	*n-mer*	$T_{g'}$ (°C)	*Structure*
Maltose	2	−29·5	Glucose 1→4 Glucose
Cellobiose	2	−29	Glucose 1→4 glucose
Isomaltose	2	−32·5	Glucose 1 ↓
Gentiobiose	2	−31·5	6 glucose
Nigerose	2	−35·5	Glucose 1 ↓
Laminaribiose	2	−31·5	3 glucose
Maltulose	2	−29·5	Glucose 1→4 fructose
Cellobiulose	2	−32·5	Glucose 1→4 fructose
Isomaltulose	2	−35·5	Glucose 1 ↓ 6 fructose
Lactose	2	−28	Glucose 4←1 galactose
Melibiose	2	−30·5	Glucose 6 ↑ 1 galactose
Maltotriose	3	−23·5	Glucose 1→4 glucose 1→4 glucose
Panose	3	−28	Glucose 1→4 glucose 1 ↓ 6 glucose
Isomaltotriose	3	−30·5	Glucose 1 ↓ 6 glucose 1 ↓ 6 glucose
Raffinose	3	−26·5	Galactose 1 2 fructose ↓ ↑ 6 glucose 1
Maltotetraose	4	−19·5	Glucose 1→4 glucose 1→4 glucose 1→4 glucose
Stachyose	4	−23·5	Galactose 1 ↓ 6 galactose 1 2 fructose ↓ ↑ 6 glucose 1
Nystose	4	−26·5	Glucose 1 ↓ 2 fructose 1 ↓ 2 fructose 1 ↓ 2 fructose

Reproduced from Levine and Slade (1988).

oligomer led us to suggest that the ring of α-cyclodextrin apparently has a much larger hydrodynamic volume in solution (due to its relative rigidity (Beesley, 1985)) than does the linear chain of maltohexaose, which is relatively flexible and can assume a more compact solution conformation (Levine and Slade, 1987*b*).

The above comparisons pointed up the subtleties of structure/property analysis of SHPs. We reached the obvious conclusion, as others have (Dziedzic and Kearsley, 1984; Medcalf, 1985), regarding the choice of a suitable maltodextrin for a specific application, that one SHP is not necessarily interchangeable with another of the same nominal DE, but from a different commercial source. We advised that basic characterization of structure/property relations, e.g. in terms of $T_{g'}$ (rather than DE, which can be a less significant (Dziedzic and Kearsley, 1984) and even misleading quantity), be carried out before one selects such food ingredients (Levine and Slade, 1986).

PREDICTED FUNCTIONAL ATTRIBUTES OF SHPs AND POLYHYDROXY COMPOUNDS

We have demonstrated how further insights into structure/function relationships may be gleaned by treating Fig. 10 as a predictive map of regions of functional behavior for SHP samples (Levine and Slade, 1986). For example, polymeric SHPs which fall on the entanglement plateau demonstrate certain functional attributes, some of which have been reported in the past, but not quantitatively explained from the theoretical basis of the entanglement capability revealed by our studies. The plateau region defines the useful range of gelation, encapsulation, cryostabilization, thermomechanical stabilization, and facilitation of drying processes. The lower end of the $\bar{M}_n$ range corresponds to the region of sweetness, hygroscopicity, humectancy, browning reactions, and cryoprotection. The intermediate region at the upper end of the steeply-rising portion represents the area of anti-staling ingredients (Krusi and Neukom, 1984). We have described how the map (labeled as in Fig. 10) can be used to choose individual SHPs or mixtures of SHPs and other carbohydrates (targeted to a particular $T_{g'}$ value) to achieve desired complex functional behavior for specific product applications. Especially for applications involving such mixtures, use can also be made of the data for the polyhydroxy compounds represented in Fig. 4, in combination with Fig. 10. One will recognize that the area represented by the left-hand third of Fig. 4 (and the low-MW sugars and polyols included therein) corresponds to the sweetness/hygroscopicity/humectancy/browning/cryoprotection region of Fig. 10. Likewise, the tri- through hepta-saccharides occupying the right-hand portion of Fig. 4 have been predicted to function similarly to the SHPs in the anti-staling region of Fig. 10 (Levine and Slade, 1987*b*).

As a specific example, the production of SHPs capable of gelation from solution should be designed to yield materials of $DE \leq 6$ and $T_{g'} \geq -8°C$ (Levine and Slade, 1986). This prediction agreed with results reported (Richter *et al.*, 1976*a,b*; Braudo *et al.*, 1979) for 25 w% solutions of potato starch maltodextrins of 5–8 DE, which produced thermoreversible, fat-mimetic gels, and for tapioca SHPs of $DE < 5$, which also form fat-mimetic gels from solution (Lenchin *et al.*, 1985). We also tested the accuracy of this prediction, as illustrated by the experimental results shown in Table 1 for the gelling ability of 20 w% solutions of many of the commercial SHPs. These results (Levine and Slade, 1987*b*) demonstrated a clear line of demarcation between gelling ($T_{g'} \geq -7{\cdot}5°C$ and $DE \leq 6$) and non-gelling SHPs. Thus, thermoreversible gels produced by refrigerating 20 w% solutions of SHPs of $DE \leq 6$ and $T_{g'} \geq -7{\cdot}5°C$ appear to form by a mechanism involving gelation-via-crystallization-plus-entanglement (Levine and Slade, 1987*a*) in concentrated solutions undercooled to $T < T_m$ (Wunderlich, 1976, 1980). As suggested previously, for SHPs of sufficiently long linear chain length ($\overline{DP}_n \gtrsim 18$), a partially-crystalline gel network is formed which apparently contains random interchain entangelements in the amorphous regions and microcrystalline intermolecular junction zones (Slade, 1984). The latter accounts for the thermoreversibility of such SHP gels, via a crystallization (on undercooling)–melting (on heating) process (Slade and Levine, 1984*b*).

Other experimental evidence, which supported these conclusions about the gelation mechanism for partially-crystalline polymeric SHP gels, included the following (Levine and Slade, 1987*a*): (1) DSC analysis of 20 w% SHP gels, after setting by overnight refrigeration, revealed a small crystalline melting endotherm at $T_m \simeq 60°C$, similar to the characteristic melting transition observed in retrograded B-type starch gels (Slade, 1984); and (2) the relatively small extent of crystallinity in these SHP gels was increased significantly by a 2-step temperature-cycling gelation protocol (12 h at 0°C, followed by 12 h at

40°C), adapted from one originally developed by Ferry (1948) for gelatin gels, and subsequently applied by Slade (Slade and Levine, 1987*a*) to retrograded starch gels. We suggested (Levine and Slade, 1987*a*) that, in some fundamental respects, the thermoreversible gelation of concentrated aqueous solutions of polymeric SHPs appears to be analogous to the thermoreversible gelation and crystallization of synthetic homopolymer and copolymer–organic diluent systems (Boyer *et al.*, 1985; Domszy *et al.*, 1986). For partially-crystalline gels of the latter type, the possibly-simultaneous presence of random interchain entanglements in the amorphous regions (Boyer *et al.*, 1985) and microcrystalline junction zones (Domszy *et al.*, 1986) has been reported. However, controversy still exists, even in these recent publications, over which of the two conditions (if either alone) might be the necessary and sufficient one primarily responsible for the structure/property relations of such partially-crystalline polymeric gelling systems. The controversy could be resolved by a simple dilution test; entanglement gels can be dispersed by dilution, microcrystalline gels cannot be.

We have suggested that maltodextrins to be used for encapsulation of volatile flavors/aromas and oxidizable lipids, an application requiring superior barrier properties (i.e. relative impermeability to gases and vapors), should likewise be capable of entanglement and network formation ($T_{g'} \geq -8$°C) (Levine and Slade, 1986). It has been reported (To and Flink, 1978; Flink, 1983; Karel and Flink, 1983) that effectiveness of encapsulation increases with increasing T_c, which in turn increases with increasing $\overline{DP}_n$ within a series of SHPs, although a quantitative relationship between T_c and MW had not been established (To and Flink, 1978), prior to our work. Recently, Karel (1986) noted that oxidation of lipids can be inhibited by encapsulation in stable, low-moisture carbohydrate glasses, in which, at $T < T_c = T_g$, mobility is restricted and diffusion of gases and vapors is consequently retarded. Karel stated that lipids so encapsulated will not oxidize until plasticization by moisture and/or heat, to $T_c = T_g < T$, allows sufficient mobility and permeability to permit penetration by oxygen. Reineccius (1986) described some recent results, concerning encapsulation (to retard oxidation) of orange oil in amorphous spray-dried SHP matrices, that seemed to confirm our suggestion regarding entanglement. He measured shelf-life in terms of extent of oxygen uptake and subsequent oil oxidation, and reported that exceptionally long shelf-life (i.e. essentially no oxyen uptake) resulted for a substrate made from ARD 2326, an SHP which had been predicted by our analysis to be an effective encapsulator and had been shown by our results to be capable of entanglement and gelation (Levine and Slade, 1987*b*). However, Reineccius also tested a homologous series of SHPs (Maltrins M040–M365), and reported (Anandaraman and Reineccius, 1986) that shelf-life increased with *increasing* DE of the encapsulating SHP (although it did not approach that of the ARD 2326 sample). This finding, which contradicts what we would have predicted and also the result for ARD 2326, might be explained as a consequence of pre-existence of entrapped oxygen (rather than oxygen uptake) in the spray-dried matrices (samples were spray dried in air, and not de-gassed), a possibility that Karel (1986) appears to have recognized. If true, the SHP matrices would have shown increasing permeability to release of entrapped oxygen with increasing DE, thus explaining these confusing and potentially-misleading results.

There is a substantial recent patent literature describing uses of SHPs for encapsulating/coating applications in various food products, exemplified by the following: (a) maltodextrins of 4–20 DE to produce stable amorphous spray-dried substrates for encapsulation of acetaldehyde (Saleeb and Pickup, 1985); (b) SHPs of ≤20 DE to encapsulate flavor oils in stable extruded glassy melts (Miller and Mutka, 1986); (c) a heat-stable sweetening composition produced by encapsulating APM in maltodextrin (Colliopoulos *et al.*, 1984); (d) maltodextrins of DE ≤ 10 as amorphous coatings for encapsulation of crystalline salt-substitute particles (Meyer, 1985); and (e) maltodextrins or dextrins as coating agents for roasted nuts candy-coated with honey (Green and Hoover, 1979).

With regard to the freezer-storage (in our terminology, 'cryo') stabilization of fabricated frozen foods (e.g. desserts such as ice cream, with smooth/creamy texture) against ice crystal growth over time, inclusion of low-DE maltodextrins elevates the composite $T_{g'}$ of a mix of soluble solids, which is typically dominated by low-MW sugars (Levine and Slade, 1986). In practice, a retarded rate of ice recrystallization ('grain growth') at the characteristic freezer temperature (T_f) results, along with an increase in observed T_r. Such behavior has been documented in several soft-serve ice cream patents (Cole *et al.*, 1983, 1984; Holbrook and Hanover, 1983). In such products, ice recrystallization is known to involve a diffusion-controlled maturation process with a mechanism analogous to 'Ostwald ripening', whereby larger crystals grow with time at the expense of smaller ones which eventually disappear (Maltini, 1977; Bevilacqua and Zaritzky, 1982; Harper and Shoemaker, 1983). The rate of such a process (at T_f), and thus also ΔT ($T_f - T_{g'}$), is reduced by formulating with low-DE maltodextrins of high $T_{g'}$. The fact that ice recrystallization exemplifies a collapse process whose kinetics can be described by the WLF equation is illustrated in Fig. 11 (Levine and Slade, 1987*a*). Because typical ice creams and frozen novelties have $T_{g'}$ values in the range −30 to −43°C, they would exist as rubbery fluids (with embedded ice and fat crystals) under typical freezer storage at −18°C, and WLF (rather than Arrhenius) kinetics would describe the rate of ice crystal growth. Figure 11 contains two WLF plots of log rate of iciness development vs ΔT, which show that iciness increases with increasing ΔT, with $r = 0{\cdot}68$ for a set of ice cream prototypes (Cole *et al.*, 1983, 1984) and 0·83 for a set of frozen novelties. Considering that iciness scores were obtained by sensory evaluation, we described this unprecedented experimental demonstration of WLF behavior in a frozen system as remarkable (Levine and Slade, 1987*a*). However, we mentioned that this non-Arrhenius behavior of freezer-stored ice cream is apparently recognized, at least empirically, by the British frozen foods industry. We have seen on ice cream packages in England the following shelf-life code: in a * home freezer (15°F), storage life = 1 day; ** freezer (0°F), 1 week; and *** freezer (−10°F), 1 month. In practice, the technological utility of $T_{g'}$ and $W_{g'}$ results for sugars, polyols, and SHPs (in Tables 1, 2, and 5), in combination with corresponding relative sweetness data, has been demonstrated by the successful formulation of fabricated products (e.g. Cole *et al.*, 1983, 1984) with an optimum combination of stability and softness at 0°F freezer storage. Low-DE maltodextrins are also used to stabilize frozen dairy products against lactose crystallization (another example of a diffusion-controlled collapse phenomenon) during storage (Kahn and Lynch, 1985).

Low-DE maltodextrins and other high-MW polymeric solutes (e.g. see Table 5 (Levine and Slade, 1987*b*)) are well known as drying aids for processes such as freeze, spray, and drum drying (Flink, 1983; Karel and Flink, 1983; MacKenzie, 1981; Nagashima and Suzuki, 1985; Szejtli and Tardy, 1985). Through their simultaneous effects of increasing the composite $T_{g'}$ and reducing the unfrozen water fraction ($W_{g'}$) of a system of low-MW solids (with regard to freeze drying) or increasing the relative vapor pressure (for spray or drum drying), maltodextrins raise the observed T_c (at any particular moisture

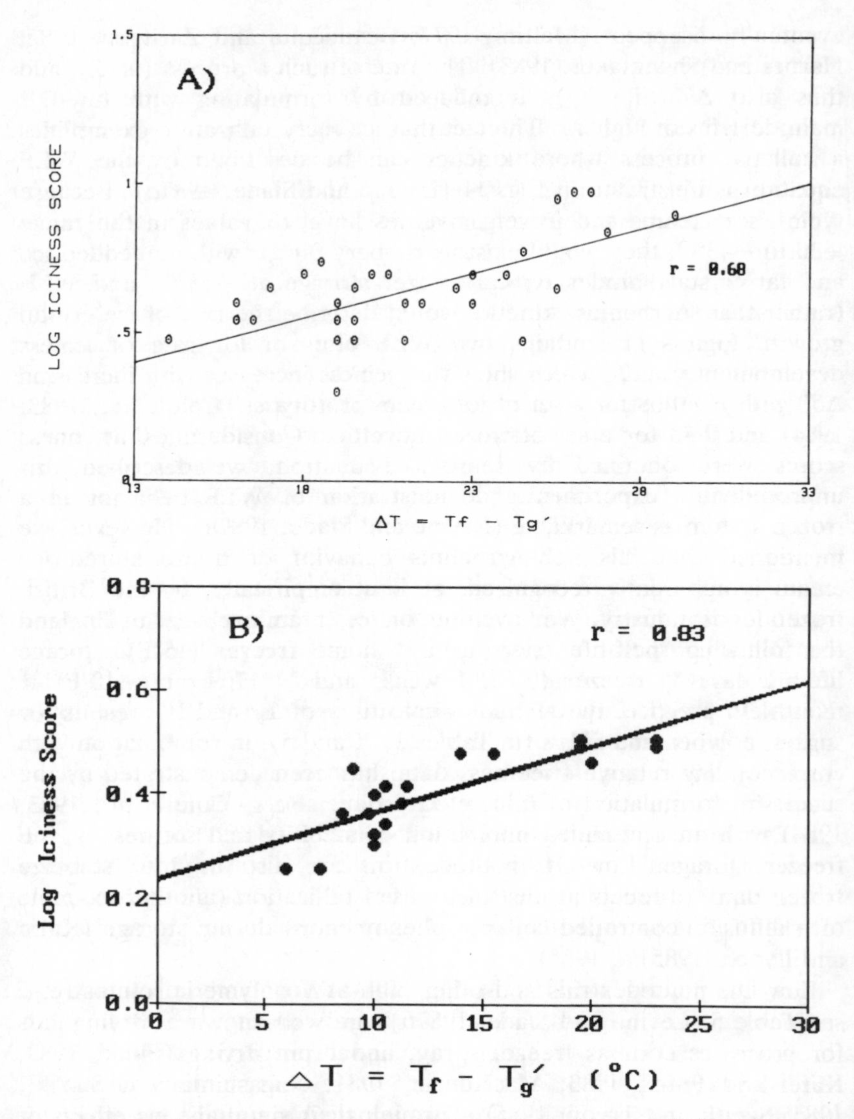

FIG. 11. Log iciness score (determined organoleptically, on a 0–10 point scale in (A) and a 0–5 scale in (B)) as a function of $\Delta T(=T_f-T_{g'})$, for experimental (A) ice cream products and (B) frozen novelties, after two weeks of deliberately-abusive (temperature-cycled) frozen storage in a so-called 'Brazilian Ice Box'. Updated from Levine and Slade (1987*a*).

content) relative to the drying T, thus stabilizing the glassy state and facilitating drying without collapse or 'melt-back' (Levine and Slade, 1986). By reducing the inherent hygroscopicity of a mixture of solids being dried, maltodextrins decrease a system's propensity to collapse (from the rubbery state) due to plasticization at low moisture. These attributes are illustrated (Nagashima and Suzuki, 1985) by recent findings on the freeze-drying behavior of beef extract with added dextrin.

Thermomechanical stabilization is a term we use to describe the stabilization of low-moisture glasses such as boiled candies (which are typically very hygroscopic and notoriously sensitive to plasticization by water (White and Cakebread, 1966; Cakebread, 1969)) against such collapse phenomena as recrystallization of sugars ('graining'), mechanical deformation, and stickiness (Levine and Slade, 1986). Incorporation of low-DE maltodextrins in low-MW sugar glasses (to increase average $\bar{M}_w$ of the solutes) has been shown (White and Cakebreak, 1966; Cakebread, 1969; Lees, 1982; Vink and Deptula, 1982) to increase T_g and thus storage stability at $T < T_g$. Even when such a candy 'melt' is in the unstable rubbery state at $T_g < T_{storage}$, maltodextrins are known to function as inhibitors of the diffusion-controlled propagation step in the sugar recrystallization process (White and Cakebread, 1966; Cakebread, 1969; Dziedzic and Kearsley, 1984). Low-DE maltodextrins are also used frequently to stabilize other amorphous solids (e.g. food powders) (Dziedzic and Kearsley, 1984) against various collapse phenomena that are exacerbated by plasticization at low moisture. For example, 'dextrins' (SHPs ≥ tetrasaccharides of DE ≤ 25) are used as anti-caking/anti-browning agents in low-moisture powders (Ogawa and Imamura, 1985), and 10-DE maltodextrin is used as an additive in production of shelf-stable amorphous juice solids of unusually low hygroscopicity (Miller and Mutka, 1985).

For the lower-T_g' SHPs in Fig. 10 (as for many of the low-T_g' reducing sugars in Fig. 4), sweetness, hygroscopicity, humectancy, and browning reactions are salient functional properties (Dziedzic and Kearsley, 1984; Levine and Slade, 1987*b*). A less familiar one involves the potential for cryoprotection of biological materials, for which the utility of various other low-MW, glass-forming sugars and polyols is well known (Franks, 1982, 1985*a*; Forsyth and MacFarlane, 1986; Thom and Matthes, 1986; Vassoille *et al.*, 1986). The map of Fig. 10 led us to predict, and DSC experiments confirmed, that such SHPs and other low-MW carbohydrates, in sufficiently concentrated solution, can be quench-cooled to a completely-vitrified state, so that all the water is immobilized in the solute/unfrozen water glass (Levine and Slade, 1986). Such vitrification has also been suggested as a natural intracellular cryoprotective mechanism in winter-hardened poplar trees (Hirsh *et al.*, 1985), and demonstrated as a means of cryoprotecting whole body organs and embryos (Rall and Fahy, 1985). The essence of this cryoprotective activity, indefinite avoidance of ice formation and solute crystallization in concentrated solutions of low-MW, non-crystallizing solutes which have high $W_{g'}$ values, also has a readily-apparent relationship to food applications involving soft, spoonable, or pourable-from-the-freezer products. One example is Rich's patented 'Freeze-Flo' beverage concentrate formulated with high fructose corn syrup (Kahn and Eapen, 1982). We have reported model-system experiments with 60 w% solutions of e.g. fructose and mannose (3·3 M), methyl fructoside (3·1 M), ethyl fructoside, ethyl mannoside, and ethyl glucoside (2·9 M), in which these samples have remained pourable fluids (completely ice and solute-crystal free) under kinetic control during 0°F freezer storage (i.e. at $T <$ the theoretical freezing point due to colligative depression) for over four years to date (Levine and Slade, 1987*a*).

The literature on SHPs as anti-staling ingredients for starch-based foods (reviewed elsewhere (Slade, 1984; Slade and Levine, 1987*a*)), including the work of Krusi and Neukom (1984), reports that (non-entangling) SHP oligomers of $\overline{DP}_n$ 3–8 are effective in inhibiting, and not participating in, starch recrystallization.

We have also postulated from the map of Fig. 10 that addition of a low-MW sugar to a gelling maltodextrin could produce a sweet and softer gel, while addition of a glass-forming sugar to an encapsulating maltodextrin should promote limited collapse of the entangled network around the absorbed species (if collapse were desirable), but decrease the ease of spray drying (Levine and Slade, 1986). The latter postulate seems consistent with results of Saleeb and Pickup (1985), who reported improved encapsulation of volatiles in dense, amorphous matrices composed of a majority of maltodextrin (4–20 DE) plus a minority of a mono- or disaccharide glass-former. The map of Fig. 10 led us to two other intriguing postulates (Levine and Slade, 1986). A freeze-concentrated glass at $T_{g'}$ of an SHP cryostabilizer (of DE $\leq$ 6) would contain entangled polymer molecules, while in a glass at $T_{g'}$ of a low-MW SHP cryoprotectant, solute molecules could not be entangled. By analogy, various high-MW polysaccharide gums are believed to be capable of improving freezer-storage stability of ice-containing fabricated foods, in some poorly-understood way. In the absence of direct evidence of an effect on ice crystal size, and despite recent work which showed definitively that such stabilizers have no significant effect on either ice nucleation or propagation rates (Muhr and Blanshard, 1986; Muhr *et al.*, 1986), the effect has been attributed to increased viscosity (Keeney and Kroger, 1974; Harper and Shoemaker, 1983). Such gums may owe their limited success not only to their viscosity-increasing ability, which would be common to all glass-formers, but to their possible capabilty to undergo entanglement in a freeze-concentrated, non-ice matrix of a frozen food. Entanglement might enhance a mouthfeel which masks their limited ability to inhibit diffusion-controlled processes. In a related vein, effects of entanglement coupling on viscoelastic and rheological properties of random-coil polysaccharide concentrated solutions (Morris *et al.*, 1981) and gells (Mitchell, 1980; Braudo *et al.*, 1984), at $T > 0°C$, have been reported.

THE ROLE OF SHPs IN COLLAPSE PROCESSES AND THEIR MECHANISM OF ACTION

In the remainder of this chapter, we shall review the often critical role of polymeric SHPs in preventing structural collapse, within a context of the various collapse phenomena listed in Table 4 (Levine and Slade, 1986). These phenomena include ones pertaining to processing and/or storage at $T > 0°C$ of low-moisture foods as well as ones involving the frozen state, all of which are governed by the particular T_g relevant to the system and its content of plasticizing water. While for frozen systems, $T_{g'}$ of the freeze-concentrated, 'low-moisture' glass is the relevant T_g for describing the T_g/MW relationship (as illustrated by Figs 4, 5(B), and 10); for amorphous dried powders and candy glasses, the relavent T_g pertains to a higher temperature/very low moisture state. It has been assumed that a plot of T_g vs $\bar{M}_n$ for dry SHPs would reflect the same fundamental behavior as that shown in Figs 4, 5(B), and 10, since T_c for low-moisture SHP samples (To and Flink, 1978) represents a good quantitative approximation of dry T_g (Levine and Slade, 1986). We have verified this assumption for the homologous series of pure malto-oligomers, as shown in Fig. 5(A). The plot of dry

TABLE 4
'Collapse'-related phenomena which are governed by T_g and involve plasticization by water (reference numbers in square brackets correspond to italic numbered references within the reference list)

	References
A. Processing and/or storage at $T > 0°C$	
1. Cohesiveness, sticking, agglomeration, sintering, lumping, caking, and flow of amorphous powders $\geq T_c^a$	*[5, 6, 10, 15, 16, 23, 24, 28, 29, 30, 36, 39, 40, 41, 44]*
2. Plating of, e.g. coloring agents or other fine particles on the amorphous surfaces of granular particles $\geq T_g$	*[1, 45]*
3. (Re)crystallization in amorphous powders $\geq T_c^a$	*[6, 14, 15, 16, 24, 40, 44]*
4. Structural collapse in freeze-dried products (after sublimation stage) $\geq T_c^a$	*[6, 15, 16, 40, 41]*
5. Loss of encapsulated volatiles in freeze-dried products (after sublimation stage) $\geq T_c^a$	*[5, 6, 15, 16, 38, 40]*
6. Oxidation of encapsulated lipids in freeze-dried products (after sublimation stage) $\geq T_c^a$	*[6, 14, 15, 40]*
7. Enzymatic activity in amorphous solids $\geq T_g$	*[2, 25, 32, 33]*
8. Maillard browning reactions in amorphous powders $\geq T_g^a$	*[28]*
9. Sucrose inversion in acid-containing amorphous powders $\geq T_g$	*[46]*
10. Stickiness in spray drying and drum drying $\geq T$ sticky point[a]	*[5, 6, 15, 16, 40, 41]*
11. Graining in boiled sweets $\geq T_g^a$	*[3, 6, 11, 12, 17, 22, 37, 42, 44]*
12. Sugar bloom in chocolate $\geq T_g$	*[4, 27]*
B. Processing and/or storage at $T < 0°C$	
1. Ice recrystallization ('grain growth') $\geq T_r^a$	*[7, 8, 9, 19, 21]*
2. Lactose crystallization ('sandiness') in dairy products $\geq T_r^a$	*[6, 8, 13, 44]*
3. Enzymatic activity $\geq T_{g'}^a$	*[18, 25, 35]*
4. Structural collapse, shrinkage, or puffing (of amorphous matrix surrounding ice crystals) during freeze drying (sublimation stage) = 'melt-back' $\geq T_c^a$	*[6, 8, 16, 19, 20, 26, 40, 41, 43, 44]*
5. Solute recrystallization during freeze drying (sublimation stage) $\geq T_d$	*[31]*
6. Loss of encapsulated volatiles during freeze drying (sublimation stage) $\geq T_c^a$	*[6, 16, 38, 40]*
7. Reduced survival of cryopreserved embryos, due to cellular damage caused by diffusion of ionic components $\geq T_{g'}$	*[34]*

[a] Examples exist in the food science and technology literature of stabilization against collapse through use of low-DE SHPs.
Updated from Levine and Slade (1986).

T_g vs MW shows the same qualitative curvature (and absence of an entanglement plateau) as the corresponding $T_{g'}$ plot, and the plot of dry T_g vs 1/MW the same linearity and r value as the $T_{g'}$ plot in Fig. 5(B). Further verification is illustrated in Fig. 12, by a plot of dry T_g vs w% composition for a series of spray-dried, low-moisture powders (~2 w% water) prepared from solution blends of commercial SHPs, Lodex 10 and Maltrin M365 (Saleeb, 1987). Here again, the characteristic monotonic increase of T_g with $\bar{M}_w$ (≡increasing composition as w% Lodex 10), and curvature expected for $\bar{M}_w$ values below the plateau limit, are evident.

All the collapse phenomena mentioned in Table 4 (as well as the glass transition itself (Ferry, 1980)) are translational diffusion-controlled (many are also nucleation-limited) processes, with a mechanism involving viscous flow in the rubbery liquid state (Flink, 1983), under conditions of $T > T_g$ and $\eta < \eta_g = 10^{11}$–10^{14} Pa s (Downton *et al.*, 1982). These kinetic processes are controlled by the variables of time, temperature, and moisture content (Tsourouflis *et al.*, 1976). At the relaxation temperature, % moisture is the critical determinant of collapse and its concomitant changes (Karel and Flink,

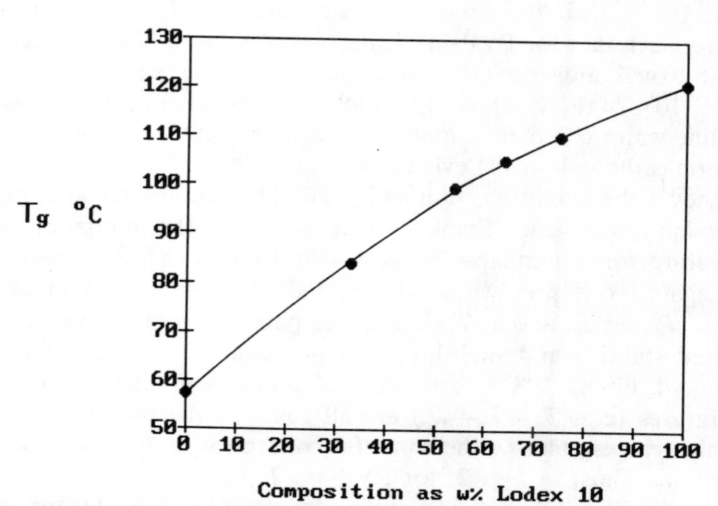

FIG. 12. Variation of the glass transition temperature, T_g, against weight % composition for spray-dried, low-moisture powders prepared from aqueous solution blends of Lodex 10 and Maltrin M365 SHPs. (Data from Saleeb (1987), personal communication.)

1983), through water's effect on T_g. This plasticizing effect of increasing moisture content at constant temperature (which is identical to the effect of increasing temperature at constant percentage moisture) leads to increased segmental mobility of polymer chains in the amorphous regions of both glassy and partially-crystalline polymers (Levine and Slade, 1987*a*). This in turn leads to the occurrence of the glass transition at decreased temperature (Cakebread, 1969).

The state diagram for PVP–water, shown in Fig. 2 (Levine and Slade, 1987*b*), illustrates the extent of this T_g-depressing effect of water. PVP is a typical water-soluble, completely-miscible, non-crystallizable polymer (Franks, 1982), whose behavior represents a good model of the analogous behavior of polymeric SHPs. The state diagram for water–PVP ($\bar{M}_n = 10\,000$, 44 000, and 700 000) in Fig. 2, compiled from several sources (MacKenzie and Rasmussen, 1972; Franks *et al.*, 1977; MacKenzie, 1977; Olson and Webb, 1978; Franks, 1982; Levine and Slade 1986), is the most complete one presently available for this polymer. PVP manifests a smooth T_g curve from about 100°C for dry PVP-44 to about −135°C, the value commonly stated (but never yet measured) for glassy water (Franks, 1982; Kanno, 1987). The dramatic effect of water on T_g is seen at low moisture, such that for PVP-44, T_g decreases about 6°C/w% water for the first 10 w% moisture. We have found that values in the range of about 5–10°C/w% water apply widely to amorphous and partially-crystalline water-compatible materials, including many monomeric and polymeric carbohydrates (Levine and Slade, 1987*a*).

Whenever the glass transition and resultant structural collapse occur on the same time scale (Franks, 1982), T_g equals the minimum onset temperature for the collapse processes in Table 4. Thus, a system is stable against collapse, within the period of experimental measurements of T_g and T_c, at $T < T_g$. Increasing percentage moisture leads to decreased stability and shelf-life, at a particular storage temperature (Karel and Flink, 1983). The various phenomenological threshold temperatures (e.g. $T_c = T_r = T_{sp}$) are all equal to the particular T_g (or $T_{g'}$) which corresponds to the solute(s) concentration for the situation in question. Thus, in Fig. 2, for PVP-44, $T_{g'} = T_r = T_c \simeq -21{\cdot}5$°C and $C_{g'} \simeq 65$ w% PVP ($W_{g'} \simeq 0{\cdot}54$ g unfrozen water/g PVP) (MacKenzie and Rasmussen, 1972; MacKenzie, 1977; Franks, 1982; Levine and Slade, 1986); while for PVP-700, $T_g = T_c = T_{sp} \simeq 120$°C at $\simeq$5% residual moisture (Olson and Webb, 1978; Levine and Slade, 1986). The equivalence of T_r for ice or solute recrystallization, T_c for collapse, and concentration-invariant $T_{g'}$ for an ice-containing system explains why T_r and T_c have always been observed in the past to be concentration-independent for all initial solute concentrations lower than $C_{g'}$ (Franks, 1982; Forsyth and MacFarlane, 1986), as illustrated in Fig. 2. Other T_g data for anhydrous polyhydroxy compounds we have analyzed are shown in Table 2. From results for $T_{g'}$ and corresponding dry T_g, 3-point T_g curves can be drawn, as a first step toward characterizing structure/property relations by constructing state diagrams for such low-MW carbohydrates. Coupled with T_m data also shown in Table 2, and resulting values of T_m/T_g ratio (the significance of which we have reviewed elsewhere (Levine and Slade, 1987*a*; Slade and Levine, 1987*b,d*), and will discuss briefly here with regard to Fig. 13), such results have been utilized to explain and predict functional behavior (Levine and Slade, 1987*a*).

Our conclusion regarding the fundamental equivalence of T_g, T_c, and T_r (Levine and Slade, 1986) represented a departure from the previous literature. For example, while To and Flink (1978) acknowledged that 'the relationship between T_c and MW is identical to the equation for T_g of mixed polymers' and that 'collapse and glass transition are (clearly) phenomenologically similar events', they differentiated between T_g and T_c by pointing out that 'while glass transitions in polymeric materials are generally reversible, the collapse of freeze-dried matrices is irreversible'. We pointed out that, while the latter facts may be true, their argument is misleading. At the molecular level, the glass-to-rubber transition for an amorphous thermoplastic material is reversible. That is, a glass at $T_{g'}/C_{g'}$ can be repeatedly warmed and recooled (slowly) over a completely-reversible T/C path between its solid and liquid states. The same is true for a completely-amorphous (and non-crystallizable) freeze-dried material. The reason collapse has been said to be irreversible for a porous matrix has nothing to do with reversibility between molecular states. Irreversible loss of porosity is simply a macroscopic, morphological consequence of viscous flow in the rubbery state at $T > T_g$, whereby a porous glass relaxes to a fluid (incapable of supporting its own weight against flow), which then becomes non-porous and more dense. Subsequent recooling to $T < T_g$ yields a non-porous glass of the original composition, which can thereafter be temperature-cycled reversibly. The only irreversible aspect of T_g-governed structural collapse is loss of porosity (Levine and Slade, 1986). However, a different scenario can pertain to freeze-

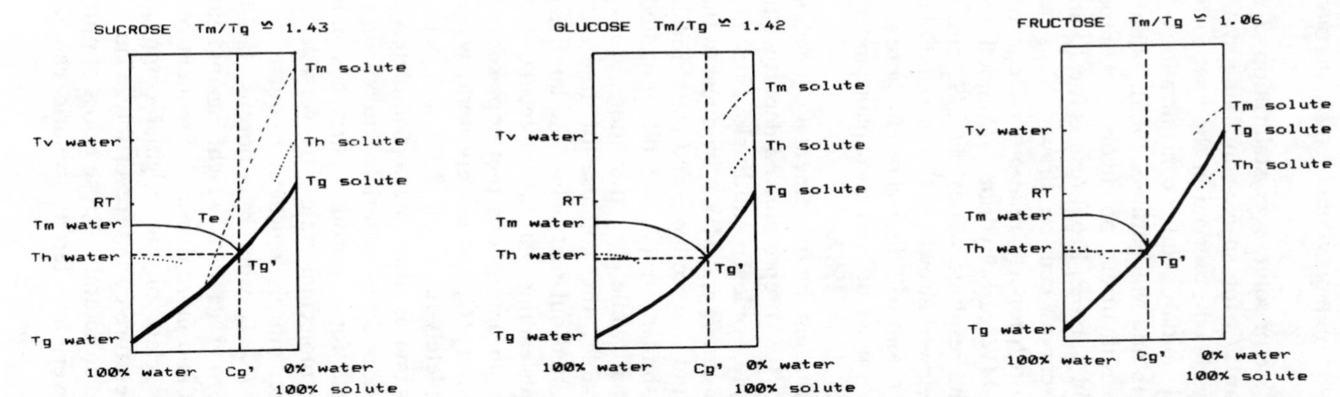

FIG. 13. Schematic state diagrams (of temperature vs weight % composition) for sucrose, glucose, and fructose, which emphasize the solute–water glass curve, and the relationship between T_m, T_g, and T_h, the estimated homogeneous nucleation temperature, for each pure solute. (T_e = eutectic melting temperature.) Adapted from Levine and Slade (1987*a*).

drying of a crystallizable solute. In this case, a second transition, representing irreversible solute (re)crystallization and concomitant disproportionation in the mobile state of the undercooled rubbery fluid at $T > T_{g'}$, can follow the glass-to-rubber transition (Phillips *et al.*, 1986).

Recently, our conclusion about the fundamental identity of $T_{g'}$ with T_r and T_c was corroborated by Reid (1985). He reported a study in which $T_{g'}$, measured by DSC, corresponded well with the temperature at which a frozen aqueous solution, viewed under a cryomicroscope, became physically mobile. Reid remarked that '$T_{g'}$, the temperature at which a system would be expected to become mobile due to appearance of the solution phase, has also been related to the T_c in freeze drying, again relating to the onset of system mobility, which presumably allows for diffusion of solution components'. Reid's study revealed another collapse-related phenomenon, governed by $T_{g'}$ of a frozen system, that was added to Table 4: slow warming of cryopreserved embryos to $T > T_{g'}$ facilitates the detrimental diffusion of ionic components (salts), resulting in cellular damage due to high ionic strength and much reduced embryo survival.

Table 5 shows a comparison of DSC-measured $T_{g'}$ values, for a variety of water-compatible monomers and polymers (including many food materials), and literature values for other collapse-transition temperatures. These results for observed T_c and T_r, which are usually measured (on an experimental time scale similar to that of our DSC method) by cryomicroscopy of frozen or vitrified aqueous solutions, are generally very close to, but almost always at a slightly higher temperature than, our values for $T_{g'}$. We have taken this fact as further support of our contention that $T_{g'}$ represents the minimum onset temperature for these subzero collapse phenomena (Levine and Slade, 1987*b*).

A Physico-Chemical Mechanism for Collapse Based on WLF Theory

We have described a universally-applicable, quantitative mechanism for collapse derived from WLF theory (Levine and Slade, 1986). The dependence of viscoelastic properties of polymers on temperature, in the rubbery range above T_g (typically from T_g to $T_g + 100\,\mathrm{K}$), is successfully predicted (Cowie, 1973) by the WLF equation derived from free volume theory (Williams *et al.*, 1955; Ferry, 1980; Soesanto and Williams, 1981). The WLF equation is shown in eqn (1):

$$\log\left(\frac{\eta}{\rho T}\Big/\frac{\eta_g}{\rho_g T_g}\right) = -\frac{C_1(T - T_g)}{C_2 + (T - T_g)} \qquad (1)$$

TABLE 5
Comparison of $T_{g'}$ values and literature values for other 'collapse' transition temperatures

Substance	T_r (°C)[a]	T_c (°C)[b]	$T_{g'}$ (°C)
Ethylene glycol	−70[c,e], −81[f]		−85
Propylene glycol	−62[e]		−67·5
Glycerol	−58, −65[c,e]		−65
Ribose	−43		−47
Glucose	−41, −38	−40	−43
Fructose	−48	−48	−42
Sucrose	−32, −30·5	−32, −34[d]	−32
Maltose		−32	−29·5
Lactose		−32	−28
Raffinose	−27, −25·4	−26	−26·5
Inositol		−27	−35·5
Sorbitol		−45	−43·5
Glutamic acid, Na salt		−50	−46
Gelatin	−11	−8	
Gelatin (300 Bloom)			−9·5
Gelatin (250 Bloom)			−10·5
Gelatin (175 Bloom)			−11·5
Gelatin (50 Bloom)			−12·5
Bovine serum albumin	−5·3		−13
Dextran		−9	
Dextran (MW 9400)			−13·5
Soluble starch	−5, −6[c]		
Soluble potato starch			−3·5
Hydroxyethyl starch	−21	−17[d]	−6·5
PVP	−22, −21, −14·5	−23, −21·6[d]	
PVP-10			−26
PVP-40			−20·5
PVP-44			−21·5
PEG	−65, −43	−13	
PEG (MW 200)			−65·5
PEG (MW 300)			−63·5
PEG (MW 400)			−61

[a] Recrystallization temperatures, from Franks (1982).
[b] Collapse temperatures during freeze-drying, from Franks (1982).
[c] From Luyet (1960).
[d] Antemelting temperatures, from Virtis Co. (1983).
[e] 'Completion of Opacity' temperatures, from Forsyth and MacFarlane (1986).
[f] Recrystallization temperature, from Thom and Matthes (1986).
Updated from Levine and Slade (1987*b*).

where η = viscosity or other diffusion-controlled relaxation process, ρ = density, and C_1 and C_2 are 'universal constants' (17·44 and 51·6, respectively, as extracted from data on numerous polymers (Williams *et al.*, 1955; Soesanto and Williams, 1981). Equation (1) describes the kinetic nature of the glass transition, and is universally applicable to any glass-forming polymer, oligomer, or monomer (including, e.g. molten glucose (Williams *et al.*, 1955)) (Ferry, 1980). The equation defines the exponential temperature dependence of any diffusion-controlled relaxation process, occurring at a temperature T, vs rate of the relaxation at a reference temperature, namely T_g below T, in terms of log η related usefully to ΔT, where $\Delta T = T - T_g$. The WLF equation is required in the temperature range of the rubbery or undercooled liquid state above T_g, and is based on the temperature dependence of free volume (i.e. temperature dependence of segmental mobility). It is not required much below T_g (i.e. in the glassy solid state) or in the very-low η liquid state (<10 Pa s (Soesanto and Williams, 1981)) more than about 100 K above T_g, where Arrhenius kinetics apply (Ferry, 1980). The WLF equation depends critically on the appropriate reference T_g for any particular glass-forming material (of any MW and extent of plasticization (Soesanto and Williams, 1981)), be it $T_{g'}$ for a frozen food system or T_g for a low-moisture one. T_g is defined as an iso-free volume state of limiting free volume for the liquid, and also approximately as an iso-viscosity state somewhere in the range of 10^{11}–10^{14} Pa s (Soesanto and Williams, 1981; Franks, 1982).

A generalized physico-chemical mechanism for collapse in low-moisture, glass-forming food systems has been described as follows (Levine and Slade, 1986). As ambient T rises above T_g, or as T_g falls below ambient T due to plasticization by water, polymer free volume increases, leading to increased segmental mobility of polymer chains (or smaller molecules). Consequently, viscosity of the dynamically-constrained solid falls below the characteristic η_g at T_g (allowing the glass transition to occur), which permits viscous liquid flow. In this rubbery state, translational diffusion can occur in practical timeframes, and diffusion-controlled relaxations (including structural collapse) are free to proceed with rates defined by the WLF equation (i.e. rates which increase exponentially with increasing ΔT above T_g).

The impact of WLF behavior on the kinetics of diffusion-controlled relaxation processes in water-plasticized foods at low moisture has been illustrated as follows (Levine and Slade, 1986). Relative relaxation rates vs ΔT, calculated from eqn (1), demonstrate the exponential relationship: for ΔT values of 0, 3, 7, 11, and 21°C, corresponding rates would be 1, 10, 10^2, 10^3, and 10^5, respectively. Such rates are dramatically different from those that would be defined by the familiar Q_{10} rule of Arrhenius kinetics for dilute solutions. Sugar recrystallization in candy glasses represents an example of a collapse process governed by WLF kinetics. The propagation step in the mechanism of recrystallization of an amorphous but crystallizable polymer (Wunderlich, 1976) (or low-MW sugar), initially quenched from the melt to a kinetically-metastable solid state, reflects a zero rate at $T < T_g$. Due to immobility in the glass, migratory diffusion of large main-chain segments (or small molecules), required for crystal growth, would be inhibited over realistic times. However, propagation rate increases exponentially with increasing ΔT above T_g (up to T_m), due to the mobility allowed in the rubbery state (Jolley, 1970). Thus, a recrystallization transition from unstable amorphous liquid to crystalline solid may occur at $T > T_g$ (White and Cakebread, 1966; Karel, 1986; Phillips *et al.*, 1986), with a rate defined by the WLF equation.

The critical message to be distilled by the reader at this point is that the structure/property relations of water-compatible food materials at low moisture are dictated by a moisture/temperature/time superposition (Flink, 1983). Referring once again to the PVP–water state diagram in Fig. 2 as a conceptual map, one sees that the T_g curve represents a boundary between physical states in which various diffusion-controlled processes (e.g. collapse phenomena) either can (at $T > T_g$, the domain of water dynamics) or cannot (at $T < T_g$, the glass dynamics domain) occur over realistic times. The WLF equation defines the kinetics of molecular-level relaxation processes that will occur above T_g, in terms of a non-Arrhenius exponential function of ΔT above this boundary condition.

Further illustration of this point is provided by the state diagrams in Fig. 13 (Levine and Slade, 1987*a*), which emphasize the glass curves for sucrose, glucose, and fructose. Comparison of these diagrams has revealed differences in the magnitudes of the kinetically-metastable domains between dry T_m and T_g (see Table 2), with respect to the homogeneous nucleation temperatures (T_h) of these sugars. In each case, T_h was estimated from the ratio of T_m/T_h (K), which, for many partially-crystalline, synthetic, pure polymers, is typically 1·25, with a reported range of 1·18–1·28 (Walton, 1969; Wunderlich, 1976). The corresponding ratio of T_m/T_g (K) is typically 1·5, with a range of

1·33–2·0 (Brydson, 1972; Wunderlich, 1980; Franks, 1982), so that, in almost all known cases (including water (Franks, 1982)), $T_h > T_g$. We have described how the relationship between T_h and T_g, relative to T_m, allows prediction about the stability towards recrystallization of concentrated and supersaturated aqueous solutions (Levine and Slade, 1987*a*). For sucrose and glucose (with typical T_m/T_g ratios of 1·43 and 1·42, respectively), which are known to crystallize readily by undercooling such solutions, $T_h > T_g$, so homogeneous nucleation can occur before vitrification on cooling from $T > T_m$. In contrast, we reported that fructose has an exceptionally-low T_m/T_g ratio of 1·06, apparently due to anomalously-high free volume and mobility, and low η, at T_g (Brydson, 1972). Thus, for fructose, which is thought to be impossible to crystallize (in a realistic time) by the same mechanism, $T_g > T_h$, so, on cooling, vitrification occurs first, thus immobilizing the system and preventing the possibility of homogeneous nucleation.

The controlled agglomeration of amorphous powders (e.g. low-MW carbohydrates) represents a specific example of a WLF-governed kinetic process related to caking, in low-moisture food systems sensitive to plasticization by water and/or heating (Levine and Slade, 1986). It has been demonstrated (Downton *et al.*, 1982; Tardos *et al.*, 1984) that spontaneous agglomeration of solid powder particles occurs when η of the liquid phase at the surface of a particle drops to $\simeq 10^7$ Pa s. This η is $\simeq 10^5$ lower than η_g. From the WLF equation this $\Delta\eta$ of 10^5 Pa s corresponds to a ΔT of ~21°C between T_g and the T_{sp} for spontaneous agglomeration. We noted that, on a state diagram of T vs w% moisture, WLF theory would predict that the T_g and T_{sp} curves should represent parallel iso-viscosity lines. The T_{sp} curve for fast agglomeration during processing would lie above the T_g curve for slow caking during storage, and the ΔT of 21°C would reflect the different time scales for these two surface phenomena (Levine and Slade, 1986). Experimental verification is demonstrated in Fig. 14, which shows actual T_g and T_{sp} data (the former ours, the latter from Masters and Stoltze (1973)) for samples of spray-dried coffee powder. Even though these data were for different coffee samples (which might explain why the two curves are not parallel), the T_{sp} curve does lie above that for T_g, with a ΔT of between 10 and 85°C for moisture contents between 12 and 2 w%. Note also that the T_g curve illustrates a typical extent of plasticization by water of about 9°C/w%. In the future, it would be informative to see T_g and T_{sp} data measured for the same sample of amorphous coffee or other food powder.

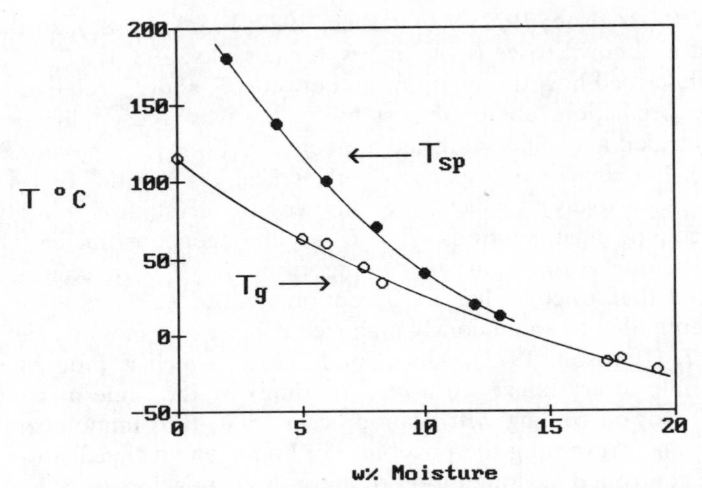

FIG. 14. Variation of the glass transition temperature, T_g, and the sticky-point temperature, T_{sp}, against moisture content for spray-dried coffee powder samples. (T_{sp} data from Masters and Stoltze (1973).)

In practice, collapse (and all its different manifestations) can be prevented, and food product quality and stability maintained, by the following three fundamental countermeasures (Levine and Slade, 1986): (1) storage at $T < T_g$ (White and Cakebread, 1966); (2) deliberate formulation to increase T_g to a temperature > the processing or storage temperature, by increasing overall $\bar{M}_w$ of the water-soluble solids in a product mixture. As described earlier, and indicated by the footnote to Table 4, this is often accomplished by adding polymeric stabilizers such as low-DE SHPs (or other polymeric carbohydrate, protein, or cellulose and polysaccharide gum stabilizers, some of which are included in Table 5) to a formulation dominated by low-MW sugars and/or polyols (Cakebread, 1969; Downton *et al.*, 1982; To and Flink, 1978; Karel and Flink, 1983; Dziedzic and Kearsley, 1984; Kahn and Lynch, 1985; White and Cakebread, 1966; Vink and Deptula, 1982; Lees, 1982; Ogawa and Imamura, 1985; Miller and Mutka, 1985; Tsourouflis *et al.*, 1976; Maltini, 1974). The effect of increased MW on T_g of PVPs is also illustrated in Fig. 2; (3) in low-moisture amorphous food powders and other hygroscopic glassy solids especially prone to detrimental effects of plasticization by water (the latter including 'candy' glasses such as boiled sweets (White and Cakebread, 1966; Cakebread, 1969; Chevalley *et al.*, 1970; Vink and Deptula, 1982; Dziedzic and Kearsley, 1984), extruded melts (Gueriviere, 1976), candy coatings (Lees, 1982), sugar in chocolate (Niediek and Barbernics, 1981), and supersaturated sugar syrups (McNulty and Flynn, 1977; Soesanto and Williams, 1981; Downton *et al.*, 1982), (a) reduction of residual moisture content to $\lesssim 3\%$ during processing, (b) packaging in superior moisture-barrier film or foil to prevent moisture pickup during storage, and (c) avoidance of high temperature/high humidity ($\gtrsim 20\%$ RH) conditions during storage (White and Cakebread 1966; Cakebread, 1969; Gueriviere, 1976; Lees, 1982; Flink, 1983). In a related vein, a 'state of the art' computer model was recently described (Marsh and Wagner, 1985), which can be used to predict shelf-life of particular moisture-sensitive products, based on moisture-barrier properties of a packaging material and temperature/humidity conditions of a specific storage environment.

On the other hand, Karel (1985) has pointed out that water plasticization (to depress T_g to a temperature below that of a phenomenon) is not always detrimental to food product quality. Examples of applications involving deliberate moisturization to produce desirable consequences include (1) controlled agglomeration or sintering (by limited heat/moisture/time treatment) of amorphous powders as described above and elsewhere (Tsourouflis *et al.*, 1976; Rosenzweig and Narkis, 1981; Flink, 1983), and (2) compression (without brittle fracture) of freeze-dried products after limited replasticization (Karel and Flink, 1983).

Prevention of Enzymatic Activity and other Chemical Reactions at $T < T_g$

One collapse-related phemonenon listed in Table 4 but not yet discussed involves enzymatic activity, in amorphous substrate-containing media, which occurs only at $T > T_g$ (Levine and Slade, 1986). Enzymatic activity represents a pleasing case study with which to close this chapter, because it is potentially important in many food applications which cover the entire spectrum of processing/storage temperatures and moisture contents, and because examples exist (Karel, 1985) which elegantly illustrate the fact that activity is inhibited in low-moisture amorphous solids at $T < T_g$, and in frozen systems at $T < T_{g'}$. Bone and Pethig (1982, 1985) studied the hydration of dry lysozyme powder at 20°C, and found that, at 20 w% water, lysozyme becomes sufficiently plasticized so that measurable enzymatic activity commences. We interpreted their results to indicate the following (Levine and Slade, 1986): a diffusion-controlled enzyme/substrate interaction is essentially prohibited in a glassy solid at $T < T_g$, but sufficient water plasticization depresses T_g of lysozyme to <20°C, allowing onset of enzymatic activity in a rubbery lysozyme solution at $T > T_g$, the threshold temperature for activity. Our interpretation was supported by results of related studies of 'solid glassy lysozyme samples' by Poole and Finney (1983*a,b*; Finney and Poole, 1984), who noted conformational changes in the protein as a consequence of hydration to the same 20 w% level, and were 'tempted to suggest that this solvent-related effect is required before (enzymatic) activity is possible'. Recently, Morozov and Gevorkian (1985) also noted a critical requirement of low-temperature, water-plasticized glass transitions for physiological activity of lysozyme and other globular proteins.

Within a context of the concept of food cryostabilization that we introduced (Levine and Slade, 1986), and the critical role of low-DE SHPs as cryostabilizers, we have verified the above conclusion in maximally-frozen biological model systems. By analogy, in such cases, the threshold temperature for onset of enzymatic activity is $T_{g'}$. Cryostabilization, as a practical industrial technology derived from the concept of glass dynamics, is a means of protecting freezer-stored and freeze-dried foods from deleterious changes in texture (e.g. grain growth of ice, solute crystallization), structure (e.g. shrinkage, collapse), and chemical composition (e.g. flavor/color degradation, fat rancidity, as well as enzymatic reactions) typically encountered. The key to this protection lies in controlling the physico-chemical properties of the freeze-concentrated matrix surrounding the ice crystals. If this matrix is maintained as an amorphous mechanical solid (at $T_f < T_{g'}$), then diffusion-controlled processes that typically result in reduced storage stability can be prevented or at least greatly retarded. If, on the other hand, a natural food is improperly stored at too high a T_f, or a fabricated product is improperly formulated, so that the matrix is allowed to exist in the freezer as a rubbery fluid (at $T_f > T_{g'}$, within the domain of water dynamics), then freezer-storage stability would be reduced. Moreover, rates of the various deleterious changes would increase exponentially with ΔT between T_f and $T_{g'}$, as dictated by WLF theory, and illustrated earlier in Fig. 11 for ice creams and frozen novelties.

The prevention of enzymatic activity at $T < T_{g'}$ was demonstrated experimentally *in vitro* in a model system consisting of glucose oxidase, glucose, methyl red, and bulk solutions of sucrose, Morrex 1910 (10-DE maltodextrin), and their mixtures, which provided a range of samples with known values of $T_{g'}$ (Levine and Slade, 1986). The enzymatic oxidation of glucose produces an acid which turns the reaction mixture from yellow to pink. Samples with a range of $T_{g'}$ values from −9·5 to −32°C were stored at various temperatures: 25, 3, −15, and −23°C. All samples were fluid at the two higher temperatures, while all looked like colored blocks of ice at −15 and −23°C. However, only samples for which the storage temperature was above $T_{g'}$ turned pink. Even after two months storage at −23°C, samples containing maltodextrin, with $T_{g'} > -23$°C, were still yellow. Frozen samples which turned pink, even at −23°C, contained a concentrated enzyme-rich fluid surrounding the ice crystals, while in those which remained yellow, the non-ice matrix was a glassy solid. Significantly, enzymatic activity was prevented by storage below $T_{g'}$ but the enzyme itself was not inactivated. When yellow samples were thawed, they quickly turned pink. Thus, cryostabilization with a low-DE SHP preserved the enzyme during storage, but prevented its activity below $T_{g'}$. As an extension of our qualitative experiment, Reid (1987) is conducting a study of quantitative enzyme–substrate reaction rates, as a function of ΔT between T_f and $T_{g'}$, in frozen model systems containing different polymeric solutes, including a series of Maltrin SHPs (M040–M250).

CONCLUSION

As emphasized in this chapter, we have seen, especially since 1980, a growing awareness among a small but increasing number of food scientists of the value of a polymer science approach to the study of food materials and systems. In this respect, food science has followed the compelling lead of the synthetic polymers field. Recognition of two key elements of this research approach, (1) the critical role of water as a plasticizer of amorphous materials and (2) the importance of the glass transition as a physico-chemical parameter which can govern food product properties, processing, quality, and stability, has also increased markedly during this decade. We have tried to illustrate here, with examples from our experiences with food materials, how one can begin to understand and explain complex behavior, design processes, and predict product quality and storage stability, based on knowledge of fundamental structure/property relationships defined by studies which employ a polymer science approach to low-moisture food systems plasticized by water. In the future, we expect to see much progress reported from this emerging, cross-disciplinary research area.

ACKNOWLEDGEMENTS

We thank our former colleagues at General Foods, Timothy Schenz and Allan Bradbury, for their contributions to our research program; Cornelis van den Berg and John Blanshard for encouragement of our work; and especially our consultant and mentor, Prof. Felix Franks of the University of Cambridge, for invaluable suggestions, discussions, encouragement, and support over the years.

Parts of this chapter, including some of the Tables and Figures, are taken from previous reports and reviews by the authors, and are updated and reproduced here with permission of the various publishers (see Reference list) holding copyrights.

REFERENCES

Ablett, S., Attenburrow, G. E. and Lillford, P. J. (1986). The significance of water in the baking process. In: *Chemistry and Physics of Baking,* J.M.V. Blanshard, P. J. Frazier and T. Galliard (Eds) Royal Soc. Chem., London, pp. 30–41.

Anandaraman, S. and Reineccius, G. A. (1986). Stability of encapsulated orange peel oil. *Food Technol.*, **40,** 88–93.

Barbosa-Canovas, G. V., Rufner, R. and Peleg, M. (1985). Microstructure of selected binary food powder mixtures. *J. Food Sci.*, **50,** 473–81 *[1].*

Beesley, T. E. (1985). Inclusion complexing. *Amer. Lab.*, May, 78–87.

Bevilacqua, A. E. and Zaritzky, N. E. (1982). Ice recrystallization in frozen beef. *J. Food Sci.*, **47,** 1410–14.

Biliaderis, C. G. (1983). DSC in food research—review. *Food Chem.*, **10,** 239–65.

Biliaderis, C. G., Page, C. M., Slade, L. and Sirett, R. R. (1985). Thermal behavior of amylose–lipid complexes. *Carbohydr. Polym.*, **5,** 367–89.

Biliaderis, C. G., Page, C. M., Maurice, T. J. and Juliano, B. O. (1986*a*). Thermal characterization of rice starches: a polymeric approach to phase transitions of granular starch. *J. Agric. Food Chem.*, **34,** 6–14.

Biliaderis, C. G., Page, C. M. and Maurice, T. J. (1986*b*). Non-equilibrium melting of amylose-V complexes. *Carbohydr. Polym.*, **6,** 269–88.

Biliaderis, C. G., Page, C. M. and Maurice, T. J. (1986*c*). Multiple melting transitions of starch/monoglyceride systems. *Food Chem.*, **22,** 279–95.

Billmeyer, F. W. (1984). *Textbook of Polymer Science*, 3rd Edn, Wiley-Interscience, New York.

Biros, J., Madan, R. L. and Pouchly, J. (1979). Heat capacity of water-swollen polymers above and below 0°C *Collect. Czech. Chem. Commun.*, **44,** 3566–73.

Blanshard, J. M. V. (1986). The significance of the structure and function of the starch granule in baked products. In: *Chemistry and Physics of Baking*, J. M. V. Blanshard, P. J. Frazier and T. Galliard (Eds), Royal Soc. Chem. London, pp. 1–13.

Bone, S. and Pethig, R. (1982). Dielectric studies of the binding of water to lysozyme. *J. Mol. Biol.*, **157,** 571–5. *[2].*

Bone, S. and Pethig, R. (1985). Dielectric studies of protein hydration and hydration-induced flexibility. *J. Mol. Biol.*, **181,** 323–6.

Boutron, P. and Kaufmann, A. (1979). Stability of the amorphous state in the water–propylene glycol system. *Cryobiol.*, **16,** 557–68.

Boyer, R. F., Baer, E. and Hiltner, A. (1985). Concerning gelation effects in atactic polystyrene solutions. *Macromolecules,* **18,** 427–34.

Braudo, E. E., Belavtseva, E. M., Titova, E. F., Plashchina, I. G., Krylov, V. L., Tolstoguzov, V. B., Schierbaum, F. R. and Richter, M. (1979). Struktur und Eigenschaften von Maltodextrin-Hydrogelen. *Starke,* **31,** 188–94.

Braudo, E. E., Plashchina, I. G. and Tolstoguzov, V. B. (1984). Structural characterisation of thermoreversible anionic polysaccharide gels by their elasto-viscous properties. *Carbohydr. Polym.*, **4,** 23–48.

Brennan, W. P. (1973). *Thermal Analysis Application Study No. 8,* Norwalk, Perkin Elmer Instrument Division.

Brydson, J. A. (1972). The glass transition, melting point and structure. In: *Polymer Science,* A. D. Jenkins (Ed.), Elsevier, Amsterdam, North Holland, pp. 194–249.

Bulpin, P. V., Cutler, A. N. and Dea, I. C. M. (1984). Thermally-reversible gels from low DE maltodextrins. In: *Gums and Stabilizers for the Food Industry 2,* G. O. Phillips, D. J. Wedlock and P. A. Williams (Eds) Pergamon Press, Oxford, 475–84.

Burghoff, H. G. and Pusch, W. (1980). Thermodynamic state of water in cellulose acetate membranes. *Polym. Engn. Sci.*, **20,** 305–9.

Cakebread, S. H. (1969). Factors affecting the shelf life of high boilings. *Manufact. Confect.*, **49**(8), 41–4. *[3].*

Chevalley, J., Rostagno, W. and Egli, R. H. (1970). A study of the physical properties of chocolate. V. *Rev. Internat. Chocolat.*, **25,** 3–6. *[4]*

Cole, B. A., Levine, H. I., McGuire, M. T., Nelson, K. J. and Slade, L. (1983). Soft, Frozen Dessert Formulation. US Patent 4,374,154.

Cole, B. A., Levine, H. I., McGuire, M. T., Nelson, K. J. and Slade, L. (1984). Soft, Frozen Dessert Formulation. US Patent 4,452,824.

Colliopoulos, J., Young, J. and Tsau, J. H. K. (1984). Aspartame Encapsulated in Glucose Polymers. European Patent application EP0100002A1.

Cowie, J. M. G. (1973). *Polymers: Chemistry and Physics of Modern Materials* Intertext, New York.

Domszy, R. C., Alamo, R., Edwards, C. O. and Mandelkern, L. (1986). Thermoreversible gelation and crystallization of homopolymers and copolymers. *Macromolecules,* **19,** 310–25.

Donovan, J. W. (1985). DSC in food research. In: *Proceedings 14th NATAS Conference,* San Francisco, pp 328–33.

Downton, G. E., Flores-Luna, J. L. and King, C. J. (1982). Mechanism of stickiness in hygroscopic, amorphous powders. *Indust. Engn. Chem. Fund.,* **21,** 447–51 *[5]*

Dziedzic, S. Z. and Kearsley, M. W. (1984). Physico-chemical properties of glucose syrups. In: *Glucose Syrups: Science and Technology,* S. Z. Dziedzic and M. W. Kearsley (Eds), Elsevier Applied Science, London, 137–68.

Eisenberg, A. (1984). The glassy state and the glass transition. In: *Physical Properties of Polymers,* J. E. Mark, A. Eisenberg, W. W. Graessley, L. Mandelkern and J. L. Koenig (Eds), American Chemical Society, Washington, 55–95.

Ellis, H. S. & Ring, S. G. (1985). A study of some factors influencing amylose gelation. *Carbohydr. Polym.,* **5,** 201–13.

Ferry, J. D. (1948). Mechanical properties of substances of high molecular weight. *J. Amer. Chem. Soc.,* **70,** 2244–9.

Ferry, J. D. (1980). *Viscoelastic Properties of Polymers,* 3rd edn, John Wiley & Sons, New York.

Finney, J. L. and Poole, P. L. (1984). Protein hydration and enzyme activity: role of hydration-induced conformation and dynamic changes in activity of lysozyme. *Comments Mol. Cell. Biophys.,* **2,** 129–51.

Flink, J. M. (1983). Structure and structure transitions in dried carbohydrate materials. In: *Physical Properties of Foods,* M. Peleg and E. B. Bagley (Eds), AVI, Westport, 473–521. *[6]*

Flory, P. J. (1953). *Principles of Polymer Chemistry,* Cornell University Press, Ithaca, New York.

Flory, P. J. (1974). Introductory lecture—gels and gelling processes. *Faraday Disc. Chem. Soc.,* **57,** 7–18.

Forsyth, M. and MacFarlane, D. R. (1986). Recrystallization revisited. *Cryo-Letters,* **7,** 367–78.

Franks, F. (1982). The properties of aqueous solutions at subzero temperatures. In: *Water: A Comprehensive Treatise,* F. Franks (Ed.), Vol. 7, Plenum Press, New York, pp. 215–338. *[8]*

Franks, F. (1985*a*). *Biophysics and Biochemistry at Low Temperatures,* Cambridge University Press, Cambridge.

Franks, F. (1985*b*). Complex aqueous systems at subzero temperatures. In: *Properties of Water in Foods,* D. Simatos and J. L. Multon (Eds), Martinus Nijhoff, Dordrecht, pp. 497–509. *[7]*

Franks, F. (1986). Unfrozen water: yes; unfreezable water: hardly; bound water: certainly not. *Cryo-Letters,* **7,** 207.

Franks, F., Asquith, M. H., Hammond, C. C., Skaer, H. B. and Echlin, P. (1977). Polymeric cryoprotectants in the preservation of biological ultrastructure. *J. Microsc.,* **110,** 223–38. *[9]*

Fukuoka, E., Kimura, S., Yamazaki, M. and Tanaka, T. (1983). Cohesion of particulate solids. VI. *Chem. Pharmaceut. Bull.*, **31**(1), 221–9. *[10]*

Fuzek, J. F. (1980). Glass transition temperature of wet fibers: its measurement and significance. In: *Water in Polymers,* S. P. Rowland (Ed.), ACS Symposium Series 127, American Chemical Society, Washington, pp. 515–30.

Graessley, W. W. (1984). Viscoelasticity and flow in polymer melts and concentrated solutions. In: *Physical Properties of Polymers,* J. E. Mark, A. Eisenberg, W. W. Graessley, L. Mandelkern and J. L. Koenig (Eds), Amer. Chem. Soc., Washington, pp. 97–153.

Green, W. M. and Hoover, M. W. (1979). Honey coated roasted nut product and method for making same. US Patent 4,161,545.

Gueriviere, J. F. (1976). Recent developments in extrusion cooking of foods. *Indust. Aliment. Agric.*, **93**(5), 587–95. *[11]*

Harper, E. K. and Shoemaker, C. F. (1983). Effect of Locust bean gum and selected sweetening agents on ice recrystallization rates. *J. Food Sci.*, **48,** 1801–6.

Herrington, T. M. and Branfield, A. C. (1984). Physico-chemical studies on sugar glasses. I. & II. *J. Food Technol.*, **19,** 409–35. *[12]*

Hirsh, A. G., Williams, R. J. & Meryman, H. T. (1985). Novel method of natural cryoprotection. *Plant Physiol.*, **79,** 41–56.

Hoeve, C. A. J. (1980). The structure of water in polymers. In: *Water in Polymers,* S. P. Rowland (Ed.), ACS Symp. Ser. 127, Amer. Chem. Soc., Washington, pp. 135–46.

Hoeve, C. A. J. and Hoeve, M. B. J. A. (1978). The glass point of elastin as a function of diluent concentration. *Organ. Coat. Plast. Chem.*, **39,** 441–3.

Holbrook, J. L. and Hanover, L. M. (1983). Fructose-containing frozen dessert products. US Patent 4,376,791.

Hoseney, R. C., Zeleznak, K. and Lai, C. S. (1986). Wheat gluten: a glassy polymer. *Cereal Chem.*, **63,** 285–6.

Jin, X., Ellis, T. S. and Karasz, F. E. (1984). The effect of crystallinity and crosslinking on the depression of the glass transition temperature in Nylon 6 by water. *J. Polym. Sci.: Polym. Phys. Ed.*, **22,** 1701–17.

Jolley, J. E. (1970). The microstructure of photographic gelatin binders. *Photogr. Sci. Engn.*, **14,** 169–77.

Kahn, M. L. and Eapen, K. E. (1982). Intermediate-Moisture Frozen Foods. US Patent 4,332,824.

Kahn, M. L. & Lynch, R. J. (1985). Freezer Stable Whipped Ice Cream and Milk Shake Food Products. US Patent 4,552,773. *[13]*

Kanno, H. (1987). Double glass transitions in aqueous lithium chloride solutions vitrified at high pressures. *J. Phys. Chem.*, **91,** 1967–71.

Karel, M. (1985). Effects of water activity and water content on mobility of food components, and their effects on phase transitions in food systems. In: *Properties of Water in Foods,* D. Simatos and J. L. Multon (Eds), Martinus Nijhoff, Dordrecht, 153–69. *[15]*

Karel, M. (1986). Control of lipid oxidation in dried foods. In: *Concentration and Drying of Foods,* D. MacCarthy (Ed.), Elsevier Applied Science, London, pp. 37–51. *[14]*

Karel, M. and Flink, J. M. (1983). Some recent developments in food dehydration research. In: *Advances in Drying,* A. S. Mujumdar (Ed.), Vol. 2, Hemisphere Publishing, Washington, 103–53. *[16]*

Keeney, P. G. and Kroger, M. (1974). Frozen dairy products. In: *Fundamentals of Dairy Chemistry,* B. H. Webb *et al.* (Eds), 2nd edn, AVI, Westport, p. 890.

Krusi, H. and Neukom, H. (1984). Untersuchungen uber die Retrogradation der Staerke in konzentrierten Weizenstarkegelen. *Starke,* **36,** 300–5.

Lees, R. (1982). Quality control in the production of hard and soft sugar confectionery. *Confect. Product.,* Feb., 50–1. *[17]*

Lenchin, J. M., Trubiano, P. C. and Hoffman, S. (1985). Converted Starches for Use as a Fat- or Oil-Replacement in Foodstuffs, US Patent 4,510,166.

Levine, H. and Slade, L. (1986). A polymer physico-chemical approach to the study of commercial starch hydrolysis products (SHPs). *Carbohydr. Polym.,* **6,** 213–44. *[18]*

Levine, H. and Slade, L. (1987*a*). Water as a plasticizer: physico-chemical aspects of low-moisture polymeric systems. In: *Water Science Reviews,* Vol. 3, F. Franks (Ed.), Cambridge University Press, Cambridge, 79–185.

Levine, H. and Slade, L. (1987*b*). Collapse phenomena—a unifying concept for interpreting the behavior of low-moisture foods. In: *Food Structure—Its Creation and Evaluation,* J. R. Mitchell and J. M. V. Blanshard (Eds), Butterworths, London, pp. 149–80.

Levine, H. and Slade, L. (1988). Thermomechanical properties of small carbohydrate–water glasses and 'rubbers': kinetically-metastable systems at subzero temperatures. *J. Chem. Soc., Faraday Trans., I,* **84,** 2619–33.

Lund, D. B. (1983). Applications of DSC in foods. In: *Physical Properties of Foods,* M. Peleg and E. B. Bagley (Eds), AVI, Westport, 125–43.

Luyet, B. (1960). On various phase transitions occurring in aqueous solutions at low temperatures. *Ann. NY Acad. Sci.,* **85,** 549–69.

MacFarlane, D. R. (1985). Anomalous glass transitions in aqueous propylene glycol solutions. *Cryo-Letters,* **6,** 313–18.

MacKenzie, A. P. (1977). Non-equilibrium freezing behavior of aqueous systems. *Phil. Trans. Roy. Soc. London B.,* **278,** 167–89. *[19]*

MacKenzie, A. P. (1981). Modelling the ultra-rapid freezing of cells and tissues. In: *Microprobe Analysis of Biological Systems,* Academic Press, New York, 397–421.

MacKenzie, A. P. and Rasmussen, D. H. (1972). Interactions in the water–PVP system at low temperatures. In: *Water Structure at the Water–Polymer Interface,* H. H. G. Jellinek (Ed.), Plenum Press, New York, 146–71.

Maltini, E. (1974). Thermophysical properties of frozen juice related to freeze drying problems. *Ann. D'Ist. Sper. Valor. Technol. Prod. Agric.,* **5,** 65–72. *[20]*

Maltini, E. (1977). Studies on the physical changes in frozen aqueous solutions by DSC and microscopic observations. *I.I.F.-I.I.R.-Karlsruhe,* 1977–**1**, 1–9. *[21]*

Marsh, K. S. and Wagner, J. (1985). Predict shelf life. *Food Engn.,* Aug., 58.

Marshall, A. S. and Petrie, S. E. B. (1980). Thermal transitions in gelatin and aqueous gelatin solutions. *J. Photogr. Sci.,* **28,** 128–34.

Masters, K. and Stoltze, A. (1973). Agglomeration advances. *Food Engn.*, Feb., 64–67.
Maurice, T. J., Slade, L., Sirett, R. R. and Page, C. M. (1983). Polysaccharide–water interactions—thermal behavior of rice starch. In: *Properties of Water in Foods*, D. Simatos and J. L. Multon (Eds), Martinus Nijhoff, Dordrecht, 211–27.
McNulty, P. B. and Flynn, D. G. (1977). Force-deformation and texture profile behavior of aqueous sugar glasses. *J. Text. Stud.*, **8,** 417–31. *[22]*
Medcalf, D. G. (1985). Food functionality of cereal carbohydrates. In: *New Approaches to Research on Cereal Carbohydrates*, R. D. Hill and L. Munck (Eds), Elsevier, Amsterdam, 355–62.
Meyer, D. R. (1985). Salt Substitute Containing Potassium Chloride Coated with Maltodextrin and Method of Preparation. US Patents 4,556,567, 4,556,568.
Miles, M. J., Morris, V. J. and Ring, S. G. (1985). Gelation of amylose. *Carbohydr. Res.*, **135,** 257–69.
Miller, D. H. and Mutka, J. R. (1985). Process for Forming Solid Juice Composition and Product of the Process. US Patent 4,499,112. *[23]*
Miller, D. H. and Mutka, J. R. (1986). Preparation of Solid Essential Oil Flavor Composition. US Patent 4,610,890.
Mitchell, J. R. (1980). The rheology of gels. *J. Text. Stud.*, **11,** 315–37.
Moreyra, R. and Peleg, M. (1981). Effect of equilibrium water activity on the bulk properties of selected food powders. *J. Food Sci.*, **46,** 1918–22. *[24]*
Morozov, V. N. and Gevorkian, S. G. (1985). Low-temperature glass transition in proteins. *Biopolymers*, **24,** 1785–99. *[25]*
Morris, E. R., Cutler, A. N., Ross-Murphy, S. B. and Rees, D. A. (1981). Concentration and shear rate dependence of viscosity in random coil polysaccharide solutions. *Carbohydr. Polym.*, **1,** 5–21.
Muhr, A. H. and Blanshard, J. M. V. (1986). Effect of polysaccharide stabilizers on the rate of growth of ice. *J. Food Technol.*, **21,** 683–710.
Muhr, A. H., Blanshard, J. M. V. and Sheard, S. J. (1986). Effects of polysaccharide stabilizers on the nucleation of ice. *J. Food Technol.*, **21,** 587–603.
Murray, D. G. and Luft, L. R. (1973). Low-D.E. corn starch hydrolysates. *Food Technol.*, **27**(3), 32–40.
Nagashima, N. and Suzuki, E. (1985). The behavior of water in foods during freezing and thawing. In: *Properties of Water in Foods*, D. Simatos and J. L. Multon (Eds), Martinus Nijhoff, Dordrecht, pp. 555–71. *[26]*
Niediek, E. A. and Barbernics, L. (1981). Amorphisierung von Zucker durch das Feinwalzen von Schokoladenmassen. *Gordian*, **80**(11), 267–9. *[27]*
Ogawa, H. and Imamura, Y. (1985). Stabilized Solid Compositions, US Patent 4,547,377. *[28]*
Olson, D. R. and Webb, K. K. (1978). The effect of humidity on the glass transition temperature. *Organ. Coat. Plast. Chem.*, **39,** 518–23.
Passy, N. and Mannheim, C. H. (1982). Flow properties and water sorption of food powders. II. *Lebensm.-Wiss. u.-technol.*, **15,** 222–5. *[29]*
Peleg, M. and Mannheim, C. H. (1977). The mechanism of caking of powdered onion. *J. Food Process. Preserv.*, **1,** 3–11. *[30]*

Phillips, A. J., Yarwood, R. J. and Collett, J. H. (1986). Thermal analysis of freeze-dried products. *Anal. Proceed.*, **23,** 394–5. *[31]*

Poole, P. L. and Finney, J. L. (1983*a*). Sequential hydration of a dry globular protein. *Biopolymers,* **22,** 255–60 *[32]*

Poole, P. L. and Finney, J. L. (1983*b*). Hydration-induced conformational and flexibility changes in lysozyme at low water content. *Int. J. Biol. Macromol.*, **5,** 308–10. *[33]*

Pouchly, J., Biros, J. & Benes, S. (1979). Heat capacities of water-swollen hydrophilic polymers above and below 0°C. *Makromol. Chem.*, **180,** 745–60.

Rall, W. F. and Fahy, G. M. (1985). Ice-free cryopreservation of mouse embryos by vitrification. *Nature,* **313,** 573–5.

Rasmussen, D. and Luyet, B. (1969). Complementary study of some non-equilibrium phase transitions in frozen solutions of glycerol, ethylene glycol, glucose and sucrose. *Biodynamica,* **10,** 319–31.

Reid, D. S. (1985). Correlation of the phase behavior of DMSO/NaCl/water and glycerol/NaCl/water as determined by DSC with their observed behavior on a cryomicroscope. *Cryo-Letters,* **6,** 181–8. *[34]*

Reid, D. S. (1987). Personal communication. *[35]*

Reineccius, G. A. (1986). Personal communication.

Reuther, F., Damaschun, G., Gernat, C., Schierbaum, F., Kettlitz, B., Radosta, S. and Nothnagel, A. (1984). Molecular gelation mechanism of maltodextrins investigated by wide-angle X-ray scattering. *Coll. & Polym. Sci.,* **262,** 643–7.

Richardson, M. J. (1978). Quantitative DSC. In: *Developments in Polymer Characterization-1,* J. V. Dawkins (Ed.), Applied Science, London, 205–44.

Richter, M., Schierbaum, F., Augustat, S. and Knoch, K. D. (1976*a*), Method of Producing Starch Hydrolysis Products for Use as Food Additives. US Patent 3,962,465.

Richter, M., Schierbaum, F., Augustat, S. and Knoch, K. D. (1976*b*), Method of Producing Starch Hydrolysis Products for Use as Food Additives. US Patent 3,986,890.

Rosenzweig, N. and Narkis, M. (1981). Sintering rheology of amorphous polymers. *Polym. Engn. Sci.,* **21,** 1167–70. *[36]*

Rowland, S. P. (1980). *Water in Polymers.* ACS Symposium Series 127, American Chemical Society, Washington.

Saleeb, F. Z. and Pickup, J. G. (1985). Fixing Volatiles in an Amorphous Substrate and Products Therefrom. US Patent 4,532,145.

Saleeb, F. Z. (1987). Personal Communication.

Schenz, T. W., Rosolen, M. A., Levine, H. and Slade, L. (1984). DMA of frozen aqueous solutions. In: *Proceedings 13th Annual Conference, North American Thermal Analysis Society,* A. R. McGhie (Ed.), NATAS, Philadelphia, 57–62.

Sears, J. K. and Darby, J. R. (1982). *The Technology of Plasticizers,* Wiley-Interscience, New York.

Slade, L. (1982). *Industrial Experience in Aw Measurement and Control.* 96th AOAC Program, Washington, DC, p. 23 (abs, no. 66).

Slade, L. (1984). Starch properties in processed foods: staling of starch-based products. *American Association of Cereal Chemists Annual Meeting*, Minneapolis, abs. no. 112.
Slade, L. and Levine, H. (1984*a*). Thermal analysis of starch and gelatin. *Amer. Chem. Soc. NE Reg. Meeting*, Fairfield, CT, abs. no. 152.
Slade, L. and Levine, H. (1984*b*). Thermal analysis of starch and gelatin. In: *Proceedings 13th Annual Conference, North American Thermal Analysis Society*, A. R. McGhie (Ed.), NATAS, Philadelphia, p. 64.
Slade, L. and Levine, H. (1987*a*). Recent advances in starch retrogradation. In: *Industrial Polysaccharides—The Impact of Biotechnology and Advanced Methodologies*, S. S. Stivala, V. Crescenzi and I. C. M. Dea (Eds), Gordon and Breach Science, New York, pp. 387–430.
Slade, L. and Levine, H. (1987*b*). Structural stability of intermediate moisture foods—a new understanding? In: *Food Structure—Its Creation and Evaluation*, J. R. Mitchell and J. M. V. Blanshard (Eds), Butterworths, London, pp. 115–47.
Slade, L. and Levine, H. (1987*c*). Polymer-Chemical properties of gelatin in foods. In: *Advances in Meat Research, Vol. 4—Collagen as a Food*, A. M. Pearson, T. R. Dutson and A. Bailey (Eds), AVI, Westport, pp. 251–66.
Slade, L. & Levine, H. (1987*d*). Non-equilibrium behavior of small carbohydrate–water systems. *Pure Appl. Chem.*, **60**, 1841–64.
Soesanto, T. and Williams, M. C. (1981). Volumetric interpretation of viscosity for concentrated and dilute sugar solutions. *J. Phys. Chem.*, **85**, 3338–41. *[37]*
Starkweather, H. W. (1980). Water in nylon. In: *Water in Polymers*, S. P. Rowland (Ed.), ACS Symposium Series 127, ACS, Washington, 433–40.
Szejtli, J. and Tardy, M. (1985). Honey Powder Preserving its Natural Aroma Components. US Patent 4,529,608. *[38]*
Tardos, G., Mazzone, D. and Pfeffer, R. (1984). Measurement of surface viscosities using a dilatometer. *Can. J. Chem. Engn.*, **62**, 884–7. *[39]*
Thom, F. and Matthes, G. (1986). Ice formation in binary aqueous solutions of ethylene glycol. *Cryo-Letters*, **7**, 311–26.
To, E. C. and Flink, J. M. (1978). 'Collapse', a structural transition in freeze dried carbohydrates. I.–III. *J. Food Technol.*, **13**, 551–94. *[40]*
Tomka, I. (1986). Thermodynamic theory of polar polymer solutions. *Polym. Prep.*, **27**(2), 129.
Tsourouflis, S., Flink, J. M. and Karel, M. (1976). Loss of structure in freeze-dried carbohydrate solutions. *J. Sci. Food Agric.*, **27**, 509–19. *[41]*
Turi, E. A. (1981). *Thermal Characterization of Polymeric Materials*, Academic Press, Orlando, Florida.
van den Berg, C. (1981). Vapor sorption equilibria and other water–starch interactions; a physico-chemical approach. Doctoral Thesis, Agricultural University, Wageningen.
van den Berg, C. (1986). Water activity. In: *Concentration and Drying of Foods*, D. MacCarthy (Ed.), Elsevier Applied Science, London, pp. 11–36.
van den Berg, C. and Bruin, S. (1981). Water activity and its estimation in food systems: theoretical aspects. In: *Water Activity: Influences on Food*

Quality, L. B. Rockland and G. F. Stewart (Eds), Academic Press, New York, pp. 1–61.
Vassoille, R., El Hachadi, A. and Vigier, G. (1986). Study by internal friction measurements of vitreous transition in aqueous propylene glycol solutions. *Cryo-Letters,* **7,** 305–10.
Vink, W. and Deptula, R. W. (1982). High Fructose Hard Candy. US Patent 4,311,722. *[42]*
Virtis Company, Inc. (1983). Virtis SRC Sublimators Manual, Gardiner, New York. *[43]*
Walton, A. G. (1969). Nucleation in liquids and solutions. In: *Nucleation,* A. C. Zettlemoyer (Ed.), Marcel Dekker, New York, p. 225.
White, G. W. and Cakebread, S. H. (1966). The glassy state in certain sugar-containing food products. *J. Food Technol.,* **1,** 73–82. *[44]*
Williams, M. L., Landel, R. F. and Ferry, J. D. (1955). Temperature dependence of relaxation mechanisms in amorphous polymers and other glass-forming liquids. *J. Amer. Chem. Soc.,* **77,** 3701–6.
Wright, D. J. (1984). Thermoanalytical methods in food research. *Crit. Revs. Appl. Chem.,* **5,** 1–36.
Wuhrmann, J. J., Venries, B. and Buri, R. (1975). Process for Preparing a Colored Powdered Edible Composition. US Patent 3,920,854. *[45]*
Wunderlich, B. (1973). *Macromolecular Physics, Vol.* 1—*Crystal Structure, Morphology, Defects,* Academic Press, New York.
Wunderlich, B. (1976). *Macromolecular Physics, Vol.* 2—*Crystal Nucleation, Growth, Annealing,* Academic Press, New York.
Wunderlich, B. (1980). *Macromolecular Physics, Vol.* 3—*Crystal Melting,* Academic Press, New York.
Wunderlich, B. (1981). The basis of thermal analysis. In: *Thermal Characterization of Polymeric Materials,* E. A. Turi (Ed.), Academic Press, Orlando, Florida, pp. 91–234.
Yannas, I. V. (1972). Collagen and gelatin in the solid state. *J. Macromol. Sci.-Revs. Macromol. Chem.,* **C7,** 49–104.
Yost, D. A. and Hoseney, R. C. (1986). Annealing and glass transition of starch. *Starke,* **38,** 289–92.
Zeleznak, K. J. and Hoseney, R. C. (1987). The glass transition in starch. *Cereal Chem.,* **64,** 121–4.
Zemelman, V. (1987). Personal communication. *[46]*

Chapter 4

PROTEIN–WATER INTERACTIONS

MARILYNN SCHNEPF

Department of Human Nutrition and Foods, Virginia Polytechnic Institute and State University, Blacksburg, Virginia, USA

SYMBOLS AND ABBREVIATIONS

a_w	Water activity
ATP	Adenosine-5′-triphosphate
BET	Brunauer, Emmett and Teller
%EM	Percent expressible moisture
G	Free energy
H	Enthalpy
IR	Infrared
NMR	Nuclear magnetic resonance
NSI	Nitrogen solubility index
pI	Isoelectric point
S	Entropy
T_1	Spin-lattice relaxation time
WBC	Water binding capacity
WHC	Water holding capacity
π	Surface pressure
Γ	Surface concentration

INTRODUCTION

The interaction between protein and water is a complex and important relationship. This interaction plays an important role in determining the three dimensional structure of the protein molecule as well as determining many of the functional properties of proteins in foods.

The first part of this paper will cover the types of bonding between water and protein molecules, followed by the methods used to determine this interaction. A discussion of how protein–water interaction affects the functional properties of protein in foods will follow.

STRUCTURE OF WATER

A detailed discussion of the structure of water is beyond the scope of this paper. However, a few relevant points should be emphasized before discussing the association of protein and water. The many unique properties of water are known to be related to the structure of water. Compared to other molecules of similar molecular weight, water has larger values for heat capacity, melting point, boiling point, surface tension, heats of fusion, vaporization, and sublimation than would be expected from its components. These higher values are related to the extra energy needed to break intermolecular hydrogen bonds in water (Fennema, 1977). Because each water molecule has two areas of positive charge (hydrogen-bond donor sites) and an equal area of negative charge (hydrogen-bond acceptor sites), water is able to engage in three dimensional bonding in a tetrahedral arrangement. The extent of this three dimensional network or the exact structure of water is a complex problem. The fluid nature of water means that bonds are going to exist for only a relatively short time compared to the bonds found in ice. The momentary nature of the bonds gives rise to difficulty in studying the bonds which leads to different theories on the structure of water. Fennema (1977) has divided these theories into two major categories: (1) continuum theories which are also called uniform or homogeneous theories and (2) mixture theories. Those who espouse continuum theories propose that all molecules in cold liquid are totally hydrogen-bonded (four bonds/molecule) but that the bonds differ in angle, length, and energy. Supporters of mixture theories contend that water consists of two or more distinguishable species which exist in dynamic equilibrium and differ in the degree of hydrogen bonding. An example of a mixture theory is the 'interstitial model'. Proponents of this model hold that liquid water exists partly in the form of bulky framework structures that evolve from a 4-coordinated, approximately tetrahedral arrangement of water molecules. Within the cavities formed by their framework, entrapped single water molecules exist.

TYPES OF BOUND WATER

One of the first problems encountered in attempting to understand protein–water interaction is one of definition and methodology. There are numerous methods used to study the interaction of water with macromolecules. The results obtained from different methods often describe different types of bound water. Chou and Morr (1979), in their discussion of protein–water interactions, have divided the three types into six. The first is structural water which engages in hydrogen bonding to the protein molecule. Structural water stabilizes the native structure and is found inside the macromolecule. The water molecules are often engaged in two or more hydrogen bonds and are not available for chemical reaction. Only a small portion of the total water associated with the protein molecule is structural, but the water is very important in determining the three dimensional conformation of the protein in the native state.

The second type of water is monolayer water which fills the first adsorbed layer around the protein. Monolayer water is attached to specific water-binding sites through hydrogen bonding or dipole interaction. About 4–9% of the water associated with the protein is adsorbed at the surface of the protein. This water is not available as a solvent, but may be available for certain reactions.

The third type of bound water is unfreezable water which represents the total water clustered around each polar group. This water does not freeze at a sharp transition temperature and may include both structural and monolayer water. The amount of unfreezable water depends on the amino acid composition of the protein and the polar side chains. Researchers have reported that one gram of protein may have 0·3–0·5 g of water associated with it as unfreezable. This corresponds to a water activity (a_w) of up to 0·9 so reactivity is a function of a_w (Bull and Breese, 1968*b*).

The remaining three types of water associated with the protein molecule are not as well defined as the first three types. The fourth type of water defined by Chou and Morr (1979) is that water associated with the protein molecule via hydrophobic hydration. This type of water has been described as clathrate-type or ice-like structured water but the real nature of this water is not clear. A fifth type is imbibition or capillary water which is held physically or by surface forces. This water is available for chemical reactions and acts as a solvent but can only be removed by force. It is the major type of water found in meat and cheese curd. A sixth type of water, hydrodynamic hydration, is that water which is transported along with the molecule. This water is independent of a_w and has normal physical properties of water.

Other researchers define only three types—constitutional, interfacial (vicinal and multilayer), and bulk phase water (free or entrapped) (Kuntz, 1975; Fennema, 1977). Constitutional water corresponds to structural water and, in a 20% protein solution, would make up about 0·1% of the total water. Interfacial water is made up of vicinal water, the first one or two molecules adjacent to the protein, and multilayer water, the next few layers of water molecules. Interfacial water would correspond to monolayer, unfreezable, and water of hydrophobic hydration. In a 20% protein solution the vicinal water would constitute about 5–10% of the total water content or about 0·3–0·5 g water/g protein. Multilayer water is harder to characterize but it does differ from normal water in that it is more structured. Bulk phase water, which is the remaining water associated with the protein, constitutes the major portion of water. It may be physically entrapped as in a gel or held in dilute protein dispersions as in a sol. The degree to which this water differs from normal water is disputed.

Trying to define and categorize types of bound water is necessary but one should not lose sight of the fact that bound water exists as a continuum. It is difficult to know precisely where one type of water ends and another type begins.

ROLE OF WATER IN PROTEIN CONFORMATION

Information about the structure of water molecules further removed from the protein surface may still be debated, but information about the water molecules tightly bound to the surface or held in the interior of the protein molecule is more certain.

The number of different conformations which a protein could assume is statistically very large. The final conformation is determined by both peptide–peptide and peptide–solvent interactions.

The net free energy of stabilization of native protein is equivalent to no more than three to five hydrogen bonds (Franks, 1985). This clearly demonstrates that the solvent plays a central, rather than a peripheral, role in protein conformation which in turn is essential to protein function.

Kauzmann (1959) studied the role water plays in determining the native conformation of proteins. Three types of bonds determine the pattern of folding of the protein. The oxygen atoms of carbonyl groups and hydrogen atoms of amide groups participate in hydrogen bonding which is an important stabilizer of the native structure. Hydrophobic bonds are also very important. Normally 20–30% of the amino acids in a protein are nonpolar, that is valine, leucine, isoleucine, and phenylalanine. If proline, alanine, and tryptophane are included the percentage rises to 35–45%. The third type of bond is salt linkages. Charged groups that are separated are surrounded by strongly oriented water molecules which are highly compressed by electrostrictive forces arising from the large electric fields in the vicinity of the charges. When these oppositely charged groups come into contact, the electric fields no longer pass so extensively through the water, so the solvent molecules are much less strongly oriented and compressed. A large increase in entropy and volume will result. Salt linkages and hydrophobic bonds are stabilized by entropy effects rather than by energy effects. The addition of electrolytes will strengthen hydrophobic bonds but will weaken salt linkages. The addition of nonpolar solvents will weaken hydrophobic bonds but will strengthen salt linkages by lowering the dielectric constant of the medium. Because of this relationship solvents such as alcohols and acetone will denature proteins while salts act as inhibitors of denaturation.

The traditional view of protein conformation has attributed much of the stability of the native structure to hydrophobic energy. Nonpolar groups were considered to be buried on the inside of the molecule. The loss of entropy on folding was compensated by a gain in 'hydrophobic energy' resulting from a reduction in the number of contacts between the nonpolar groups and water. Polar groups were considered to contribute little to the molecules' stability. Now that more protein structures have been elucidated it is clear that the true picture is more complex. After a review of the literature on protein structure, Finney (1979) made the following generalization about protein structures and the role of water.

1. Formal charges on protein side groups are almost invariably

exposed to the solvent. Where they are not, salt bridges are usually formed.
2. Polar groups tend to be more evenly distributed between internal and external positions. Polar groups which are buried are generally involved in hydrogen bonding, with efficiency generally over 80%.
3. As many as 40–50% of the apolar groups may be accessible to the solvent. The simple picture of buried hydrophobic side chains may be an oversimplification. The actual protein conformation may be more complex.

X-ray and neutron scattering have made possible the identification of discrete water molecules within protein structures. A limitation of this procedure is that the protein must be in a crystalline state and only the position of the oxygen atom is depicted. Most of the water molecules that can be located by these methods are found linking main chain C—O groups to NH groups which are too well separated from one another for direct hydrogen bonds. The water molecules may be hydrogen bonded to three or four different residues. Other water molecules are found linking main chain atoms to side chain hydrogen bond-donors and acceptors. Phenylalanine, serine, methionine, and tryptophane would participate in this type of interaction with water molecules. Another function which water molecules may play is in the stabilization of reverse β-turns in regions where there are few intrapeptide hydrogen bonds. Water bridges may also be formed between amino acid residues belonging to different protein molecules in a crystal or to different subunits of the same protein (Franks, 1985).

Using models, Warner (1981) demonstrated that peptides form regular hexagonal or honeycomb patterns. The distance between peptide oxygen is about 4·8 Å which is the same as the 'second-neighbor' oxygen distances in the ice lattice. The ice-like lattice could lie above a polypeptide layer and exact colinear hydrogen bonds could be formed at each position to satisfy all of the bonding requirements of the peptide and water. This bound water could be important in stabilizing the polypeptide backbone.

Water molecules have been found to form hydrogen-bond bridges between polar groups too far apart to otherwise link. Water mediated salt bridges are also found. Sometimes small groups of water molecules are found near internal charged groups. Here the water molecules may perform a stabilizing role by spreading the buried charge through a region of higher dielectric constants. Each water molecule found internally will make generally three or four hydrogen-bond contacts (Finney, 1979).

Surface exposed polar groups are generally solvated as one would expect. Often short surface bridges are found between polar groups in the same molecule. There is no direct crystallographic evidence for the existence of localized 'clathrate cage' structures about exposed apolar groups. The time scale of X-ray and neutron experiments may be so long that the time-averaged results of such interactions may not be detected (Finney, 1979).

Computer simulation studies suggest that water molecules can form a larger variety of cavity hydration structures which are energetically almost equivalent. This would also make their detection by scattering methods almost impossible (Franks, 1985).

However, water molecules have been found in hydrophobic clefts of some proteins such as trypsin and myoglobin. The water molecules may play an important role in the activity of these proteins. Water molecules may be released as the enzyme binds to its substrate (Finney, 1979).

TECHNIQUES USED TO STUDY WATER BINDING BY PROTEIN

Numerous methods have been used to study water binding by protein. Each method may describe different properties of water–protein interaction which may then give rise to conflicting data. Since this interaction is complex, each method should be viewed as providing one piece of a complex puzzle with no method able to provide complete information on all types of protein–water interactions. Bull and Breese (1968*a*) have listed 12 different methods available, Labuza (1977) described nine methods, and Franks (1975) three groups of methods. Chou and Morr (1979) have grouped the methods into four categories depending on the main properties of the protein–water interaction being studied. The first group includes methods related to the thermodynamic properties of water. These properties include changes in enthalpy (H), entropy (S), free energy (G), water activity (a_w), the freezing point, and the boiling point. The simplest technique used to study these properties is sorption–desorption isotherms. Graphs of a_w versus moisture content can be obtained. Values for a_w

between 0 and 0·3 are related to the water in the monolayer. As moisture content increases with a corresponding increase in a_w between 0·3 and 0·9, the water is said to be unfreezable. At a_w 0·90–0·99 water content increases sharply with increases in a_w. This water is capillary or imbibition water. Isotherm equations correlate with the number of binding sites on protein molecules and the strength of the binding.

In a critique of thermodynamic measurements of protein and water interaction, Kuntz (1975) states that the most serious difficulty with adsorption isotherms is that changes in protein surface are ignored. Protein conformational changes are coupled with the degree of hydration of the protein. The same problem exists when measuring the enthalpy, entropy, and heat capacity. When dry, the protein may be trapped in an unfavorable energy state and any measurement taken will reflect these conformational changes. Kuntz (1975) states that the following conclusions can be reached from thermodynamic studies. Some water molecules are held to protein molecules more tightly than to other water molecules. No simple relationship exists between the amount of water taken up at any arbitrary humidity and temperature and the amount of 'bound' water determined by other methods. Hysteresis and the lack of crystallization of protein at low humidities suggest that one is dealing with a metastable system below a water content of 0·5 g water/g protein. From thermodynamic measurements that use a two component phase diagram such as melting point or freezing point depression, one can conclude that some water molecules are sufficiently altered in their characteristic properties that their activity is less than that of ice. These molecules retain considerable mobility in the presence of ice. The existence of at least two distinct water environments can be demonstrated from thermodynamic studies. One type of water has properties similar to bulk water and the other interacts more closely with the protein molecule.

The second group of techniques measure kinetic properties or the change in the mobility of water as it associates with protein and changes in the protein molecule as affected by its interaction with water. Specific techniques used to study this interaction include nuclear magnetic resonance (NMR), dielectric dispersion, laser light scattering, and intrinsic viscosity. As water and protein associate, changes in the density, volume, and shape of the protein influence the viscosity. Changes in the relaxation rates of protein can be measured as a function of hydration.

Kinetic studies measure the motion of water molecules and molecule measurements of the dielectric constant and NMR relaxation dispersion. The dielectric constant changes rapidly in the region which can be related to different types of water binding by protein. The 10^5–10^7 Hz region is assigned to the tumbling motion of the macromolecule and is sensitive to the molecular weight of the polymer. It is not possible to distinguish the motion of water molecules from the tumbling motion of the protein molecule. At 10^8–10^9 Hz the partially hindered rotation of some hundreds of water molecules per protein molecule is observed. At 2×10^{10} Hz the relatively unperturbed motion of most of the water molecules in the system is noted. From the study of NMR relaxation times, good evidence exists to show that some water molecules move in a way that is strongly coupled to the slow protein rotational motion. The number of such molecules is uncertain but could be in the range of 10–100 water molecules per protein molecule. From kinetic studies, one can conclude that three water environments exist, 'bulk', 'bound', and 'irrotationally bound', with approximate rotational correlation times in the ps, ns, and μs range, respectively. These three types of water are referred to as type I, II, and III and support a discrete distribution rather than a continuous distribution. Type II is the same water detected by low temperature NMR and calorimetric experiments (Kuntz, 1975).

The third group of techniques are spectroscopic methods which measure the nature and strength of hydrogen bonds. These techniques include infrared (IR) and RAMAN spectroscopy as well as NMR which can determine unfreezable water.

In his classic studies using high resolution proton NMR Kuntz (1971) was able to detect the amount of water which does not freeze when an aqueous macro-molecular solution is rapidly frozen and then held at −20 to −60°C. The spectrum shows a single, broad band (0·2–2 kHz) whose area is a direct indication of frozen water.

An analysis of the IR spectra of water on protein films detects the OH band at 3410–3420 cm^{-1}. Comparison of maxima may be unrevealing since IR deals with some type of average environment. Band width may better reflect band length and angles. Protein films do show a 10–20% increase in OH line width compared to liquid water. Low temperature NMR of frozen biopolymer solutions indicate 0·3–0·4 g bound water/g protein. The amount of water bound depends upon the amino acid composition. Using the data obtained from homopolypeptides, Kuntz (1975) developed the following formula to predict the amount of water bound to the protein molecule, $A = f_e + 0{\cdot}4f_p + 0{\cdot}2f_n$ where A is the amount of bound water in g water/g protein, f_e is the fraction of charged side chains, f_p is the fraction of polar side chains, and f_n is the fraction of non-polar side chains. Problems with the formula include the overestimation of the extent of hydration since all side chains are counted, even those buried in the native protein structure. Hydration will increase slightly upon denaturation and decrease slightly upon hydrophobic aggregation. At a pH below 4, dehydration will occur since the number of charged carboxylate groups will decrease sharply. The equation counts carboxylate and cationic side chains equally even though carboxylate groups appear to be somewhat more hydrated. Neutralization of the nitrogen charge at a high pH does not cause dehydration so there is a moderate underestimate (10–15%) of hydration above pH 10.

The fourth group of techniques are diffraction methods which provide information on the average position and orientation of water molecules with respect to each other and to the protein molecule. These methods include light scattering and small angle X-ray scattering as well as high resolution X-ray and neutron diffraction. These methods can be used to locate regions of structural water in the interior of the protein. X-ray measurements are an accurate means of determining highly ordered water molecules within protein crystal structure but cannot be used to define the nature of the solvent region away from the immediate protein–solvent interface. Crystal structure analysis can give direct information concerning the internally bound fraction and, in some cases, the molecules strongly bound at the molecular surface (Finney, 1979).

Chou and Morr (1979) conclude their list of techniques by pointing out that since the techniques measure different properties of water–protein interaction the data may seem to be conflicting but should really be viewed as giving supporting information about a complex interaction.

Many of the techniques discussed above are only of academic interest to the food scientist who is concerned with water as it relates to complex food systems. Water holding capacity (WHC) or water binding capacity (WBC) is often the ultimate concern. Water binding capacity can be determined by four different types of tests which also may give conflicting values (Chen *et al.*, 1984). The first method involves the application of some kind of external stress such as compression, centrifugation, or suction. The second method is based on the equilibrium uptake of water vapor which results in sorption isotherms. The third method is based on liquid diffusion into a capillary swelling system which is achieved with the Bauman apparatus. This method measures the capillary, or imbibing, water which is the ability of the food to spontaneously take up water. The fourth method is based on the determination of a colligative property of water as a change in the freezing point. Chen and coworkers (1984) concluded by stating the WBC will strongly depend on the physico-chemical properties and composition of the test materials as well as on the experimental conditions.

PROTEIN–WATER INTERACTION: POLAR, NONPOLAR, AND IONIC EFFECTS

Since the structure of water itself is not completely elucidated, its interactions with other molecules is also not clearly understood. It is known that when large nonpolar to weakly polar molecules are put in aqueous solutions, they cause structuring of the water around them to some distance. This structuring protects the macromolecule from thermal denaturation and gives water a polymeric nature (Labuza, 1977).

The number of polar amino acids, cationic, anionic, or nonionic, affects the amount of water bound to the protein (Chou and Morr, 1979). The location of the polar amino acids is also a factor since it will be easier to bind water if the charge is located at the surface rather than buried inside the molecule. Early workers believed that the amount of water absorbed depended upon the number and availability of two types of hydrophilic groups found in the protein molecule. The polar chains and the carboxyl and imido groups of peptide bonds are capable of binding water through hydrogen bonds. It was thought that each polar group sorbs one molecular of water followed at higher humidities by multi-molecular absorption (Leeder and Watt, 1974).

Researchers now seem to agree that not all sorption sites have equal degrees of hydrophilicity. Carboxyl and amino groups seem to be mainly responsible for the binding of water (Table 1) with other groups showing less of an ability to bind water (Leeder and Watt, 1974).

TABLE 1
Water associated with hydrophilic groups in proteins

Sorption site	*Moles H_2O per mole of sorption site at RH of:*						
	5%	*10%*	*20%*	*35%*	*50%*	*65%*	*80%*
Carboxyl (—COOH)	0·7	0·92	1·2	1·63	2·0	2·3	2·5
Amino (—NH_2)	0·6	0·83	1·2	1·63	2·1	2·4	2·7
Guanidino (—NH·C(=NH)·NH_2)							
Aliphatic hydroxyl (OH)	0·05	0·09	0·17	0·27	0·34	0·46	0·60
Phenolic hydroxyl (—OH)	0·16	0·25	0·5	0·75	1·0	1·3	1·8
Peptide (—CO·NH—)							
Amide (—CO·NH_2)	0·04	0·06	0·11	0·17	0·25	0·36	0·56
Heterocyclic imino (—NH—)							

Reference: Leeder and Watt (1974).

Kuntz and coworkers (1969) used NMR to determine the amount of water bound to several purified proteins (Table 2). Using polypeptides, Kuntz (1971) was able to determine the amount of water bound to acidic, basic, hydrophobic, and hydrophilic groups (Table 3). When applied to real proteins, the difference between the calculated and experimental results was in the range of 10%. Good agreement exists between the theoretical values and the calculated values at low humidities. At relative humidities over 65% there is a sharp increase in the actual amount of water bound compared to calculated values (Leeder and Watt, 1974). The deviation from calculated values seems to be a reflection of protein solubility. Leeder and Watt (1974) concluded that at higher humidities when multilayer formation becomes more evident the structure of the protein will determine the extent of water uptake. Soluble proteins lack cohesive forces which restrict swelling while, for insoluble proteins, sorption is restricted by swelling constraints.

The calculated results tended to overestimate the experimental results. This overestimation could be due to the fact that some charged groups are buried on the inside of the protein and not exposed to the solvent. Kuntz (1971) reached several conclusions about the interaction between water and protein. He found that the ionized side chains of amino acids are heavily hydrolyzed with ionic group binding between 25–30% of the water. Because of this, the hydration of

TABLE 2
Protein hydration at −35°C and 60 MHz

Sample	*Normalized conc.* *(mg ml^{-1})*	*Hydration*[a] *(g water per g of protein)*
Gelation in distilled water	100	0·4–0·5
	50	0·4–0·5
	25	0·4–0·5
Lysozyme	100	0·36
α-Chymotrypsin	100	0·37
Bovine serum albumin	100	0·43
Oxyhemoglobin	100	0·45
Ovalbumin	100	0·31
tRNA, denatured	100	1·7
α-Chymotrypsin, denatured	100	0·4
Ovalbumin, denatured	100	0·3
Bovine serum albumin		
pH 4·3–5·3	100	0·37
pH 3·4–4·0	100	0·31
pH 2·0–3·0	100	0·28

[a] Absolute values may be uncertain by ±20%; relative values are more reproducible (±10% or better).
Reference: Kuntz *et al.* (1969).

proteins is sensitive to acid titration, but should be insensitive to change in pH above 6. Exposing the hydrophobic core of globular proteins should produce a small increase in hydration. Aggregation of proteins, which involves hydrophobic sites, should result in a slight decrease in hydration while aggregation, which protects ionic sites, should cause large decreases. Hydration is independent of polymer concentration and is not greatly affected by conformation.

The amount of water bound is also a function of pH. Ionized proteins tend to bind more water. When the pH of the protein is at its isoelectric point with no net charge there is minimal hydration and swelling. The protein matrix becomes shrunken. Below the isoelectric point, carboxyl groups became nonionized which also reduced the water binding properties of the protein molecule.

In a study of protein hydration using egg albumin, Bull and Breese (1968*b*) found that the hydrophilic residue became saturated at a relative humidity of 0·92 which corresponded to a water content of about 0·3 g water/g protein. As water was added to the tightly

TABLE 3
Polypeptide hydration

Polypeptide	*pH*	*Moles H_2O per mole of amino acid*		
		−25°C	*−35°C*	*−45°C*
L-Glu	7–12	8·3	7·7	6·3
L-Glu	4·5		1·8	
L-Asp	8–12	8·1	6·0	4·8
L-Asp	4·5	2·1		
L-Tyr	11·5–12		8·5	6·5
L-Tyr	11·3		5·5	5·1
DL- or L-Lys	3–9	5·0	4·3	3·8
DL- or L-Lys	10–12	5·0	4·5	3·7
L-Orn	1·5–9	4·0	3·4	3·5
L-Orn	10–12	4·5	3·7	3·5
L-Arg	3–8	3·1	2·7	
L-Arg	10	3·0		
L-Pro		3·1	2·8	
L-Asn		2·0		
DL-Ala		1·4		
L-Val		0·9		
Gly		0·9		
Polymers				
$Lys^{40}Glu^{60}$	2–4	2·5	2·4	
$Lys^{40}Glu^{60}$	11–12	7·8	7·5	
$Lys^{50}Phe^{50}$	2–9	2·6	3·8	
$Lys^{50}Phe^{50}$		1·2		

Reference: Kuntz (1971).

complexed water, the complex tends to relax and the apparent density of the complexed water tends to decrease.

While admitting that hydration of nonpolar groups is difficult to study, Noguchi (1981) did reach some conclusions about hydrophobic hydration.

(1) The change in volume due to hydrophobic hydration around the methylene group is −1 to −2 ml $mole^{-1}$. Because of this small change, it is reasonable to conclude that an 'iceberg' block around hydrophobic groups does not exist.
(2) The diabatic compressibility of water of hydrophobic hydration is larger than electrostricted water and smaller than normal water.

(3) With increasing temperature, water molecules participating in hydration are released more rapidly in the order of electrostrictional hydration < hydrogen-bonded hydration < hydrophobic hydration.

The unfolding or relaxation of the compact globular protein to a more relaxed random coil must satisfy thermodynamic requirements. A change in enthalpy is not acceptable unless a compensating change occurs in entropy so that the total free energy of the system is not changed. As the protein molecule unfolds, buried nonpolar groups are exposed. The thermodynamically unstable situation is compensated for by the formation of hydrogen-bonded water clusters (ice-like structures) around the exposed nonpolar groups. Change in the solvent accommodates the nonpolar groups of the unfolded polypeptide (Labuza, 1980). Fontan and coworkers (1982) showed that the enthalpy/entropy compensation was highly correlated (correlation coefficient 0·988) in a wide range of protein and starch foods.

PROTEIN–WATER–ION INTERACTION

Many workers have studied the three way interaction of protein, water, and ions (Von Hippel and Wong, 1962; Bull and Breese, 1970; Damodaran and Kinsella, 1982). The 'salting in' and 'salting out' effect of ions on proteins have been used to isolate and characterize many proteins. 'Salting in', which causes an increase in solubility, usually occurs at lower ion concentration and lower electrostatic interactions are involved. At concentrations in excess of 0·1 molar, precipitation or 'salting out' of the protein occurs.

Since proteins are colloidal particles, they are surrounded by an electric double layer (Eagland, 1975*b*). The interaction depends on both the nature of the protein colloid as well as the concentration of the electrolyte. Hydrophobic colloids are compact in shape. Neutral electrolytes, at a concentration below 0·1 mol dm^{-3}, increase the thickness of the double layer and, hence increase the stability of the colloidal particle. Reactions with hydrophilic colloids, which are long and threadlike, are more complex. Low electrolyte concentrations increase the electric double layer which will bring about changes in the conformation of the protein. Conformational changes may be due to repulsion between adjacent charged groups and to the exposure of previously buried hydrophobic groups (Eagland, 1975*a*). At concentrations of 0·1–0·15 mol dm^{-3}, the electric double layer is effectively suppressed. At concentrations greater than 0·1 mol dm^{-3}, changes in protein conformation are most likely to occur due to possible interactions. At these higher concentrations there is more solvent–ion interaction and also an exchange between the ions in the solvent and ions more firmly held by the macromolecule.

Anions decrease the polarity of water more than cations (Damodaran and Kinsella, 1982). Negatively charged anions attract both of the positively charged hydrogen poles of water while the cations attract the slightly negatively charged oxygen. Both anions and cations tend to arrange themselves in a pronounced series in their effect upon proteins. The ions generally follow the lyotropic series. With cations, the series is closely related to the ionic radii with Cs^+ having the greatest effect followed by $Rb^+ > K^+ > Na^+ > Li^+$. The larger the ionic radii of the cation the greater the tendency of the ions to dehydrate the protein by binding water. Ions may have an effect on protein by preferentially binding the protein molecules itself in addition to its affect on water. Bull and Breese (1970) studied this three way interaction using egg albumin. They reported that the size of the cation studied had little effect on the binding of the ion to the protein, with about 5·7 moles of LiCl, RbCl, and CsCl bound per mole of egg albumin. Very little NaCl and KCl were bound to the protein. The cations bound water according to ionic radii as expected. In contrast to cations, there was a relationship between the size of the monovalent anion and its ability to bind to protein. The larger the monovalent anion the greater the tendency to bind to the protein and the greater is its dehydrating effect on the protein. The practical application of the interaction of solutes with protein and water will be discussed later.

HYDRATION OF DENATURED PROTEINS

One way to help understand the complex interaction between water and protein is to study the changes in protein hydration which occur when proteins are denatured. While there is some disagreement, most researchers have reported a slight increase in bound water as the protein denatures. Kuntz and Brassfield (1971), using proton magnetic resonance, attributed the narrow band to bound water when a solution of bovine serum albumin at −25°C was studied. In the presence of urea, this band width increases. A 10% increase in protein hydration was observed. Water binding was occurring in the regions of the protein that were exposed by unfolding. As many as one-third of the amino acid residues may be buried. They theorized that the small size of the increase could be due to the fact that polar and charged groups already on the surface are mainly responsible for sites of water binding. The small increase may also be due to internal solvation of native proteins and incomplete unfolding of the protein in the denatured state. Denaturing agents such as urea or guanidine HCl may affect protein hydration by one of two ways. Denaturants may produce these effects by an indirect effect on water structure rather than by direct interaction with the protein (Tombs, 1985). Other researchers reported that denaturants greatly diminish hydration of protein by binding to the protein (Bull and Breese, 1970).

FUNCTIONAL PROPERTIES IN FOODS

Many of the important functional properties of proteins in foods are related to the interaction of water with proteins. These functional properties include solubility, water absorption and binding, viscosity, and gelation. As can be seen in Table 4, the relationship between water and soy protein is a major determinant in the application of soy to many different food systems (Kinsella, 1979).

The interaction of solutes–protein–water has practical application in many food products. Hardy and Steinberg (1984) studied the interaction of sodium chloride and paracasein as a function of water sorption. They concluded that as the concentration of salt increased, the amount of interacted salt also increased. Salt tended to partition itself between the protein and water. At a higher a_w, the salt tended to bind the water and interact less with the protein. At a lower water content, salt tended to increase its interaction with protein reaching a constant level at 0·7 g salt/g paracasein. These interactions are important in cheese production since the binding of salt will influence the growth of microorganisms which, in turn, will influence the hydrolysis of protein during ripening. This hydrolysis will determine many of the rheological and textural properties of the final product.

Another example of this three way interaction in foods is the effect of salts on the conversion of collagen to gelatin, also known as the collagen fold. Neutral salts affect the collagen fold by competitively

TABLE 4
Functional properties performed by soy protein preparations in actual food systems

Functional property	*Mode of action*	*Food system*
Solubility	Protein solvation, pH dependent	Beverages
Water absorption and binding	Hydrogen-bonding of HOH, entrapment of HOH, no drip	Meats, sausages, breads, cakes
Viscosity	Thickening, HOH binding	Soups, gravies
Gelation	Protein matrix formation and setting	Meats, curds, cheese
Cohesion–adhesion	Protein acts as adhesive material	Meats, sausages, baked goods, pasta products
Elasticity	Disulfide links in gels deformable	Meats, baked goods nnnnn
Emulsification	Formation and stabilization of fat emulsions	Sausages, bologna, soup, cakes
Fat absorption	Binding of free fat	Meats, sausages, donuts
Flavor-binding	Adsorption, entrapment, release	Simulated meats, baked goods
Foaming	Forms stable films to entrap gas	Whipped toppings, chiffon desserts, angel cakes
Color control	Bleaching of lipoxygenase	Breads

Reference: Kinsella (1979).

reorganizing the water involved in stabilizing the helix of collagen (Von Hippel and Wong, 1962).

The importance of ions in protein–water interaction is evident in the study of the hydration of milk proteins. Berlin (1981) concluded that the ions in milk assume a greater role in controlling water binding than the status of either the casein or whey proteins with regard to denaturation. The loss of protein solubility through thermal denaturation apparently had little effect on the capacity of the protein to bind water.

The application of the findings of basic research to actual food systems is an important and often difficult process. In an attempt to relate the mobility of water as detected by NMR with apparent viscosity, Richardson and coworkers (1985) studied wheat flour suspensions at various concentrations. At low concentrations, between 0% and 30%, wheat flour exhibits Newtonian flow characteristics, between 30% and 85%, pseudoplastic, and greater than 85%, viscoelastic properties as gluten develops. NMR response was directly related to apparent viscosity only within each region, not for all concentrations studies. This study illustrates the complexity of interaction of water and macromolecules. At different concentrations of water, different types of interactions between the water and other components of the wheat flour would occur. Not only would protein have to be considered but other components in the flour as well.

In sausage-making, water holding capacity (WHC) is an important function of the muscle protein and will often determine the textural properties of the final product. Prerigor meat exhibits superior water holding capacity especially if it is salted prior to blending with other ingredients (Poulanne and Terrell, 1983). The interaction of fat–water–protein may also be important. The use of a pre-emulsified fat was found to increase the WHC in the final sausage. Heat treatment is also important since the proteins will be denatured which will result in an overall decrease in WHC (Zayas, 1985).

The interaction of water and meat proteins is often the deciding factor in determining the quality of the product. The cross-striated muscle contains 75% water with the myofibrillar protein responsible for binding most of the water in muscle. Hamm (1985) describes the three types of water which are found in muscle tissue. Constitutional water is that water located within the protein molecule and comprises less than 0·1% of the total water associated with the muscle tissue. This tightly bound water represents about 0·3 g water/100 g protein. Another 5–15% of the total water is interfacial water which has relatively restricted mobility. This water is located on the surface of proteins in multilayers and in small crevices. Controversy exists over the exact nature of the remaining water found in muscle tissue. Many NMR studies show it is free water in the physical–chemical sense while other researchers disagree showing it to behave as bulk water in dilute salt solution (Blanchard and Derbyshire, 1975). This third type of water is the water of real importance in determining meat quality and will be discussed further. Another 10% of the water is found in the extracellular space. This water may exchange with cellular water as conditions which affect the myofibrillar proteins change.

The bulk or entrapped water, which is of real importance in meat, should not be viewed as homogeneous but rather as a more or less continuous transition. At one end of the continuum is water strongly immobilized within the tissue which can be expressed only with difficulty, while at the other end is water which can be squeezed out by very low pressure.

Because this entrapped water or, as Hamm (1985) prefers to term it, immobilized water is not homogeneous several definitions are needed to describe it. Hamm (1985) has defined the following terms:

(1) Drip loss—the exudate from meat or meat systems without the application of external force.
(2) Thawing loss—the formation of exudate from meat after freezing and thawing without the application of external force.
(3) Cooking loss—the release of fluid after heating of meat either with or without the application of external forces such as centrifugation or pressing.
(4) Expressible juice—the release of juice from unheated meat or meat systems during application of external forces such as pressing, centrifugation, or suction.

The immobilization of water in the muscle tissues is determined by the spatial molecular arrangement of the myofibrillar proteins, with myosin being mainly responsible for water binding. Swelling of the myofibrillar protein results in an increase in the amount of water that can be entrapped or immobilized. Swelling occurs by decreasing the attraction between adjacent molecules or filaments. This could be caused by increasing the electrostatic repulsion between similarly charged protein groups or by weakening of the hydrogen bonds or hydrophobic bonds between protein groups. Conditions which cause this swelling to occur which will result in an increase in the water holding capacity of the meat are numerous. Some examples include increasing the pH above 5 which is the isoelectric point of myosin. This will result in an increase in the net negative charge of the molecule. The addition of NaCl at a pH below 5 will cause the screening of positive protein charges by the preferential binding of Cl^- ions. Pyrophosphate or ATP will cause the dissociation of linkages between myosin heads and thin filaments. The interaction between hydrophobic protein groups and the hydrophobic part of smaller molecules such as lecithin will cause a weakening of interaction between hydrophobic groups. Cleavages of linkages between z-lines which occur during aging also increase water binding.

Conversely increasing the attraction between adjacent molecules will decrease the space available for the immobilization of water. This tightening of the protein network can occur by lowering the pH of the protein to pH 5, the isoelectric point of myosin. This causes an increase in the attraction between oppositely charged protein groups. The addition of NaCl at a pH less than the isoelectric point can lower the repulsion between positively charged protein groups by the screening effect of Cl^- ions. The association of actin and myosin during rigor mortis results in a tighter protein complex. The heat coagulation of protein increases the interaction between hydrophobic groups.

Protein–water interaction is also an important determinant of the ultimate nutritional value of the protein. Nonenzymatic browning, which often involves the loss of available lysine, is a function of the amount of free water available for reaction. Labuza and Saltmarch (1981) reported that a 0·1 a_w increase for food in general doubles the reaction rate constant for most reactions related to nonenzymatic browning. During the drying of foods the Maillard reaction becomes the dominant deteriorative reaction (Eichner *et al.*, 1985). This reaction between reducing sugars and amino groups of amino acids ultimately results in visible browning which may result in a change not only in color but also in nutritional quality and flavor of products. Maximum browning occurs at water activities of 0·3–0·7 (Eichner and Ciner-Doruk, 1975). At low a_w, water is tightly bound to surface polar sites by chemisorption and is generally unavailable for reaction and solution. An a_w of 0·2–0·3, which corresponds to the upper limit of the BET monolayer, is the most stable moisture content for dehydrated foods (Labuza, 1985).

The decrease in browning at high water activity is attributed to the dilution of reacting substances and to the law of mass action. During the condensation stages 3·5 ml water/mol sugar is formed during browning (Eichner and Ciner-Doruk, 1975).

The addition of humectants, like glycerol, lower the browning rate maximas to a_w of 0·41–0·55. If sugar type humectants that are reducing compounds, such as glucose or fructose, are used to control a_w, browning could be a serious problem. If liquid humectants, such as glycerol or propylene glycol, are used, problems such as taste and cost may be encountered (Labuza, 1985). Control of time and temperature when drying foods is essential to prevent the food from remaining too long at the level of water activity where maximum browning can occur.

The dehydration of high protein foods will affect the conformation of the protein. As the moisture content is reduced the distance between protein chains is diminished. This may increase the formation of cross-linkages between adjacent chains. The cross-linkages will lead to a tighter network of protein and the water holding capacity and solubility of the protein will be reduced. Thermal stability of the protein may be increased at low moisture contents since the bonds formed are tighter and less free to rotate with the addition of energy (Rustad and Nesse, 1983).

The relationship between water binding and viscosity was studied by Urbanski and coworkers (1983) using soy protein. They reported a decrease in viscosity of the soy suspensions by solutes. A high viscosity is caused by the uptake of water by polymers. When no solutes are present, the dissociation of the protein quaternary structure and the strong intramolecular repulsive forces result in molecular expansion and water uptake. Solutes which are ionic, such as NaCl, may neutralize the intramolecular repulsive forces and stabilize the quaternary structure. This results in reduced viscosity by preventing molecular expansion and water uptake. Solutes which are nonionic in nature, such as sucrose and dextrose, reduce viscosity by their affinity for water. Water will associate with the solute not the polymer. Elgedaily and coworkers (1982) used a farinograph to study the same interaction, that of salt and sugar with soy protein, in dough systems. They concluded that, with the different isolates studied, they could not generalize as to the effect of salt or nitrogen solubility index (NSI) on water absorption. Sucrose seemed to have no effect on either NSI or water absorption.

Johnson (1970) attempted to relate NSI to the functional properties of soy protein. Proteins with a NSI between 50–60% were most appropriate for use in breads, cakes, sweet doughs, cookies, donuts, dry mixes, and macaroni. When NSI was reduced to 25–35%, there was a decrease in soy flavor and protein would be best used in beverage type products, pancakes, waffles, tortillas, gravies, soup, pudding, sausage products, dietary supplements, and baby foods. A NSI of 15–25% would indicate the soy protein would be best used in calf milk replacers, crackers, beverages, cookies, cereals, baby foods, and pet foods.

PROTEIN GELS AND FOAMS

Protein–water interactions are essential to the formation of gels and foams. Gels and foams provide structure in many foods, both traditional and novel. Many proteins in foods have the ability to form gels and provide structure. These include proteins found in meat, milk, fish, eggs, and soy which give structure to products such as yogurt, gelatins, omelets, and surimi.

The acceptability of many products which rely on gel formation for their structure is determined by the capacity of the protein to bind water. Schmidt and Morris (1984) reviewed the factors which affect gelation of milk, soy, and blended protein. O'Brien and coworkers (1982) related the water holding capacity of egg proteins to the textural acceptability of precooked frozen whole egg omelets. Ziegler and Acton (1984) point out that the texture of processed meat products is a function of the properties of the protein matrix and its special interactions with the continuous aqueous phase and dispersed fat. The texture of surimi results from the thermal gelation of comminuted fish proteins (Burgarella *et al.*, 1985).

Hamm (1963) divided high water content foods into two classes, thermo-reversible secondary valence gels where hydrogen bonds cross-link macromolecules in a three-dimensional structure and principal valence gels which are thermo-irreversible and bound together by multi-valent cations or salt bridges. Syneresis occurs if macromolecules are too close and colloidal solutions occur if intermolecular cohesion becomes too weak.

Before proceeding, it is necessary to define certain terms that are used in the discussion of gelation whose meanings often overlap.

Denaturation refers to any process which causes a change in the three-dimensional structure of the native protein which does not involve rupture of peptide bonds. Protein–solvent interaction may be involved as well as changes in the physical properties of the protein.

Aggregation refers to protein–protein interactions which result in the formation of complexes of higher molecular weight.

Coagulation is the random aggregation of already denatured protein molecules in which polymer–polymer interactions are formed over polymer–solvent reaction.

Protein gels are defined as the three-dimensional network in which polymer–polymer and polymer–solvent interactions occur in an ordered manner resulting in the immobilization of large amounts of water by a small proportion of protein (Mulvihill and Kinsella, 1987). Gelation may be induced by heat or divalent cations. In gelation, polymer–polymer and polymer–solvent interaction as well as attractive and repulsive forces are balanced (Gossett *et al.*, 1984). Gelation differs from coagulation and aggregation in that a well ordered matrix is formed.

Gelation is thought to proceed by a two step mechanism. First, protein begins to unfold which changes its conformation. This may be followed by aggregation of the protein. In the second step, which proceeds more slowly, the denatured protein molecules orient themselves and interact at specific points forming the three-dimensional network (Gossett *et al.*, 1984; Mulvihill and Kinsella, 1987).

The rate of the second step is critical and may determine some of the characteristics of the gel. If the second step is slow, the protein polymer will form a fine network. The gel will be less opaque, more elastic, and exhibit less syneresis. If step two is fast, a coarser network will be set up and the gel will be opaque with more solvent expressed (Gossett *et al.*, 1984).

Many complex interactive forces will determine whether a protein will form an aggregation, coagulum, or a gel. For a protein to gel, there must be a balance between attractive and repulsive forces. Coagulation occurs if excessive attractive forces dominate and, likewise, no gel would be formed if excessive repulsive forces are present. The types of forces that hold a gel together include hydrophobic interaction, hydrogen bonding, electrostatic interaction, and disulfide crosslinks or thio-disulfide interchange (Mulvihill and Kinsella, 1987).

Using gelatin as an example of a protein gel, Labuza (1977) found that the greatest bonding of the protein was due to hydrogen bonds between the C—O groups and NH groups of the peptide linkages and not electrostatic bonds. Chemical modification of the polar groups did not greatly influence mechanical properties of the gel. Electrostatic bonds do have an effect, especially as the pH moves away from the isoelectric point (pI). At a pH lower than the pI there may be too many positively charged groups and above the pI too many negatively charged groups. As the pH of conalbumin was raised and lowered aggregation only occurred as the salt content was increased (Hegg, 1982). A change in pH or ionic strength can alter the charge distribution among amino acid side chains which will either increase or decrease protein interaction. In the case of ovalbumin, the degree of electrostatic repulsion is the main factor contributing to heat induced aggregation. As ionic strength increases the amount of water bound to albumin may decrease making it less soluble. Water–protein interactions may decrease and protein–protein interactions may increase (Gossett *et al.*, 1984).

The role of disulfide bonds in the formation of gels is controversial. Some researchers have found that heating of egg albumin causes polymerization of intermolecular sulfhydryl–disulfide exchange causing the formation of a network (Shimada and Matsushita, 1980). When studying proteins from egg white, bovine blood, and milk, Hegg (1982) found no correlation between disulfide or sulfhydryl content and gel-forming ability.

The factors that affect gelation are interrelated. Shimada and Matsushita (1980) studied the relationship between the thermo-coagulation of proteins and their amino acid compositions. Proteins may form an aggregate or coagulum (thermo-irreversible gel) or a gel (thermo-reversible gel). Proteins were divided into these two states depending on the changes they undergo upon heating when conditions such as pH, ionic strength, and the presence or absence of denaturants, are known. Most proteins can form either a coagulum or gel depending upon the conditions present. However, proteins tend to prefer one of the two states due to their inherent characteristics. Shimada and Matsushita (1980) determined that for some proteins whether a coagulum or gel is formed depends upon concentration of the protein while other proteins are concentration independent. With large proteins (>60 Kda) such as hemoglobin, egg albumin, and catalase that have a high molar percentage of hydrophobic amino acids (≥31·5%), the pH range for gelation is dependent upon concentration. With large proteins such as gelatin, soy conalbumin, and prothrombin which have a low molar percent of hydrophobic amino acids (22–31·5%), and smaller proteins, β-lactoglobulin, with a higher molar percent of hydrophobic amino acids (34·6%), the pH range for gelation is concentration independent. Hydrophobic and disulfide bonds formed at high-protein concentrations can compensate for the repulsive electrostatic forces associated with pH values well removed from the isoelectric point. Proteins showing concentration-dependent coagulation have a larger number of hydrophobic groups than those without concentration dependence.

Temperature of gel formation, pH, and the presence of salts will all influence the type of gel formed by a specific protein. These factors will interact with each other changing the gel forming temperature.

Using bovine serum albumin as an example, Yasada and coworkers (1986) reported that the gel forming temperature differs on both sides of the isoelectric point. The same forces may not be involved in gel formation at an acidic or alkaline pH. Intermolecular β-structure may be the main forces which contribute to gel formation on the acid side of the isoelectric point while disulfide bonds may contribute to gel formation on the alkaline side. At higher temperatures protein will aggregate regardless of the concentration. At lower temperatures a certain concentration is needed for gelation, usually around 1%.

Because so many factors are involved in gelation it is difficult to predict how a protein will react. Hegg (1982) studied conalbumin and lysozyme from egg white, serum albumin from bovine blood, and β-lactoglobulin from milk. Hegg concluded that it would be possible to predict the boundaries between solubility and aggregation for any globular protein by knowing the amount of salt present in the sample and simple physical data such as isoelectric point and pH-induced transitions which could be obtained from titration curves. However, to identify the special protein characteristics associated with gel formation is difficult since, as stated earlier, concentration is involved. If gel formation does occur, it is found close to the boundary between aggregation and solubility.

While much research has focused on the structure of gels, it has been difficult to relate protein–water interaction to the macrostructure of gels. The interaction between protein and water is very important in the determination of the type of gel that is formed. Substantial protein–water interaction in the system at the time of heating results in a highly hydrated viscoelastic gel. A low degree of protein–water interaction in the system results in aggregation or precipitation due to exclusion of water from the network.

Water may be held in a gel by hydrogen bonding to hydrophilic groups, dipolar interactions with ionic groups, or structured around hydrophobic groups. Protein–water interactions tend to reduce protein–protein interaction (Mulvihill and Kinsella, 1987).

Gels can be formed with only 1% solids and lose very little water on standing. The distance between macromolecules is very large so there is a tremendous amount of free space available for water. The properties of water in gels are almost the same as that of pure bulk water. Water does not leak out of the gels due to either some unusually weak, long range forces or to capillary suction in the pores formed between macromolecules (Labuza, 1977).

Goldsmith and Toledo (1985) used pulsed NMR spectrometry as a nondestructive measurement of protein–water interaction in egg albumin gels. Spin-lattice relaxation times for protons (T_1) should be shorter, the greater the degree of water binding present. As the heating temperature of egg albumin increased from 60°C to 90°C, T_1 values of the resulting gels were shorter. This would indicate an increased degree of water structuring. The T_1 values were highly correlated with the physical strength of the gels. A higher temperature is needed to form a stronger gel because of the dependence of gel structure on hydrophobic interactions (Shimada and Matsushita, 1980).

When studying gelatin gels, Labuza and Busk (1979) found that T_1 decreases with higher concentrations of gelatin, especially above 30%. The major network structure has formed and increased additions of gelatin only enhances further the helix content development without affecting the overall network pore spaces. Shorter T_1 times indicate a more highly structured proton system. Since gelatin is a highly hydrophobic molecule the water contained within the pore spaces would become more highly structured through hydrogen bonding induced by the hydrophobic shell.

However, the relationship between water binding and texture may not always be correlated. In studies using blood plasma proteins, Hermannsson and Lucisano (1982) point out that water binding and texture are not always correlated and should be treated separately. Changes in gel structure may affect texture and water binding quite differently. As gels are heated above the gelation temperature there is an increased tendency toward more protein–protein interactions. Shrinkage will occur if protein–protein interactions are uniform throughout the gel. If not, stronger protein–protein interactions will cause a partial disruption of the gel network. The gel will become more aggregated and more water will be lost since it is easily pressed out through larger capillary structures. The increased amounts of denser regions will contribute to an increase in force during the compression and penetration (Hermannsson, 1982). Likewise, the changes in blood plasma gels as the pH is increased is not reflected in water binding properties. The finer and more continuous the structure, the less is the tendency toward phase separation and moisture loss, regardless of the mobility of the polymer chains.

One of the major problems in cooked and frozen egg mixtures is the loss of water from the gel. The protein network moves closer together resulting in the expulsion of water from the gel. Albumin has been identified as the component mainly responsible for the percent expressible moisture (%EM) in precooked, frozen, and thawed whole egg mixtures. One way to help alleviate this problem is to increase the net negative charge in the protein. Gossett and Baker (1983) studied two methods used to achieve this end, increasing the pH and succinylation. When the pH of the mixture was raised to pH 9·5 or greater the %EM decreased. Succinate anions bind with two carboxyl groups causing a net gain in negative charge. Protein–protein interaction may decrease causing an increase in protein–water interaction.

Busk (1984) concludes that there is little or no correlation between gel macrostructures and the microstructures formed by polymer–water interaction. Macrostructures are assigned to the physical chemistry of the polymer itself.

Interfacial protein films can be considered as thin gel layers (Graham and Phillips, 1980). Additional components which must be considered in foam formation are surface pressure (π) and surface concentration (Γ). Foam formation by protein involves a two step operation. First, there is a slow penetration of native molecules into the surface of the film. The second step is a slow rearrangement of molecules at the surface. The first step is diffusion controlled and both π and Γ are changing. Two different modes of action are possible in the second step depending on the type of protein involved. If both π and Γ are changing, there is a positive energy barrier to both penetration and molecular rearrangement in the surface. If π is changing and Γ remains constant, conformational changes are taking place in the adsorbed layer. β-Casein represents the first type of foam formation and lysozyme the second type (Graham and Phillips, 1979*a*). The saturated monolayer coverage occurs via irreversible adsorption of 2–3 $mg\,m^{-2}$ of protein. Films generate surface pressure of about 20 $mN\,m^{-1}$ and are 5–60 Å thick. Molecules adsorbed in the first layer dominate film pressures so that further adsorption causes no change in pressure although the film thickness can increase to more than 100 Å. Molecules which give rise to the increased film thickness are reversibly adsorbed with respect to aqueous substrate exchange (Graham and Phillips, 1979*b*).

Molecular requirements for foaming include solubility, the ability of the protein to unfold at the interface, and the possession of substantial surface hydrophobicity. Denaturation improves foaming by enhancing macromolecular flexibility and surface hydrophobicity (Halling, 1981).

In dilute films, protein molecules are complete unfolded and spread out at the surface so no protein tertiary or secondary structure remains. In concentrated films, native and completely unfolded molecules coexist at the surface. The type of film formed also depends upon the conformation of the original protein. β-Casein, which is a flexible and disordered molecule, forms either trains of amino acids residues in the interface or layers and tails of residues protruding into the bulk phase. In contrast, globular proteins such as lysozyme and BSA retain elements of their native structure when adsorbed so their surface pressure curves and molecular area are relatively condensed (Graham and Phillips, 1979*c*).

Transmission electron micrographs show an electron opaque layer at the film surface. Layers appeared to be the formation of conglomerates of unfolded polypeptides. Protein clusters formed a network of parallel filaments (Johnson and Zabik, 1981).

Foam stability is enhanced if the film at the interface is a highly elastic and densely packed layer of solid particles. Globular proteins that remain soluble at their isoelectric point, can form strong films. Interfacial disulfide bonds also assist in the formation of strong films (Dickinson and Stainsby, 1987).

Measuring foam characteristics is difficult. The two main attributes of foams are foam power, or capacity, and foam stability. Foam power or capacity is a measure of the increase in foam volume upon the introduction of a gas into the protein solution. Foam stability is a measure of the rate of liquid leakage from foam or the rate of a decrease in foam volume with time. These characteristics can be measured by conductivity (Kato *et al.*, 1983). Difficulties exist in making these measurements. Foam power, or capacity, is partially dependent on the method used to introduce the gas into the protein. Foam stability depends on the thickness and strength of the adsorbed film at the air–water interface. Changes in film thickness may occur before there is any leakage from the foam or any change in volume.

Townsend and Nakai (1983) attempted to overcome difficulties of characterizing foam capacity and stability by measuring the physicochemical properties of protein in bulk solution and correlating these measures to foaming capacity and foam stability. These researchers found that there was good correlation between foaming characteristics and the flexibility of the protein, viscosity, and average hydrophobicity not just surface hydrophobicity since the protein uncoils. No good correlation was observed between foaming capacity and charge density. Ionic strength showed no significant correlation although there was a tendency toward decreased foaming capacity as the net charge increased. A negative correlation was found between foaming capacity disulfide linkages.

CONCLUSIONS

The interaction of protein and water can be studied on several levels. Very basic research has detailed the exact position of water molecules within many protein macromolecules. The important role of water in protein conformation and stability can be elucidated. As the water molecules become less tightly bound to the protein, more controversy exists as to the exact nature of the interaction. The most important and difficult task, as Karel (1975) points out, is to relate the various theoretical and empirical definitions of the state of water to predict the behavior of foods. The major problem to be resolved is the relationship between the types of bound water and the practical aspects of food technology, especially the processibility, palatability, and stability of foods.

REFERENCES

Berlin, E. (1981). Hydration of milk proteins. In: *Water Activity: Influences on Food Quality,* L. Rockland and G. Steward (Eds), Academic Press, NY, pp. 467–88.

Blanchard, J. and Derbyshire, W. (1975). Physico-chemical studies of water in meat. In: *Water Relations of Foods,* R. Duckworth (Ed.), Academic Press, NY, pp. 559–71.

Bull, H. and Breese, K. (1968*a*). Protein hydration. I. Binding sites. *Arch. Biochem. Biophys.* **128,** 488–96.

Bull, H. and Breese, K. (1968*b*). Protein hydration. II. Specific heat of egg albumin. *Arch. Biochem. Biophys.,* **128,** 497–502.

Bull, H. and Breese, K. (1970). Water and solute binding by proteins. II. Denaturants. *Arch. Biochem. Biophys.,* **139,** 93–6.

Burgarella, J., Lanier, T., Hamann, D. and Wu, M. (1985). Gel strength development during heating of surimi in combination with egg white or whey protein concentrate. *J. Food Sci.,* **50,** 1595–7.

Busk, G. (1984). Polymer–water interactions in gelation. *Food Tech.,* **38**(5), 59–64.

Chen, J., Piva, M. and Labuza, T. (1984). Evaluation of water binding capacity (WBC) of food fiber sources. *J. Food Sci.*, **49**, 59–63.
Chou, D. and Morr, C. (1979). Protein–water interactions and functional properties. *J. Am. Oil Chem. Soc.*, **56**, 53A–62A.
Damodaran, S. and Kinsella, J. (1982). Effects of ions on protein conformation and functionality. In: *Protein Structure Deterioration*, J. Cherry (Ed.), ACS Publication, Am. Chem. Soc., Washington, DC, pp. 327–56.
Dickinson, E. and Stainsby, G. (1987). Progress in the formulation of food emulsions and foams. *Food Tech.*, **41**(9), 74–81, 116.
Eagland, D. (1975*a*). Nucleic acids, peptides and proteins. In: *Water, A Comprehensive Treatise*, F. Franks (Ed.), Plenum Press, NY, pp. 305–518.
Eagland, D. (1975*b*). Protein hydration—its role in stabilizing the helix conformation of protein. In: *Water Relations of Foods*, R. Duckworth (Ed.), Academic Press, NY, pp. 73–92.
Eichner, K. and Ciner–Doruk, M. (1975). Formation and decomposition of browning intermediates and visible sugar–amine browning reactions. In: *Water Activity: Influences of Food*, L. Rockland and G. Steward (Eds), Academic Press, NY, pp. 567–603.
Eichner, K., Laible, R. and Wolf, W. (1985). The influence of water content and temperature on the formation of Maillard reaction. Intermediates during drying of plant products. In: *Properties of Water in Foods*, D. Simatos and J. Multon (Eds), Martinus Nijhoff Publishers, Dordrecht, The Netherlands, pp. 191–210.
Elgedaily, A., Campbell, A. and Penfield, M. (1982). Solubility and water adsorption of systems containing soy protein isolates, salt and sugar. *J. Food Sci.*, **47**, 806–9.
Fennema, O. (1977). Water and protein hydration. In: *Food Proteins*, J. Whitaker and S. Tannenbaum (Eds), AVI, Westport, CT, pp. 50–90.
Finney, J. (1979). Organization and function of water in protein crystals. In: *Water, A Comprehensive Treatise*, F. Franks (Ed.), Plenum Press, NY, pp. 47–122.
Fontan, C. Chirife, J., Sancho, E. and Inglesias, H. (1982). Analysis of a model for water sorption phenomena in foods. *J. Food Sci.*, **47**, 1590–4.
Franks, F. (1975). Water, ice and solutions of simple molecules in water relations of foods. In: *Water Relations in Food*, R. Duckworth (Ed.), Academic Press, NY, pp. 3–22.
Franks, F. (1985). Water in aqueous solutions: recent advances. In: *Properties of Water in Foods*, D. Simatos and J. Multon (Eds), Martinus Nijhoff Publishers, Dordrecht, The Netherlands, pp. 1–23.
Goldsmith, S. and Toledo, R. (1985). Studies on egg albumin gelation using nuclear magnetic resonance. *J. Food Sci.*, **50**, 59–62.
Gossett, P. and Baker, R. (1983). Effect of pH and of succinylation on the water retention properties of coagulated, frozen and thawed egg albumin. *J. Food Sci.*, **48**, 1391–4.
Gossett, P., Rizvi, S. and Baker, R. (1984). Quantitative analysis of gelation in egg protein systems. *Food Tech.*, **38**(5), 67–74, 96.

Graham, D. and Phillips, M. (1979*a*). Proteins at liquid interfaces. I. Kinetics of adsorption and surface denaturation. *J. Colloid Interface Sci.*, **70,** 403–14.

Graham, D. and Phillips, M. (1979*b*). Proteins at liquid interfaces. II. Adsorption isotherms. *J. Colloid Interface Sci.*, **70,** 415–26.

Graham, D. and Phillips, M. (1979*c*). Proteins at liquid interfaces. III. Molecular structure of adsorbed films. *J. Colloid Interface Sci.*, **70,** 427–39.

Graham, D. and Phillips, M. (1980). Proteins at liquid interfaces. V. Shear properties. *J. Colloid Interface Sci.*, **76,** 240–50.

Halling, P. (1981). Protein stabilized foams and emulsions. *CRC Critical Reviews in Food Sci. Nutr.*, **15,** 155–203.

Hamm, R. (1963). The water imbibing power of foods. *Recent Adv. Food Sci.*, **31,** 218.

Hamm, R. (1985). The effect of the quality of meat and meat products: problems and research needs. In: *Properties of Water in Foods,* D. Simatos and J. Multon (Eds). Martinus Nijhoff Publishers, Dordrecht, The Netherlands, pp. 591–602.

Hardy, J. and Steinberg, M. (1984). Interaction between sodium chloride and paracasein as determined by water sorption. *J. Food Sci.*, **49,** 127–31.

Hegg, P. (1982). Conditions for the formation of heat-induced gels of some globular food proteins. *J. Food Sci.*, **47,** 1241–44.

Hermannsson, A. M. (1982). Gel characteristics—structure as related to texture and waterbinding of blood plasma gels. *J. Food Sci.*, **47,** 1965–72.

Hermannsson, A. and Lucisano, M. (1982). Gel characteristics—Waterbinding properties of blood plasma gels and methodological aspects on the water-binding of gel systems. *J. Food Sci.*, **47,** 1955–9.

Johnson, D. (1970). Functional properties of oilseed proteins. *J. Am. Oil Chem. Soc.*, **47,** 402–7.

Johnson, T. and Zabik, M. (1981). Ultrastructional examination of egg albumin protein foams. *J. Food Sci.*, **46,** 1237–40.

Kato, A., Takahashi, A., Matsudomi, N. and Kobayashi, K. (1983). Determination of foaming properties of proteins by conductivity measurements. *J. Food Sci.*, **48,** 62–5.

Karel, M. (1975). Physico-chemical modification of the state of water in foods—a speculative survey. In: *Water Relations of Foods,* R. Duckworth (Ed.), Academic Press, NY, pp. 639–57.

Kauzmann, W. (1959). Some factors in the interpretation of protein denaturation. *Advan. Protein Chem.*, **14,** 1–63.

Kinsella, J. (1979). Functional properties of soy proteins. *J. Am. Oil Chem. Soc.*, **56,** 242–58.

Kuntz, I. (1971). Hydration of macromolecules. III. Hydration of polypeptides. *J. Am. Chem. Soc.*, **93,** 514–16.

Kuntz, I. (1975). The physical properties of water associated with biomolecules. In: *Water Relations of Foods,* R. Duckworth (Ed.), Academic Press, NY, pp. 93–109.

Kuntz, I. and Brassfield, T. (1971). Hydration of macromolecules. II. Effects of urea on protein hydration. *Arch. Biochem. Biophys.*, **142,** 660–4.

Kuntz, I., Brassfield, T., Law, G. and Purcell, G. (1969). Hydration of macromolecules. *Science,* **163,** 1329–31.

Labuza, T. (1977). The properties of water in relationship to water binding in foods, a review. *J. Food Proc. Preser.,* **1**(2), 167–90.

Labuza, T. (1980). Enthalpy–entropy compensation in food reactions. *Food Tech.,* **34**(2), 67–77.

Labuza, T. (1985). Water binding of humectants. In: *Properties of Water in Foods,* D. Simatos and J. Multon (Eds), Martinus Nijhoff Publishers, Dordrecht, The Netherlands, pp. 421–45.

Labuza, T. and Busk, G. (1979). An analysis of the water binding in gels. *J. Food Sci.,* **44,** 1379–94.

Labuza, T. and Saltmarch, M. (1981). The nonenzymatic browning reactions as affected by water in foods. In: *Water Activity: Influences on Food Quality,* L. Rockland and G. Steward (Eds), Academic Press, NY, pp. 605–50.

Leeder, J. and Watt, I. (1974). The stoichiometry of water sorption by proteins. *J. Colloid Interface Sci.,* **48,** 339–44.

Mulvihill, D. and Kinsella, J. (1987). Gelation characteristics of whey proteins and β-lactoglobulin. *Food Tech.,* **41**(9), 102–11.

Noguchi, H. (1981). Hydration around hydrophobic groups. In: *Water Activity: Influences on Food Quality,* L. Rockland and G. Steward (Eds), Academic Press, NY, pp. 281–93.

O'Brien, S., Baker, R., Hood, L. and Liboff, M. (1982). Water-holding capacity and textural acceptability of precooked frozen white egg omelets. *J. Food Sci.,* **47,** 412–17.

Poulanne, E. and Terrell, R. (1983). Effects of salt levels in prerigor blends and cooked sausages on water binding, released fat and pH. *J. Food Sci.,* **48,** 1022–4.

Richardson, S., Baiann, I. and Steinberg, M. (1985). Relation between oxygen-17 NMR and rheological characteristics of wheat flour suspension. *J. Food Sci.,* **59,** 1148–51.

Rustad, T. and Nesse, N. (1983). Heat treatment and drying of capelin mince. Effect of water binding and soluble protein. *J. Food Sci.,* **48,** 1320–2, 1347.

Schmidt, R. and Morris, H. (1984). Gelation properties of milk proteins, soy proteins, and blended protein systems. *Food Tech.,* **38**(5), 85–96.

Shimada, K. and Matsushita, S. (1980). Relationship between thermocoagulation of proteins and amino acid compositions. *J. Agric. Food Chem.,* **28,** 413–17.

Tombs, M. (1985). Phase separation in protein–water systems and the formation of structure. In: *Properties of Water in Foods,* D. Simatos and J. Multon (Eds), Martinus Nijhoff Publishers, Dordrecht, The Netherlands, pp. 25–36.

Townsend, A. and Nakai, S. (1983). Relationship between hydrophobicity and foaming characteristics of food proteins. *J. Food Sci.,* **48,** 588–94.

Urbanski, G., Wei, L., Nelson, A. and Steinberg, M. (1983). Rheology models for pseudoplastic soy systems based on water binding. *J. Food Sci.,* **48,** 1436–9.

Von Hippel, P. and Wong, K. (1962). The effect of ions on the kinetics of formation and stability of the collagen fold. *Biochem.*, **1,** 664–74.

Warner, D. (1981). Theoretical studies of water in carbohydrates and proteins. In: *Water Activity: Influences on Food Quality,* L Rockland and G. Steward (Eds), Academic Press, NY, pp. 435–65.

Yasada, K., Nakamura, R. and Hayakawa, S. (1986). Factors affecting heat-induced gel formation of bovine serum albumin. *J. Food Sci.*, **51,** 1289–92.

Zayas, J. (1985). Structural and water binding properties of meat emulsions prepared with emulsified and unemulsified fat. *J. Food Sci.*, **50,** 680–92.

Ziegler, G. and Acton, J. (1984). Mechanisms of gel formation by proteins of muscle tissue. *Food Tech.*, **38**(5), 77–82.

Chapter 5

THE BEHAVIOUR OF ENZYMES IN SYSTEMS OF LOW WATER CONTENT

J. E. McKAY

Procter Department of Food Science, University of Leeds, UK

SYMBOLS AND ABBREVIATIONS

a_w	Activity of water
AMP	Adenosine monophosphate
ATP	Adenosine triphosphate
ERH	Equilibrium relative humidity
ΔH^*	Enthalpy of activation
K_m	Michaelis constant
MC	Moisture content
p/p_0	Relative vapour pressure
RH	Relative humidity
ΔS^*	Entropy of activation
x_s	Mole fraction of solvent
x_w	Mole fraction of water

INTRODUCTION

The availability of water in any system is described on the one hand by the amount of water present, the moisture content (MC), and on the other by the capacity of that water, subject to the limitations imposed by its interactions with other components of the system, to participate in physical or chemical processes. This capacity is described conventionally for systems at equilibrium by the thermodynamic potential but in complex systems, for reasons of practicality, it is more often represented by the closely related thermodynamic activity of water as displayed by its vapour pressure at equilibrium with its environment in relation to the vapour pressure of pure water at the same temperature (ERH). Unfortunately the experimental systems used in practice are not necessarily at equilibrium even when steady states of long duration are observed and the apparent ERH may not portray precisely the true thermodynamic activity even if it does afford a satisfactory practical definition of the water status of the system at a given temperature (Reid, 1976).

Large numbers of foodstuffs, ingredients and raw materials have been studied and the behaviour of the materials defined in terms of water sorption isotherms which display the variation of moisture content with the apparent ERH or with the corresponding water activity (a_w) (Iglesias and Chirife, 1982). Although it is reasonably easy to construct a water sorption isotherm it is by no means easy to interpret the observations so recorded other than in the most general way, even if the activities derived are taken at face value. Indeed it provides a frustratingly enigmatic guide to the understanding of the behaviour of the water present in molecular terms, particularly with respect to its capacity to act as a vehicle for chemical and biochemical reactions.

Although it appears from spectroscopic and thermodynamic studies that sorbed water may exhibit anisotropy with respect to molecular mobility (Lechert, 1981) and binding energy (Soekarto and Steinberg, 1981) it is also evident that exchange occurs between the molecules in different categories and that the activity of water in a system expresses the resultant of the kinetic complexities thereof. As the amount of water in the system increases the proportion of relatively unrestricted water increases and the resultant activity reflects the changes by approaching a value of 1·0. In the case of sorption by an insoluble solid material the interactions with the surface profoundly affect the overall properties of water when it is present in small amount. As the quantity of water increases and consequently the thickness of the adsorbed layer the influence of molecules which interact strongly with the solid surface by dint of their close proximity to it becomes progressively smaller to the point of insignificance. Should a soluble component also be present in the system then the dissolution of this component, by mutual interaction with water molecules and the solid surface, also exerts an influence upon the measured activity and may continue to do so at water contents which are high enough to reduce the influence of the solid surface to insignificance, depending upon its concentration in the system.

The elucidation of the molecular structures of enzymes is achieved mainly by the application of X-ray diffraction techniques to hydrated crystals in which the enzyme molecules are still in an aqueous environment. The polypeptide chains appear to be folded to give an approximately globular form with a close-packed internal structure which is largely free from water. The water molecules which do occur in the interior are highly localised in the crystal structure, i.e. firmly held, singly, in pairs or even in more extended networks. These molecules appear to fulfill a structural function by bridging internal polar groups in a largely hydrophobic environment (Kuntz and Kaufmann, 1974).

In the crystal array the protein molecules are thought to touch at only a few points, the spaces between molecules being occupied with liquid water. The volume fraction devoted to this interstitial space would be 26% for close-packed rigid spheres of uniform diameter and, in practice, values of 27–63% are observed (Matthews, 1970). In both internal and external situations water molecules satisfy otherwise unrequited bonding capacity and occupy voids. When the crystal is dried and this water is removed a loss of order is observed in the X-ray diffraction pattern, probably due to enforced changes in conformation (Kuntz and Kaufmann, 1974).

Despite the relatively low degree of stability associated with the folded form, the folded structures revealed by X-ray investigations are believed to persist in solution with relatively little change, other than the conferment upon the peripheral chains of some degree of freedom to flex. The possession of enzymic activity by these folded forms has been demonstrated by Bernard and Rossi (1970) and it seems likely that water molecules form an essential part of the active conformation since the loss of order which is detected when protein crystals are dried is accompanied by the loss of activity when the proteins concerned are enzymes.

Most studies of enzyme activity are conducted in highly dilute aqueous solutions with well defined pH, substrate concentration and ionic strength for largely practical reasons. The results of such studies lead to concepts in which the role of water is emphasised strongly, being responsible for the transport of the substrate to the active site of the enzyme, the development of charge on the protein through protolysis, the moderation of electrostatic interactions, particularly hydrogen bonding, the promotion of hydrophobic interactions and, in some cases, also serving as a co-substrate. Obviously any lowering of the activity of water will affect its capacity to fulfill any of these functions which may be required of it in an enzymic reaction.

A relatively small number of studies has been conducted with the sole purpose of investigating the effect of reduced water activity on enzyme activity. Some investigations have been carried out for quite different purposes in systems of low water activity. The implications of some of these studies is discussed below.

HOMOGENEOUS MODEL SYSTEMS

Mixtures of water with a number of simple organic compounds have been used as solvents for the conduct of enzymic reactions under ostensibly well defined conditions (which are not always revealed in the published work). Simple aliphatic alcohols and polyols have been most commonly used and the enzymes studied include polyphenoloxidase (Tome *et al.*, 1978), peroxidase (Blain, 1962), trypsin (Inagami and Sturtevant, 1960), chymotrypsin (Barnard and Laidler, 1952), ATP-ase (Ethier and Laidler, 1953), and invertase (Nelson and Schubert, 1928).

The general outcome of these studies is that enzymic activity is depressed by the lowering of x_w and is finally extinguished at some critical value of x_w as shown in Table 1. Small amounts of glycerol, on the other hand, were found to increase the activities of peroxidase and polyphenoloxidase and small amounts of dioxan activated trypsin. It is also evident that different enzymes are affected differently by the same solvent mixture and that a particular enzyme may be affected differently by different co-solvents, as is polyphenoloxidase. Tome *et al.* (1978) followed their reactions to virtual completion and found that the reactions came to an end prematurely for $x_w < 1{\cdot}0$ for all the co-solvents used. The amount of substrate conversion was found to decrease with x_w over the range of x_w in which the enzyme was active.

While the various effects on enzymic activity can be associated with a_w, i.e. the effect of the co-solute on the fugacity of the water in the system, they can be attributed at least in part to the overall effect of the co-solute upon the water in terms of the modified nature of all the properties of the mixed solvent compared to the properties of water.

TABLE 1
Effect of x_w (mole fraction of water) on enzyme activity

Enzyme	*Co-solvent*	x_w *limit of detectable activity*	*Temperature (°C)*
Peroxidase	Glycerol	0·369	Not specified
Lipoxygenase (Blain, 1962)	Glycerol	0·637	
Trypsin (Inagami and Sturtevant, 1960)	Dioxan	$<0·392$	25
	Dimethyl sulphoxide	$<0·650$	
Polyphenoloxidase (Tome *et al.*, 1978)	Methanol	0·61	30
	Ethanol	0·71	
	Ethylene glycol	0·44	
	Diethylene glycol	0·65	
	Propylene glycol	0·68	
	Glycerol	0·43	
	Sorbitol	0·75	

The modification of properties is demonstrated by the examples given in Table 2 which show some of the properties of the mixtures possessing the lowest a_w values at which enzymic activity was studied. Although it can be seen from the corresponding values of x_w and p/p_0 that these are not ideal solutions the departures from ideality are relatively small compared to the variations in the other properties in which quite large specific effects are shown. These changes in property, of course, are merely reflections of the changes in structure produced by the inclusion of the co-solute in water. Nevertheless a separate consideration of a particular aspect of the overall change is necessary for practical purposes.

Changes in dielectric constant are probably the most significant in terms of affecting enzymic activity. Indeed Laidler and co-workers (Barnard and Laidler, 1952; Ethier and Laidler, 1953) studied enzymic activity in the presence of methanol and dioxan in order to determine the effect of solvent dielectric constant upon the formation of the enzyme–substrate complex. Laidler (1955) concluded that an enzymic reaction in which the activated complex is more polar than the reactants would be accelerated by an increase in dielectric constant while one in which the complex is less polar conversely would be slowed down and vice versa. He was obliged to admit, however, that

TABLE 2
Some physical properties of aqueous solvents

Co-solvent (S)	*S%*	x_w	x_s	η/η_0 (20°C)	D/D_0 (20°C)	p/p_0 (30°C)
Glycerol	89·7	0·369	0·631	270	0·58	0·32
Methanol	46·8	0·61	0·39	1·8	0·70	0·67
Ethanol	48·9	0·71	0·29	2·8	0·63	0·81
Dioxan	84·8	0·47	0·53	—	0·13	—
Sucrose	60·0	0·927	0·073	49·3	0·76	0·90

x_w and x_s are the mole fractions of water and co-solvent respectively.
η/η_0 represents the ratio of the viscosity of the mixture to the viscosity of pure water (Weaste, 1984).
D/D_0 represents the ratio of the dielectric constant of the mixture to that of pure water (Åkerløv, 1932).
p/p_0 represents the water vapour pressure of the mixture compared with that of pure water (Money and Born, 1951; Tome *et al.*, 1978).

complications could arise from 'specific solvent–solute interactions' which appeared to include the inactivation of the enzyme at higher concentrations (>25% for methanol) of the co-solvent.

The presence of the co-solvent affects other electrostatic phenomena, notable among which is the dissociation of the buffer salts which are used to define and maintain the pH of the system. Because of the effect on ion-pair dissociation the activities of all ions in the system will be reduced so that not only will the conventional pH scale become invalid but also the effectiveness of all ionic species will become impaired. These effects are demonstrated particularly well by the observations of Inagami and Sturtevant (1960) working with water–dioxan mixtures of greatly lowered dielectric constant.

Changes in viscosity also would seem to have some potential for affecting enzymic activity by influencing the rates of diffusion of reactants and products. Several workers (Ruchti and McLaren, 1964, Bowski *et al.*, 1971) have attempted to investigate the role of viscosity in the anomalous behaviour of the invertase–sucrose system at high (weight) concentrations of substrate. The conclusion emerging from these investigations was that the increase in viscosity even in 60% sucrose solutions had little effect on the rate of reaction. This view was confirmed by experiments in which pectin was used to change the viscosity of the medium (Kertesz, 1935*a*). It is difficult to believe that viscosity has no effect but significant changes may only appear at viscosities much greater than one or two poise.

It seems very likely that general effects of the presence of a solute or co-solvent are greatly outweighed by specific effects either on water itself or on the other components of the system. This is particularly evident in the case of the invertase–sucrose system in which the reaction rate is depressed by substrate concentrations greater than 0·3 M (Nelson and Schubert, 1928). The fall in the rate of hydrolysis is attributed essentially to enzyme–substrate interactions, substrate inhibition being the major of these but some reverse reaction (transferase) also being observed (Bowski *et al.*, 1971). This appears to be an unusual case with respect to the inhibitory effect of the substrate but reversal of reaction is not unknown in systems of high substrate concentration, e.g. the plastein reaction.

Although the work in this particular area is sparse it serves to show that enzyme action is mainly affected by specific interactions with co-solutes which are largely independent of general solute effects such as the lowering of water vapour pressure. The phenomena of inhibition and activation obviously come into this category. Even solute–solvent interactions have some degree of specificity as shown by changes in viscosity and dielectric constant and effects on dissociation processes including apparent pH.

HETEROGENEOUS MODEL SYSTEMS

Simple Mixtures

These so-called simple mixtures consist essentially of enzyme and substrate, the substrate predominating. Cooled solutions of the components are mixed together, in some cases with a solid support, and the mixture usually freeze-dried. The solid mixture is then allowed to 'equilibrate' with a salt solution of appropriate water vapour pressure and at the required temperature. The moisture content is recorded at suitable time intervals along with the amount of product formed by any enzymic reaction which has occurred.

The simplicity of the systems lies firstly in the absence of other constituents, excepting the small amounts of salts imported into the system with the enzyme preparation, and secondly in the (presumably) even distribution and intimate mixing of the potential reactants. The activity of water in the partially hydrated mixtures is determined in the early stages of sorption solely by its interaction with the surfaces present, i.e. predominantly that of the supporting solid. As sorption progresses and components begin to dissolve then the water is also subject to solvent–solute interactions which in turn are reflected in the resultant a_w. If enzymic activity occurs the composition of the system will be subject to modification as long as net reaction persists so that true equilibration and the corresponding moisture content will be established only if reaction equilibrium has been attained.

As conditions are changed to allow greater amounts of water to be sorbed the effects of both surface and solute interactions with water diminish and the a_w increases in a manner determined by the nature of the components of the system and the temperature. Where the components are soluble the system will progress towards complete homogeneity. At low a_w values however the heterogeneous system will differ very significantly from a homogeneous counterpart of the same a_w in terms of the extent of Brownian motion of the constituent molecules and the consequent limitations imposed upon the processes of solubilisation and transport.

Despite such fundamental differences the general pattern of behaviour observed for heterogeneous systems is similar to that of the homogeneous systems. Enzymic activity declines as a_w decreases and is finally extinguished at a critical value of a_w as shown in Table 3. The limiting a_w for enzymic activity is more or less as high for urease (Skujins and McLaren, 1967) as for the amylases (Drapron and Guilbot, 1962) notwithstanding the significant differences in diffusibility of the respective substrates. The sorption behaviour of the substrate (urea) on its own, however, showed that significant dissolution occurred only at $a_w > 0{\cdot}7$ (Skujins and McLaren, 1967) so that the constraint on enzyme action may arise from restricted solution of the substrate rather than restriction on its subsequent transport to the active site. The observations of Acker and Wiese (1972) on the action of lipases on individual triglycerides emphasise that diffusibility is nevertheless important. The liquid triolein was susceptible to lipolysis at much lower a_w than the crystallisable trilaurin which was hardly hydrolysed at all in mixtures in which it could be guaranteed to be crystalline. The authors found, in fact, that the susceptibility of trilaurin to hydrolysis varied considerably with the method of deposition on the cellulose support but did not investigate the degree of crystallinity of the various deposited forms. The increase in reaction rate with increasing a_w for both triglycerides suggests that either the

TABLE 3
Effect of water activity on enzyme activity in simple hydrated solids

Enzyme	*Effective adsorbent*	*a_w limit of detectable activity*	*Temperature (°C)*
Diastase (= α-amylase)	Potato starch	c. 0·53[a]	20 (Kiermeier and Coduro, 1954)
	Filter paper	0·46	
β-Amylase	Soluble starch (freeze-dried)	0·75–0·80	21 (Drapron and Guilbot, 1962)
		0·65–0·70	31
α-Amylase	Soluble starch (freeze-dried)	0·75–0·80	31
Urease	Urea	0·65	20 (Skujins and McLaren, 1967)
Lipase	Cellulose	0·21 (trilaurin) <0·02 (triolein)	25 (Acker and Wiese, 1972)

[a] Estimated from the data of Wolf *et al.* (1984).

enzyme molecules are progressively converted into the fully active form or that they are able to expedite the reaction by their increasing diffusion rates, the surface coverage of the substrate (5% w/w) being limited even when completely liquid.

The inclusion of 'crystalloids' was found to have a marked effect on the behaviour of mixtures of β-amylase and soluble starch (Drapron and Guilbot, 1962) on hydration (see Table 4). It was also found that

TABLE 4
Effect of added components on the a_w limit of detection of enzyme activity in freeze-dried mixtures of β-amylase and soluble starch at 31°C (Drapron and Guilbot, 1962)

Component	*wt% in system*	*% water sorbed*	*a_w*
Glycerol	83·5	10	0·35
Lithium chloride	48·7	47·0	0·11
Sodium bromide	37·1	35·0	0·57
Potassium bromide	40·6	73·0	0·82
None	—	20·0	0·73
None[a]	—	26·9	0·80

[a] Dialysed enzyme.

the dialysed enzyme required a significantly higher a_w (0·80–0·85) for the limit of detectable activity than did the undialysed. This suggests that the difference in behaviour shown by diastase (Kiermeier and Coduro, 1954) (see Table 1) which was used undialysed may have been due to some extent to the amount and nature of salts etc. included in the preparation as well as to the different method of drying the mixture.

Kiermeier and Coduro (1954) found that diastase showed activity only at MC > 14% in the starch–enzyme mixtures studied but when the mixture was dried on filter paper activity could be detected at a MC as low as 5·5% corresponding to 46% RH. The RH corresponding to a MC of 14% in the starch–enzyme mixture was not quoted by the authors but the sorption isotherm obtained for potato starch by Wolf *et al.* (1984) indicates a value of *c.* 53% RH (microcrystalline cellulose had MC 5·5% at 50% RH (see Fig. 1)). In other words the a_w limit is roughly the same for the two systems but the amounts of water required to produce this level of activity in the water vicinal to the respective solid phases are clearly different.

In the course of their thorough investigation of the behaviour of mixtures of β-amylase and soluble starch, Drapon and Guilbot (1962)

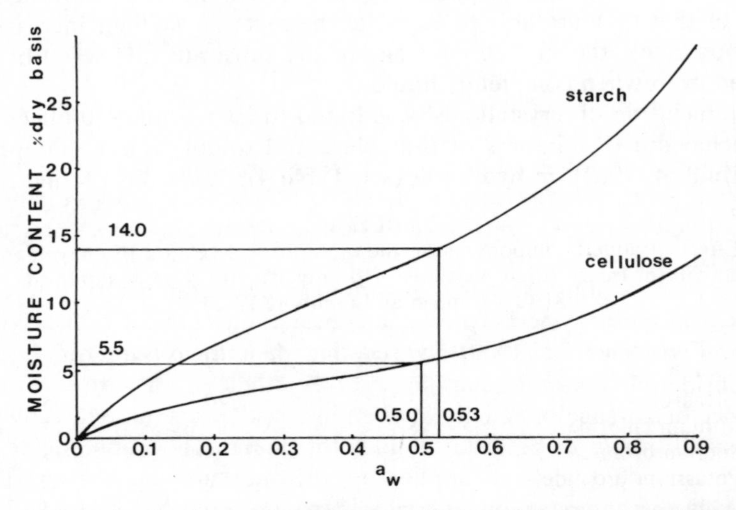

FIG. 1. Sorption isotherms for microcrystalline cellulose and purified potato starch at 25°C and drawn from the data of Wolf *et al.* (1984). Ordinate: moisture content, per cent dry weight. Abscissa: a_w.

observed that the final extent of enzyme action achieved decreased with a_w over the range in which activity could be detected. This is a pattern of behaviour similar to that observed by Tome *et al.* (1978) for homogeneous model systems. The course of events in the heterogeneous systems appeared to be complex, particularly in the early stages of reaction at a_w values at the lower end of the active range, with some indication of reverse reaction taking place. Similar indications were found in the investigation of the behaviour of mixtures of diastase and potato starch (Kiermeier and Coduro, 1954) but Drapron and Guilbot (1962) observed no sign of reverse reaction in the case of mixtures of α-amylase and soluble starch. They did find evidence of some abnormality even here, however, in the early stages of hydrolysis.

The latter authors were led to suspect that the restricted availability of water induced 'topochemical' processes as a result of limited contact between enzyme and substrate and justify their suspicion by demonstrating that enzymic activity could be restored in mixtures in which it had become quiescent by drying, grinding and rehydrating to the same RH. The effect of the addition of glycerol (Table 4) strengthens the suspicion that restricted diffusion is a powerful limiting factor in reducing enzyme activity and that activity is determined by total liquid content rather than a_w. The presence of added salts is also effective in causing the total liquid (water) content to increase at given a_w relative to that sorbed in their absence. In the light of subsequent experience with aqueous organic solvents, however, Tome and his colleagues (1978) suggested that the lack of reactant mobility was not the only factor involved in solid systems. Enzyme inactivation by interaction with other solutes, particularly of a time-dependent nature, is also likely to play a part in both homogeneous and heterogeneous systems.

It is worth noting that the quantitative measurement of enzymic activity in these systems presents considerable difficulty such that Drapron and Guilbot (1962) advised that their measurements should be considered as semi-quantitative. Nevertheless, in terms of the presence or absence of activity or even increase or decrease in activity, there would seem to be little grounds for doubt. The same degree of reservation could also be applied to the measurement of enzymic activity in the homogeneous model systems.

In summary it can be concluded that enzymic activity will be initiated during the rehydration of dry systems only when the conditions allow suitable kinetic encounters between enzyme and substrate along with any other species required to participate in the events of the overall reaction scheme. Not suprisingly the capacity of the solvent to free potential solute molecules from the forces of association which confine them within existing aggregates, either regular or irregular, is a function of energy i.e. activity. This is demonstrated by the work of Duckworth and his co-workers (Duckworth *et al.*, 1976; Duckworth, 1981) who have shown that, under controlled conditions, different molecular species are mobilised (into solution) at characteristic a_w values and that, when more than one solute species is present, the activity threshold of solubility of one species is modified by the presence of another.

Once the threshold of solubility has been reached reaction will proceed within the thin film of concentrated solution subject to the constraints imposed upon it by the influence of the adsorbing surface, on the one hand, and by the presence or absence of the various solute species, on the other. The latter may include those species which are essential for particular stages of reaction as well as those which exercise inhibitory effects. As sorption progresses and a_w increases, the influences of surface and solutes upon the free energy of the solvent molecules diminish and the latter will become increasingly effective in fulfilling the various roles that they play in the enzymic reaction. There will be effects of specific nature, affecting the enzyme, or possibly other reactants, directly or, alternatively, affecting the reaction indirectly through changes in such properties of the system as the dielectric constant or viscosity. (It should be borne in mind that these properties may be affected by the adsorbent surfaces in thin adsorbed films as well as by solute molecules.) These specific effects will, of course also diminish as the amount of water in the system increases.

Complex Mixtures

The complexity here lies in the complete or partial retention of its original biological organisation by the solid phase. This is usually a ground or milled tissue of plant origin which has been selected because it contains the particular enzymic activity to be investigated. The comminution of the tissue will inevitably cause some, possibly extensive, degree of disruption to its organisation entailing the chance of damage or redistribution at the molecular level. Although the survival of enzymic activity can easily be demonstrated, the degree to which enzymes retain their native condition and distribution is uncertain. In addition, the ground tissue will retain its non-volatile components unless steps are taken to remove them and these may influence the subsequent observations.

The substrate used may be natural or artificial and in most cases is mixed with the dry solid, either mechanically or by deposition from solution in a volatile non-aqueous solvent. In consequence the mixing of enzyme and substrate may not be so efficient as in the freeze-dried mixtures described previously. The dry mixtures are allowed to sorb moisture under controlled conditions of RH and temperature and the moisture content and amount of product formed by enzyme action measured at suitable intervals of time.

Despite the differences in the structural complexity of the solid phases and the different enzymes involved the results obtained are qualitatively similar to those observed in simple mixtures as Table 5 shows. The limiting a_w values are lower than those shown in Table 3 probably for reasons connected with the nature of the enzymes. This is certainly true in the case of lipase (Drapron, 1972) which is believed to act characteristically at the oil–water interface rather than in solution.

The invertase system studied by Kertesz (1935*b*), on the other hand,

TABLE 5
Effects of water activity on enzyme activity in complex hydrated systems

Enzyme	*Adsorbent (substrate)*	*a_w limit of detectable activity*	*Temperature (°C)*
Phospholipase B	Barley malt	0·55	30
Phospholipase D (Acker and Luck, 1958)	(lecithin)	0·45	
Acid phosphatase (Acker and Kaiser, 1961)	Barley malt (phenyl phosphate)	<0·35	30
Lipase (Drapron, 1972)	Wheat germ (olive oil)	<0·05	30
Yeast invertase (Kertesz, 1935*b*)	Sucrose/apple pomace (sucrose)	*c.* 0·7[a]	25
Pork extract (Matheson, 1962)	Cellulose (glycogen)	<0·11	37

[a] Estimated on the basis of MC from the data of Lewicki and Lennart on apple tissue presented by Iglesias and Chirife (1982).

is intermediate between simple and complex mixtures in that the complex solid component did not contain the enzyme and the substrate, up to its dissolving, was the major solid component, albeit an easily soluble one. Experimental samples were defined by their water contents and were allowed to come to vapour pressure equilibrium in small closed containers. Consequently the a_w limit (*c.* 0·7) quoted in the table, which is more like those observed in systems of simple mixtures, is based on the assumption that the MC (4·75%) corresponding to the limit of detectable activity consists mainly of water sorbed by the apple pomace (20% w/w of the total solids). At this RH crystalline sucrose would have sorbed very little water but would be at the point of incipient dissolution.

The meat model contrived by Matheson (1962) differs in quite another way. An enzyme extract of formidable complexity was applied to a simple solid adsorbent along with glycogen solution and the mixture freeze-dried. The fall in glycogen content was measured at suitable time intervals for samples stored at a variety of RH at 37°C. Detectable loss of glycogen was observed at RH as low as 11% in contrast to the behaviour described for starch in the previous section and in spite of the even more intractable nature of the substrate. Details of the MC are not given by the author but were probably greatly in excess of the 2% or so expected of cellulose alone.

The observations are again consistent with the idea of enzymic activity being limited mainly by the obstacle to encounter imposed by restricted transport of substrate and/or enzyme. The relative success of lipase at low moisture levels serves to emphasise this point since here the transport of substrate molecules to the interface is achieved by the Brownian motion of these molecules themselves in the liquid film deposited upon the solid phase. This has been demonstrated by Acker and Wiese (1972). Presumably a similar but less efficient transport mechanism may have operated in the case of the phospholipases (Acker and Luck, 1958) where the lecithin substrate was probably deposited as a semi-solid. The diffusion of this substrate would be unlikely to occur in any aqueous phase other than an emulsion (Naessens *et al.*, 1982). The use of a crystalline phosphatidyl choline as substrate in such a system would be of interest.

The similarity in behaviour in simple and complex mixtures extends to the curtailment of the extent of reaction at low a_w. This behaviour does not appear to have been observed in the lipase. Drapron (1972) mentions only the relationship between the rate of lipolysis and a_w and suggests that at low a_w, i.e. below 0·25, the activity of the enzyme may be restricted by the unavailability of water molecules to occupy sites in the enzyme structure which are essential to the active conformation and from some of which the water may participate in the hydrolytic reaction. The lack of data on the extent of lipolysis at low a_w leaves the connection with restricted diffusion in other systems largely as a matter of conjecture.

Among its unusual features the lipase system showed maximum activity at $a_w = 0{\cdot}8$ and a minimum K_m at a_w between 0·3 and 0·4. The author reasoned that the presence of excess water could be a disadvantage, interfering with the formation of the highly hydrophobic enzyme–substrate complex. A similar effect has been observed in the behaviour of some enzymes in aqueous organic solvents.

ENZYMIC ACTIVITY IN FOOD MATERIALS OF LOW MOISTURE CONTENT STORED AT AMBIENT TEMPERATURES

The fact that food materials of low moisture content can be stored successfully for long periods of time must be regarded as one of the prerequisites for the development of human civilisation. Surprisingly, in the light of its long existence in practice, the understanding of the phenomena involved is still essentially superficial. Perhaps this is a tribute to the degree of success with which the technology has been conducted even without depth of understanding.

The realisation that the degree of availability of water required to support the growth of spoilage microorganisms in foods can be related to the ERH values appropriate to their MC (Scott, 1957) has led to high hopes that ERH and therefore a_w, may hold the key to the control of all forms of spoilage in stored foods, chemical and biochemical as well as microbial. Indeed Rockland (Rockland and Nishi, 1980) has embodied this concept in general terms in the form of a 'stability isotherm' in which the enzymic activity is associated with $a_w > 0{\cdot}4$. It is evident from the studies on model systems, however, even though these are far from comprehensive, that a_w is not of itself a primary determinant of enzyme behaviour. Nevertheless, ERH/a_w provide a good working guide to the water status, and the aspects of behaviour related to it at a given temperature, of types of food material whose compositions are unlikely to vary to any great extent.

Acker (1962, 1969) has reviewed comprehensively the investigation of enzymic activity in foods of low moisture content up to 1969 and a later and briefer overview has been given by Schwimmer (1981). Because the earlier work still provides the main basis for current thinking on the subject, a reappraisal will be included here, notwithstanding the excellence of Acker's account of 1962.

The experimental procedure used in these investigations is closely similar to that required for the complex heterogeneous model systems. The food material is brought to the required RH (or MC) and temperature and the formation of the products expected to arise from the enzymic reaction is measured by sampling at suitable time intervals. The fundamental difference from the model systems lies in the use of endogenous substrate instead of added artificial or natural exogenous substrate. This, in turn, may lead to restriction of reaction due to the nature, quantity and distribution of the endogenous substrate. The rates of reaction may be rather slow in consequence and long reaction times may be required to allow sufficient product to accumulate in order to provide decisive analytical data.

The early workers in the field were confronted, not only with the technical difficulties of measuring the slow rates of increase in initially small amounts of product, but also with the problem of establishing, firstly, that the changes in composition were biochemical rather than chemical in origin and, secondly, whether the enzymes responsible were those of the material, itself, or those introduced to the system by invading microorganisms.

Stored Cereals and Derived Products

Cereals and products derived from them have been subjects of extensive study as befits their economic importance and this part of the field has been reviewed in more depth than others. In addition to the general coverage given by Acker (1962, 1969) the specific topics of lipid-hydrolysing and -oxidising enzymes have been discussed by Morrison (1978), Galliard (1983) and others in the course of more general accounts of the role of cereal lipids and the enzymes acting on them in the context of cereal technology. This emphasis on enzymes with specific reactivities towards lipids reflects the significance of this group of substances in cereal technology but there has been some element of practical strategy also in the use of the relatively easily detectable lipases as indicators of potential activity on the part of other enzymes.

Stored Intact Grains

The undamaged mature cereal grain is capable of remaining physiologically quiescent at MC above 20% due to the regulation of its complex system by phytohormones. The breaking of dormancy and the inception of germination involve an elaborate sequence of developmental processes (Meredith and Pomeranz, 1984) so that, although they are associated with the imbibition of water by the seed they are in no way to be compared with the much simpler events involved in the rehydration of model systems of instantaneous reaction potential. The seed may be brought, under suitable conditions, to much lower MC levels without loss of viability and therefore presumably without damage to its internal organisation.

On rehydration the tissues are capable of being restored to their original state of integrity, the elements of the control system and the condition of dormancy included. Sorption isotherms show that layers of water of slowly increasing a_w are built up, presumably over large areas of accessible internal surface, as the MC increases, the manner varying significantly with temperature. The sorption behaviour of the intact grain, in fact, varies only very slightly from that of the derived flour (Iglesias and Chirife, 1982).

The conditions required to set in motion the process of germination do not appear to be attained in water sorption experiments, due either to the large amount of imbibed water needed (45–46%) (Meredith and Pomeranz, 1984) or the discretion of the experimenters. The pioneering investigators discovered, however, that it was by no means impossible to promote mycotic growth and the changes arising from it.

As mentioned above, the enzymic release of free fatty acid was much studied by the earlier workers, partly because of the obvious connection with off-flavour, partly because of the suspected connection with impaired baking performance and partly because of its practicality. Significant levels of fatty acid production were observed in wheat grain stored at high RH/MC at ambient temperature and these were coupled all the more significantly with enhanced rates of respiration. Both effects were found subsequently to be greatly reduced by fungistatic agents (Milner *et al.*, 1947) and decisive experiments confirmed the involvement of contaminating moulds (Bottomley *et al.*, 1950; Hummel *et al.*, 1954; Golubchuk *et al.*, 1955; Sorger-Domenigg *et al.*, 1955).

A minimum critical ERH of 75% (MC 14–15%) was shown to exist for mould growth in wheat, in the absence of which respiration and fatty acid release were insignificant. Understandably the curtailment of established mould growth by reducing the MC of the grain did not necessarily prevent the continuing action of the lipolytic enzymes then present (Sorger-Domenigg *et al.*, 1955).

This demonstration of the persistent inactivity in the intact, uninfected grain is all the more impressive on consideration of the presence of acyl lipids (Morrison, 1983), the presence of various acyl hydrolases (Galliard, 1983) and the facility of lipases, in particular, for reaction at extremely low a_w. Acker (1962) sensibly attributed the lack of reactivity to the maintenance of the tissue organisation in the grain. Later investigations have revealed that the major lipid component of the cereal grain, the triglyceride fraction, is concentrated in the germ and aleurone tissues in the form of liquid droplets or spherosomes within the cells (Morrison, 1983). The location of lipase activity is less certain but there is evidence that it is not closely associated with the spherosomes (Galliard, 1983). When germination begins lipolysis first occurs in the aleurone layer, the endosperm and the embryo axis and only later in the scutellum, triggered in the first two (non-germ) regions by some agent produced within the germ tissues (Clarke *et al.*, 1983).

The fact that the sorption of water vapour, even at relatively high RH, appears to be unable to initiate lipolysis, a vital component in the early stages of germination, suggests that it is unlikely that any other enzymic activity would occur under the same conditions.

Milled Cereal Products

This condition of structural integrity and physiological control is comprehensively disturbed by the operations of milling. During the reduction of wheat or maize to flour, or the polishing of rice, despite the removal of the oil- and enzyme-rich germ tissue, there is sufficient redistribution of lipid (Morrison and Barnes, 1983) and possibly also of enzyme to create a situation in which lipolysis can be shown to occur even at very low moisture levels. The amounts of lipid present in the primary milled products are usually low and variable with the percentage extraction but even in white flours the levels of triglyceride owe much to transfer from germ or aleurone tissues during the milling process either of liquid oil or fine particles of tissue, scutellum in particular (Morrison and Barnes, 1983).

Even on the basis of the total lipid content (1–2%) it is impossible that a continuous lipid film could cover fully the surface of the flour particles thereby providing a liquid medium for the diffusion of substrate to the oil–water interface throughout the material. In any case the low degree of volatility of triglycerides would impede the distribution of individual molecules over the surfaces such as is achieved by water. Interfacial surfaces will be formed, however, on a scale that will be limited by the amount of water present at low MC. How these interfaces may develop as the water content increases and what possibility may exist for emulsification are intriguing questions.

Although the detailed circumstances may be as yet unrevealed there is no doubt that lipolysis occurs in flour at rates of reaction that increase with temperature, RH, percentage extraction (Cuendet *et al.*, 1954) and decrease with flour strength (Warwick *et al.*, 1979). The MC threshold of reactivity is also dependent on these factors but is essentially low, for example below MC 6% ($a_w = 0{\cdot}1$–$0{\cdot}2$) for wheat flour at 37·5°C. The susceptibility of stored flours of various origins to lipolysis is mainly determined by the lipid and lipase contents which are in turn the result of a combination of genetics and processing conditions.

The developing thin layer of water with its putative interfacial facilities provides opportunities for reaction between any enzyme species capable of diffusing through the aqueous phase to the interface and any substrate diffusing through the oil phase, provided that the enzymic activity is not lost due to unfolding which may occur at the boundary and that essential cofactors are available. Since cereals contain a variety of lipid acyl hydrolases, phospholipases and lipoxygenases it is inevitable that lipase activity will be accompanied by other degradations and transformations of lipids.

Although model experiments have shown that phospholipases are capable of activity at low a_w there is evidence that such activity in flour may be restricted by the low degree of availability of the substrate. The considerable amount of triglyceride redistributed during milling does not appear to contain any significant amount of phospholipid, according to Morrison *et al.* (1982). Since there is no significant transport from the germ tissue it seem unlikely that any redistribution of phospholipid as solute in the oil phase will occur within the endosperm tissue either. Some contact can be made between enzyme and substrate, however, since Warwick *et al.* (1979) report the disappearance of both galacto- and phospholipid during the long-term storage of wheat flours.

Because of the effective aeration of the system flour would seem to be an ideal subject for the oxidation by lipoxygenase (also for autoxidation) of the polyunsaturated fatty acid constituents of the various acyl lipids. The mono- and dihydroxy fatty acids which have been found to accumulate in long-stored wheat flours are not specific indicators of enzyme-catalysed oxidation, since no information on isomer ratios is available. The decline observed in carotenoid (Warwick and Shearer, 1980) and thiol (Cuendet *et al.*, 1954) contents could also arise from autoxidation. The contribution of lipoxygenase to the oxidative development and eventual deterioration of wheat flour in storage is presumed rather than proven. In the case of the mill products of higher oil content, such as wheat germ, maize germ and rice bran, there is no doubt about the involvement of lipoxygenase in the development of oxidative rancidity (Galliard, 1983, 1986).

In addition to the lipid-specific enzymes proteolytic enzymes are capable of causing deterioration in stored flour and, like the former, may be present as endogenous forms or as exogenous forms introduced by microbial infestation. Some evidence has been presented for the modification of the solubility properties of wheat proteins during storage of the grain and flour under conditions (25°C; 55% RH) unlikely to produce mould growth (Jones and Gersdorf, 1941) but the current view appears to be that indigenous proteases do not cause any significant damage during storage. Some interest has been shown in the effects of proteolytic action in grain which has germinated prior to or during storage but the extent of breakdown even under these circumstances is limited and not obviously detrimental (Lukow and Bushuk, 1984).

The importance of enzyme action during the storage only becomes of consequence at later stages of production because of the effect on functional properties. Indeed some enzymes may continue their activities into such later stages at a higher level, where more favourable conditions for action occur, while others, previously inactive, may be able to act for the first time. The technological consequences of enzymic modification of lipid constituents are described in detail by Hammond (1983), Nicolas and Drapron (1983) and Shearer and Warwick (1983).

Stored Oil-Seeds

The situation of stored oil-seeds is essentially the same as that of stored cereal grains, viz. one of dormant physiology maintained by physical segregation and hormonal control. The deterioration of the stored material can be caused by mechanical damage, microbial infestation, or a combination of the two and stems mainly from the free fatty acids liberated by lipolysis but also possibly from products of lipoxygenase action. The stability of some oil-seeds is further enhanced by the fact that they contain little or no lipase activity prior to germination. Rapeseed (Theimer and Rosnitschek, 1978), cottonseed, sunflowerseed and peanuts (Huang and Moreau, 1978) fall into this category as does maize (Huang and Moreau, 1978) which, like rice, contains sufficient oil to qualify as an oil-seed.

As in the case of cereals, a threshold of stability towards mould infestation has been discerned at 75% RH (10–11% MC) (Milner, 1950). The only difference lies in the corresponding MC values which tend to be lower for the oil-seeds than for the cereal grains due to differences in water sorption behaviour. Below this threshold value undamaged seeds and nuts show a high degree of stability in storage. The slighly reduced stability of shelled seeds and nuts may be due to superficial damage sustained by the kernels during the shelling process. Rockland (1957) observed that little enzymic damage occurred in walnuts at RH values above 50% while oxidative and browning processes of a purely chemical nature became increasingly effective at lower RH.

Dried Fruits and Vegetables

Mature leguminous seeds are harvested, like other edible seeds, in a state of partial dehydration achieved by natural means and are subjected to further dehydration for storage. Enzymes have been shown to survive in dried beans and it is very likely that the integrity of cellular and tissue organisation also survives. The situation is, in fact, very similar to that of the cereal and the relatively small amount of information on the subject suggests that the similarity extends to the behaviour during storage. Beans show a similar tendency to accumulate free fatty acid above a critical MC (13–14%) along with other signs of deterioration (Morris and Wood, 1956).

The dehydration of fruits and other vegetables is rather a different matter. Here water is removed from tissues which are for the most part mature and permanent and composed of fully hydrated physiologically active cells. There is no question of retention of cell and tissue organisation in the dehydrated product to be restored by rehydration. However, the efficiency of the thermal disruption of cellular organisation is not necessarily matched by that of the thermal inactivation of the enzymes present and enzymes have long been suspected to be the causes of the rather ostentatious symptoms of deterioration observed in storage of unblanched dehydrated vegetable products (Joslyn, 1951).

Oxidative enzymes became prime suspects (Aylward and Haisman, 1969) and the practice of blanching was regarded as an essential pretreatment to successful dehydration. Although timely and suitably controlled blanching proved to be beneficial doubts continued with respect to its effectiveness against the notoriously heat-stable peroxidase enzymes (Burnette, 1977). Even less durable enzymes were known to survive sun-drying (Whittaker, 1958) and the introduction of milder heat treatments in the interests of flavour retention in dehydrated products increased the chances of survival of peroxidase activity (Farkas *et al.*, 1956, Burnette, 1977). Indeed testing for residual peroxidase activity has been suggested as a check on blanching efficiency. Even when inactivation has been achieved there remains the possibility of progressive partial regeneration of the activity on subsequent storage (Burnette, 1977).

While it is relatively easy to detect the presence of peroxidase activity in dried products by using a convenient exogenous substrate it is by no means easy to follow the action of peroxidase on endogenous substrates because of the highly indiscriminate nature of the process. Consequently the involvement of peroxidase in the deterioration of stored dehydrated products still remains as a conclusion based on indirect evidence.

Deterioration caused by lipases does not seem to affect dried products, other than the dry mature seeds, proabably because of the effectiveness of blanching. Lipoxygenases, however, although not known to be particularly heat-stable (Aylward and Haisman, 1969) can cause deleterious effects in stored products due to the decomposition during storage of hydroperoxides formed by the enzymes prior to their thermal inactivation.

The survival of enzymic activity in some dehydrated products has its advantages, notably in those vegetables, such as Brussels sprouts and onions, with flavours which are developed by enzyme action (Schwimmer, 1981) following any breach of cellular integrity. Indeed in the production of any dehydrated product the level of heat treatment applied must represent compromise between the achievement of an acceptable degree of flavour quality, on the one hand, and a reasonable shelf-life on the other.

Stored Dried Products of Animal Origin

Dried Egg

Commenting on the possible connection between the deterioration of dried foodstuffs and the action of enzymes surviving the rigours of the dehydration process, Joslyn (1951) expressed the view that dried whole egg products could well suffer from the effects of endogenous enzyme activity during storage. Reviewing the subject some ten years later Acker (1962) was able to conclude that, in spite of the lack of severity of the processing conditions and the abundance of lipid in eggs, there was no firm evidence which could implicate the endogenous enzymes. Indeed the conclusion was encouraged at least in part by the fact that enzymes of the types capable of causing spoilage, particularly lipases and phospholipases, were difficult to find in the uninfected, unincubated egg. Enzymes introduced through bacteria infecting the liquid egg prior to drying were considered to be far more likely to contribute to the instability of the dried product and a combination of hygienic handling and storage at suitable RH (<75%) has been suggested as an important factor in successful storage (Bergquist, 1986).

As in the case of grain storage, the conditions used for processing and storing dried egg products have to take account of the need to protect protein functionality from thermal damage. There is, therefore, an obvious reluctance to use any more than the minimum thermal severity. The phosphatase test appears to be a suitable guide to the efficiency of pasteurisation.

Dried Milk

At the same time endogenous enzymes were suspected of contributing to the deterioration of dried milk products during storage (Acker, 1962). This suspicion was better founded at least on the grounds of the potential of endogenous enzymes (Kitchen, 1985) to cause damage, particularly lipases, proteases and lactoperoxidase, and the improved opportunity for enzyme–substrate encounter afforded by redistribution of fat during processing. The potential to cause damage is further increased by the introduction of exogenous enzymes during the development of bacterial growth, particularly of psychrotrophs (Stead, 1984) at any stage of holding liquid milk prior to drying.

Acker (1962) points out the critical importance of pasteurisation and suggests that, while the use of the disappearance of phosphatase activity as the criterion for pasteurisation efficiency ensures the loss of lipase (Coulter *et al.*, 1951; Deeth and Fitz-Gerald, 1983), it does allow lactoperoxidase to persist in subsequently derived products. It can be argued, however, that, at the MC recommended for storage (<5%) and the corresponding a_w (<0·3), the only enzymes likely to cause damage are lipases and that peroxidases and other enzymes acting in solution rather than interfacially are unlikely to be effective.

Dried Meat

Excepting the unlikely event that a sudden popular demand should arise for pemmican, biltong and beef jerky, this is an area for the scientifically inquisitive. Such was the motivation claimed by Potthast and his colleagues (1975) for investigating the possible enzyme activity in freeze-dried beef stored at various RH, i.e. to compare changes in the freeze-dried material with those observed *post mortem*. They were no doubt encouraged by the observation that catalase and ATP-ase activities appeared to survive drying (Hunt and Matheson, 1959).

The findings of this investigation (Potthast *et al.*, 1975) were consistent with the behaviour to be expected of enzymes in such a system. ATP appeared to be broken down slowly in samples stored at RH 40% and above at 25°C, the rates increasing with RH. The final extent of breakdown was partial at the lower RH but increasing with RH and becoming complete at RH 70%. The breakdown of glucose, fructose and their 6-phosphates showed a similar threshold RH and pattern of behaviour as did the liberation of fatty acids but this occurred at RH as low as 10%.

The investigators were satisfied that these changes were caused by enzyme action, mainly because ATP was not broken down in precooked freeze-dried beef and partly because of the rapidity of the process at high RH. The fact that glycogen levels remained unchanged even at high RH was attributed to its failure to dissolve at the low MC involved despite Matheson's (1962) experience to the contrary. Prolonged storage times at very high RH (97·5%) proved to be impractical due to microbial growth.

The levels of lactate and AMP were found to remain effectively constant during storage. While no explanation was offered for the former observation, the latter was attributed to the high activity of AMP-deaminase in converting AMP to inosine monophosphate. While the constancy of the lactate level seems curious, its elucidation would have entailed the scrutiny of the entire glycolytic pathway which might well be interrupted by the destructive events of freezing. Later work by the same group (Potthast *et al.*, 1977) showed that glycogen was broken down on rehydration of the stored material but with the efficiency of breakdown decreasing with increasing time of storage and RH. The same pattern of deterioration was found for the formation of lactate on rehydration and both effects were ascribed to the possible damage to enzymes by Maillard reactions during storage.

SYSTEMS OF LOW MOISTURE CONTENT AT TEMPERATURES ABOVE AND BELOW AMBIENT

Throughout the preceding discussion no account has been taken of the effect of temperature. The experimental work involved was conducted largely within the temperature range 20–31°C, but even within this narrow range differences in temperature can cause differences in behaviour (Drapron and Guilbot, 1962).

Such effects can be seen in the manner in which water is sorbed and desorbed. Comparing the sorption isotherms in Fig. 2 it can be seen

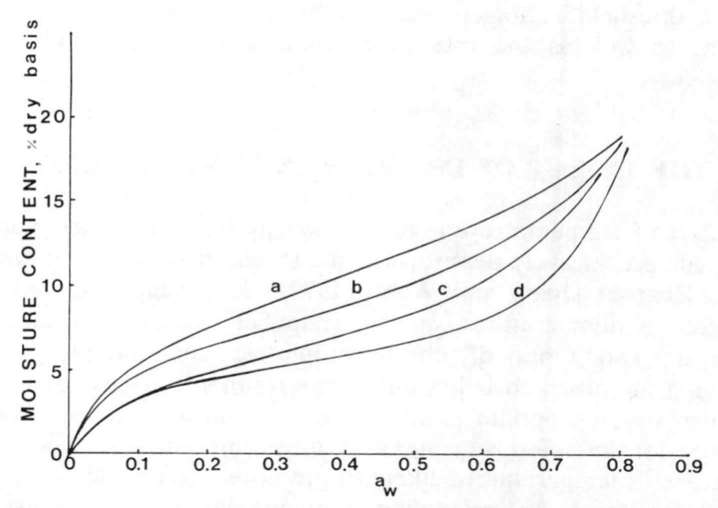

FIG. 2. Sorption isotherms for cardamom seeds at various temperatures drawn from the data of Wolf *et al.* (1973): (a) 5°, (b) 25°, (c) 45°, (d) 60°C. Ordinate: moisture content, per cent dry weight. Abscissa: a_w.

that they converge in the regions below $a_w = 0{\cdot}1$, where all the water molecules are very strongly adsorbed, and above $a_w = 0{\cdot}8$, where some molecules are becoming virtually free from the influence of the surfaces and are constrained only by interaction with the solutes present. In the intervening region the curves become flatter with rising temperature as the increasingly energetic water molecules escape from the attractive forces of the solid phases to demonstrate progressively greater mobility and fugacity. As the adsorptive power wanes its effective range diminishes so that the thickness of the adsorbed film which can be maintained at a given a_w decreases and thereby the MC.

Although a_w is a good guide to the condition of water in various situations at a given temperature it provides a much poorer basis for comparison of situations at different temperatures. It is essentially a measure of effective concentration, after all, and is only related to free or available energy in that capacity. In providing a comparison with pure water under the prevailing conditions a_w does not include any indication of the change in energy conferred on pure water with change in temperature or of the changes in behaviour resulting from it. In other words the film of water in a sorption system at 5°C and $a_w = 0{\cdot}5$ has nothing like the reactivity of water in a film at 80°C and the same a_w, despite its greater thickness. It is not surprising, then, that the threshold a_w for enzymic activity in systems of low moisture content, as well as the rate of reaction, varies significantly with temperature.

THE EFFECT OF INCREASE IN TEMPERATURE

The effect of temperature on enzyme activity has been much studied and is comprehensively described in the standard works on enzymes, such as that of Dixon and Webb (1979). The complexities of the influence of temperature can be simplified to some extent by identifying two types of effect, promotive and inhibitory, and envisaging the observed behaviour as the resultant of these two. This is particularly appropriate in systems of low moisture content where the scope for inhibition is greater than for promotion.

Increase in temperature is likely to promote enzymic action in low moisture systems by promoting the dissolution and diffusional distribution of enzyme, substrate and cofactors, thus improving the probability as well as the effectiveness of contact. This applies equally well to interactions with inhibitors. The fall in MC which inevitably accompanies a rise in temperature may be appreciable and may, on the one hand, assist interaction by the concentration effect yet, on the other, impede it through the loss of liquid continuity in the system.

At all levels of temperature, however, the inactivation of the enzymes must be taken into account as demonstrated by Stearn (1949) for the hydrolysis of starch, using the change in the temperature coefficient as an indication of the increasing influence of the inactivation process. An understanding of the mechanism(s) of thermal inactivation of enzymes has been sought over the years within the more general context of protein denaturation. Despite the mechanistic complications it is relatively easy to observe the effective denaturation of enzymes in terms of the loss of activity and, indeed, the loss of technological functionality has been used for non-enzymic proteins in a similar way (Becker and Sallans, 1956). From the data obtained by quite simple procedures it is possible to characterise the process of denaturation in thermodynamic terms (Stearn, 1949) or in the terminology of the thermal processing of food (Aylward and Haisman, 1969) according to the investigator's purpose.

The experimental approach to the investigation of enzyme inactivation at low moisture levels is essentially the same as that used for enzymic activity. Model systems, usually freeze-dried mixtures, or dried biological materials are adjusted to the appropriate MC/RH and a thermal shock is administered. The residual enzymic activity is then assayed under standard conditions, which may or may not entail the extraction of the enzyme, and compared with the activity present before the thermal treatment. Some care is necessary here because the preliminary treatment may itself cause some loss of activity.

Drapron and Guilbot (1962) found that β-amylase was partly inactivated by freeze-drying alone although the loss of activity could be prevented by including maltose prior to drying. Roozen and Pilnik (1970, 1971) observed similar behaviour on the part of the notably thermoresistant horseradish peroxidase which was partially inactivated when mixtures of the enzyme with water-soluble polymers (carboxymethyl cellulose, starch and polyvinylpyrrolidone) were freeze-dried and stored at various RH at 25°C and to a lesser degree at 2°C. The loss of activity appeared to be irreversible and depended upon the nature of the polymer, the presence or absence of buffer salts and the initial pH of the system.

With all experimental studies of thermal inactivation some pains have to be taken to define the time/temperature characteristic of the thermal shock exactly. Very small volumes of solutions can be heated in thin-walled glass capillary tubes so that the time of immersion in the heating bath and the time at the nominal temperature are virtually the same. Larger containers and volumes of sample show distinct lag phases in temperature rise and the rapidity of thermal equilibration is further hindered by the presence of air when the sample is a solid. When the effect of moisture content is to be studied the sample container must enclose a headspace of minimum volume to keep the MC from changing appreciably during heating but, of course, there is always some evaporative loss to the atmosphere during the equilibration process.

Studies conducted on enzymes in solution show that inactivation is coincident with the denaturation of the polypeptide chain(s) in many if not most cases. Enzymes with labile prosthetic groups could be expected to depart from this pattern of behaviour. The kinetics of denaturation processes have been examined and analysed in terms of the thermodynamic characteristics of the activated state (in the Arrhenius context), the enthalpy (ΔH^*) and entropy (ΔS^*) of activation. The experimental values obtained for the latter correspond to the number of intramolecular bridges broken and the degree of unfolding sustained, respectively, in the formation of the activated intermediate and are consequently large and positive (Stearn, 1949). Horseradish peroxidase (haemoprotein), on the contrary, showed a negative ΔS^* indicating the formation of an intermediate which was actually more rigid than the native state (Joffee and Ball, 1962). It is probable that both conformational and constitutional changes can arise during the thermal inactivation of enzymes, the causes and extent of which vary with environmental conditions.

In systems of low moisture content the efficiency of the various processes which contribute to enzyme inactivation will be impaired by restriction of the availability of water, with the exception of autoxidation and dehydration, so that the contributions of these processes may be weighted differently for different MC at a given temperature. Indeed the contributions of some may virtually disappear at low MC.

Few investigators have ventured into this area of enzymology and the experience of some of those who have may have been such as to discourage others. While Sisler and Johnson (1965) observed the effect of heat and moisture on the activity of *o*-diphenol oxidases in tobacco with apparent success Multon and Guilbot (1975) in a much more extensive study of the thermal inactivation of ribonuclease in wheat at various temperatures and MC, obtained results that defied interpretation. Both groups found, as would be expected, that the rate of inactivation increased with temperature and MC. From the values calculated for ΔH^*, Sisler and Johnson found signs of a greater extent of structural change in the intermediate formed at high (66%) MC compared to that formed at low (5–24%) MC. Using their data the corresponding ΔS^* can be calculated and these are in accord with different degrees of unfolding at high and low MC.

The data of Multon and Guilbot also show differences in behaviour at high (35–45%) and low (4·5–30%) MC but the behaviour in the low MC range appears to be complex, with considerable variation in numerical value of ΔH^* and ΔS^* and changes of sign in the latter as well. Although the investigators would appear to have had some technical problems with their experimental system is it only fair to point out that similar anomalies have been reported for the denaturation of wheat proteins (Becker and Sallans, 1956). This is clearly a matter that requires further investigation.

Despite the problems in understanding these events in detail there would seem to be little doubt that the thermal inactivation of enzymes becomes increasingly difficult as the moisture content of the system decreases. This is demonstrated very clearly, albeit indirectly, by Hutchinson's study of the effects of temperature and moisture content on the germination capacity of wheat (1944). This showed that, at a given temperature, longer heating was required to cause the same degree of damage in the grain sample of lower MC and, for given time of heating a higher temperature was required for the sample of lower MC. For example in wheat of 3% MC germination was totally suppressed by heating for 60 min at 114°C while the same result could be achieved by heating a sample of 35% MC at 54°C for the same time.

The survival of enzyme activity at elevated temperatures is of considerable importance in the baking process where a final enzymic contribution is added to those which have been made during the storage of whole grain and flour and the dough mixing process (Nicolas and Drapron, 1983). The enzymic processes which occur or continue during baking have been discussed by Linko and Linko (1986) who suggest that some transformations may continue even after baking. Although a number of enzymes play a part in determining the final quality of baked products, e.g. proteases, oxidoreductases and pentosanases, α-amylase, either endogenous or added, must make its main contribution during the baking process. It is only then that the starch is rendered vulnerable by progressive gelatinisation to attack by the enzyme as it in its turn loses activity through the rise in temperature. The requisite degree of breakdown of the starch is achieved through a subtle interplay of events which could only take place in a system of low and diminishing water content.

THE EFFECTS OF LOW TEMPERATURE AND FREEZING

It is expedient at this point to draw a distinction between systems of low and intermediate a_w. Solids are capable of sorbing water vapour at low temperatures and it is possible, with sufficient ingenuity, to display the sorption behaviour in the form of the traditional sorption isotherm for temperatures as low as −50°C (Mackenzie, 1975). In situations at equilibrium in which an ice phase exists the water vapour pressure is that of ice at that temperature (and pressure) irrespective of the compositional properties of the system. The activity of the liquid water phase is given by the ratio of the vapour pressure of ice to that of pure liquid (supercooled) water at the given temperature provided by extrapolation and is thereby a function of temperature (and pressure) independent of composition. The attainment of low activities in sorbed water is achieved by equilibration of the sample at a given temperature with ice at a much lower temperature and requires much more elaborate equipment than is used for most sorption work conducted above 0°C (Mackenzie, 1975).

Since foods of low moisture content are designed to be stored at ambient temperatures and since frozen foods are likely to be in equilibrium with ice at the same temperature with resultant a_w in the intermediate range (>0·65) enzyme behaviour will be considered only in the context of the latter. It is highly improbable that any significant amount of investigation has been carried out on enzyme activity at low a_w and temperature, due to the technical difficulty in conducting such work coupled with the very low levels of activity to be expected.

MODEL SYSTEMS

In general terms there has been considerable interest in the stability and activity of enzymes at low temperatures resulting in a number of reviews including those by Tappel (1966), Douzou (1973), Fennema (1975) and Schwimmer (1981). Most of the information available has come from experiments conducted in mixtures of low freezing points produced from water and an organic solvent in suitable proportions. The design of these experiments was concerned primarily with homogeneity at the low temperatures and the role of a_w was not a matter for consideration. Douzou (1973) points out the consequences of temperature and composition on the properties of the solvent in terms of the dielectric constant and the resultant pH produced by various buffers without comment on a_w. The disadvantages incurred in using these solvents to study the influence of low temperature have been reviewed by Douzou (1973) and are essentially those incurred in using them to study the effect of a_w, viz. the specific effects of the organic solvent, particularly in denaturing the enzymes.

The results of such investigations have shown that most enzymes are capable of continuing to operate at low temperatures, subject to a rather greater decrease in rate than required by the Arrhenius relationship with temperature. This phenomenon may be seen to reflect a particular aspect of the general state of equilibrium between active and inactive enzyme forms since some enzymes exhibit instability at temperatures around 0°C (Dixon and Webb, 1979) even in the absence of organic solvent. Suggested causes for the behaviour of individual enzymes have been proffered by various workers and have been listed by Schwimmer (1981). The suggestions concern mainly possible changes in conformation.

The behaviour of enzyme systems can be quite different at a given temperature depending upon whether or not ice is formed. This is perhaps not surprising in the sense that the separation of a solid ice phase can cause considerable compositional change in the residual liquid, depending upon the extent to which it occurs. This aspect of behaviour has also attracted much attention and is discussed at some length by Tappel (1966) and Fennema (1975) among the reviews quoted earlier. Of particular interest is the fact that some enzymes show accelerated rates below the freezing point as a result of the favourable conditions created within the concentrated residual aqueous phases.

By considering the effects of freezing on buffer solutions, Fennema (1975) demonstrated the changes in pH and ionic strength to which model systems are liable upon freezing. The separation of ice and salts as solid phases will also provide opportunities for adsorption phenomena to occur under such circumstances. Freezing of solutions containing organic cryoprotectives, such as glycerol or sucrose, can cause appreciable changes in viscosity, as Fennema (1975) also points out but viscosity changes of moderate degree appear to have little effect on the rates of enzyme reactions as shown by Kertesz (1935*a*). Even in well defined model systems the resultant effect of freezing on enzyme activity can only be predicted by the wisdom of hindsight because of the highly specific nature of the interaction of the enzyme with its environment.

The activities of enzymes in frozen systems are significantly affected by the conditions of freezing, particularly the rate of temperature reduction and the temperature nadir (minimum) reached during the freezing process. Again some degree of individuality is displayed in this respect, the effects varying with the nature of the enzyme, the presence of other substances and the extent to which they are present. The nadir of the freezing process has a fairly uniform effect, the extent of inactivation being greater the lower the nadir. The conditions of thawing also influence the retention of activity, particularly the rate of rise in temperature. Presumably in influencing the rates of formation and removal of ice the rates of freezing and thawing control the variations in concentration and especially the conditions of temperature and time under which the enzyme may be exposed to them. Predictably the effects of freezing and thawing become intensified with repetition.

When plant or animal tissues are frozen another factor comes into play, viz. the disruptive effect of freezing conditions on the integrity of the membranes of cells and organelles (Partmann, 1975). Fennema (1975) sees cellular systems as being distinguished from non-cellular enzyme systems because the enzymic activity of the former is increased by freezing in the great majority of cases studied whereas such activation is exceptional in the latter. The behaviour of enzymes in frozen cellular materials is also strongly influenced by rates of freezing and thawing which are likely to exercise an influence through the extent of membrane damage in addition to the concentration effect operating in non-cellular materials.

There appears to be a general tendency for enzymic activity to decline in stored, frozen materials, although exceptions inevitably occur, and for the survival of activity to improve as the storage temperature is lowered (Fennema, 1975). The temperature range, −2 to −10°C, appears to be the least suitable for storing frozen enzyme samples and −196°C the best.

ENZYMIC ACTIVITY IN FROZEN FOODS

The success of freezing as a method of preserving foodstuffs depends upon its ability to achieve the maximum retention of those attributes which determine the quality of the product, especially the flavour and texture. From the discussion above it is obvious that the process is unavoidably physically destructive to plant and animal cells and tissues and that enzymic reactions are likely to occur in the stored product at rates which may even be enhanced by the freezing process.

In both cellular and non-cellular materials the separation of ice creates a concentrated liquid aqueous phase in which the water is hindered from freezing by the influences of the sorbing surfaces and by the interactions with other solutes in the solution phase. The situation created at this stage is not necessarily one at thermodynamic equilibrium but were it so then the a_w would be determined by the temperature and would be the same for any foodstuff at that temperature. The proportions of ice and liquid phase and the composition of the latter would, of course, vary with the initial composition of the material and would determine the course of subsequent events.

Poulsen and Lindeløv (1981) give figures for the percentages of the water in some foodstuffs which is converted to ice at various temperatures. The percentage of ice rises rapidly as the temperature falls to −15°C and the levels off at a characteristic, virtually constant level. The effective MC values corresponding to these final levels ranged from 20% for bread to about 4% for strawberries, with beef (10%) and shellfish (7%) at intermediate values. Under equilibrium conditions at −30°C the a_w for all of these materials would be 0·75 (Fennema, 1981).

Long term storage would seem to offer the chance for true thermal equilibrium to be approached and indeed changes such as the recrystallisation of ice do occur (Partmann, 1975). However, even during storage a true equilibrium is unlikely in a system which is subject to slow changes in chemical composition and, more to the point, cycles of rising and falling temperature (Schwimmer, 1981).

Subject to the general decline in the enzymic activity of frozen samples over long periods of storage, frozen foodstuffs are certainly vulnerable to deterioration during storage through the action of enzymes, particularly those likely to damage flavour, such as peroxidases, lipases and lipoxygenases. Attempts to reduce the destructive physical effects of the freezing through the use of cryoprotective substances may reduce the degree of enhancement of enzymic activity concomitant with the process but at the same time may achieve an increased stabilisation of the enzymes present.

Although the measures taken to control undesirable enzyme activity in frozen foodstuffs are not highly sophisticated in relation to the complexities of the problem they seem, nevertheless, to be remarkably successful. Blanching and chemical treatments of fruits and vegetables prior to freezing have made possible the creation of stable, frozen products of acceptable flavour quality while judicious technology, particularly with respect to the time interval between death and freezing, provide frozen meat and fish of acceptable quality. There is, no doubt, room for improvement in the technology of preserving food by freezing, especially in the control of rates of freezing. However, it has to be borne in mind that freezing techniques of increasing delicacy will tend to favour the preservation of enzymic potency and the need for the preservation of the integrity of the cellular compartmentation could become essential. It is possible that the apparently crude procedures of the present time may strike an optimum in a compromise between damage and preservation.

SUMMARY

It is highly improbable that any enzyme could function in the total absence of water. Nevertheless some enzymes can function in the presence of remarkably small amounts of water, the cauliflower esterase observed by Duden (1971) being a strong contender for the record low threshold of activity.

Most enzymes studied require fairly substantial amounts of water to be present in the system before any detectable level of activity can occur. The limiting factor appears to be the MC/a_w required to solubilise the substrate (along with any other participating molecules) and enable it (them) to reach the active site of the enzyme. In the case of a sufficiently volatile substrate transport may be achieved through the vapour phase (Duden, 1971) to make a solvent unnecessary as a vehicle for that particular reactant. In addition the efficiency of reaction will be affected by the ability of the products formed to escape from the vicinity of the active site and by specific effects exercised by other components of the system.

The activity of enzymes in the concentrated aqueous phases retained in foods of low moisture content is determined: firstly by their ability to survive undenatured in that environment through the temperature effects of processing; secondly by MC/a_w in as far as they determine, in turn, the specific interactions between the enzymes and the other components of the system. In foodstuffs of consistent composition the threshold conditions for enzymic activity at a given temperature may be related to a particular a_w.

The duration and rate of enzymic action during storage will be determined in the main by the storage temperature for a given MC. The results of such enzymic action may in some cases be beneficial within limits and the enzymes surviving in stored ingredients may continue to function beneficially during some or even all stages of subsequent processing.

REFERENCES

Acker, L. (1962). Enzymic reactions in foods of low moisture content. *Adv. Fd Res.*, **11,** 263–330.

Acker, L. (1969). Water activity and enzyme activity. *Fd Technol.*, **23,** 1257–70.

Acker, L. and Luck, E. (1958). Über den Einfluss der Feuchtigkeit auf den Ablauf enzymatischer Reaktionen in wasserarmen Lebensmittel. *Z. Lebensmittelunters u.-Forsch.*, **108,** 256–69.

Acker, L. and Kaiser, H. (1961). Über den Einfluss der Feuchtigkeit auf den Ablauf enzymatischer Reaktionen in wasserarmen Lebensmittel. III. Mitteilung: Das Verhalten der Phosphatase in Modellgemischen. *Z. Lebensmittelunters. u.-Forsch.*, **115,** 201–10.

Acker, L. and Wiese, R. (1972). Über das Verhalten der Lipase in wasserarmen Systemen. II. Mitteilung: Enzymatische Lipolyse im Bereich extrem niedriger Wasseraktivität. *Z. Lebensmittelunters. u.-Forsch.*, **150** 205–11.

Åkerløf, G. (1932). Dielectric constants of some organic solvent mixtures at various temperatures. *J. Am. Chem. Soc.*, **54,** 4125–39.

Aylward, F. and Haisman, D. R. (1969). Oxidation systems in fruits and vegetables—their relation to the quality of preserved products. *Adv. Fd Res.*, **17,** 1–76.

Barnard, M. L. and Laidler, K. J. (1952). Solvent effects in the α-chymotrypsin–hydrocinnamic ester system. *J. Am. Chem. Soc.*, **74,** 6099–101.

Becker, H. A. and Sallans, H. R. (1956). A study of the relation between time, temperature, moisture content and loaf volume by the bromate formula in the heat treatment of flour. *Cereal Chem.*, **33,** 254–65.

Bergquist, D. H. (1986). Egg dehydration. In: *Egg Science and Technology* 3rd Edn, W. J. Stadelman and O. J. Cotterill (Eds), AVI, Westport, Conn., pp. 285–323.

Bernard, S. A. and Rossi, G. L. (1970). Are the structure and function of an enzyme the same in aqueous solution and in the wet crystal? *J. Molec. Biol.*, **49,** 85–91.

Blain, J. A. (1962). Moisture levels and enzyme activity. In: *Recent Advances in Food Science. Vol. 2. Processing,* J. Hawthorn and M. J. Leitch (Eds), Butterworth, London, pp. 41–5.

Bottomley, R. A., Christensen, C. M. and Geddes, W. F. (1950). Grain storage studies. IX. The influence of various temperatures, humidities and oxygen concentrations on mold growth and biochemical changes in stored yellow corn. *Cereal Chem.*, **27,** 271–96.

Bowski, L., Saini, R., Ryu, D. Y. and Vieth, W. R. (1971). Kinetic modelling of the hydrolysis of sucrose by invertase. *Biotechnol. Bioeng,* **13,** 641–56.

Burnette, F. S. (1977). Peroxidase and its relationship to food flavour and quality: a review. *J. Fd Sci.*, **42,** 1–6.

Clarke, N. A., Wilkinson, M. C. and Laidman, D. L. (1983). Lipid metabolism in germinating cereals. In: *Lipids in Cereal Technology,* P. J. Barnes (Ed.), Academic Press, London, pp. 57–92.

Coulter, S. T., Jeness, R. and Geddes, W. F. (1951). Physical and chemical aspects of the production, storage and utility of dry milk products. *Adv. Fd Res.*, **3,** 47–118.

Cuendet, L. S. Larson, E., Norris, C. G. and Geddes, W. F. (1954). The influence of moisture content and other factors on the stability of wheat flours at 37·8°C. *Cereal Chem.*, **31,** 363–9.

Deeth, H. C. and Fitz-Gerald, C. H. (1983). Lipolytic enzymes and hydrolytic rancidity in milk and milk products. In: *Developments in Dairy Chemistry—2. Lipids,* P. F. Fox (Ed.), Applied Science Publishers, London and New York, pp. 195–239.

Dixon, M. and Webb, E. C. (1979). *Enzymes,* 3rd Edn, Longmans, London.

Douzou, P. (1973). Enzymology at subzero temperatures. *Mol. Cell. Biochem.*, **1,** 15–27.

Drapron, R. (1972). Réactions enzymatiques en milieu peu hydraté. *Annls Technol. agric.*, **21,** 487–99.

Drapron, R. and Guilbot, A. (1962). Contribution à l'étude des réactions enzymatiques dans les milieux biologiques peu hydratés: la dégradation de l'amidon par les amylases en fonction de l'activité de l'eau et de la température. *Annls Technol. agric.*, **11,** 175–218; 275–371.

Duckworth, R. B. (1981). Solute mobility in relation to water content and water activity. In: *Water Activity: Influences on Food Quality,* L. B.

Rockland and G. F. Stewart (Eds), Academic Press, New York, pp. 295–317.

Duckworth, R. B., Allison, J. Y. and Clapperton, H. A. A. (1976). The aqueous environment for chemical change in intermediate moisture foods. In: *Intermediate Moisture Foods,* R. Davies, G. G. Birch and K. J. Parker (Eds), Applied Science Publishers, London, pp. 89–99.

Duden, R. (1971). Zum Problem enzymatischer Reaktionen in Lebensmitteln bei sehr niedriger Wasseraktivität. *Lebensmitt.-Wiss. u. Technol.,* **4,** 205–6.

Ethier, M. C. and Laidler, K. J. (1953). Molecular kinetics of muscle adenosine triphosphatase. II. Solvent and structural effects. *Archs Biochem. Biophys.,* **44,** 338–45.

Farkas, D. F., Goldblith, S. A. and Proctor, B. E. (1956). Stopping off-flavors by curbing peroxidase. *Food. Eng.,* **28,** 52–3.

Fennema, O. (1975). Activity of enzymes in partially frozen systems. In: *Water Relations of Foods,* R. B. Duckworth (Ed.), Academic Press, London, pp. 397–413.

Fennema, O. (1981). Water activity at subfreezing temperatures. In: *Water Activity: Influences on Food Quality,* L. B. Rockland and G. F. Stewart (Eds), Academic Press, New York, pp. 713–32.

Galliard, T. (1983). Enzymic degradation of cereal lipids. In: *Lipids in Cereal Technology,* P. J. Barnes (Ed.), Academic Press, London, pp. 111–48.

Galliard, T. (1986). Wholemeal flour and baked products; chemical aspects of functional properties. In: *Chemistry and Physics of Baking: Materials, Processes and Products,* J. M. V. Blanshard, P. J. Frazier and T. Galliard (Eds), Special Publication No. 56, The Royal Society of Chemistry, London, pp. 199–215.

Golubchuk, M., Sorger-Domenigg, H., Cuendet, L. S., Christensen, C. M. and Geddes, W. F. (1955). Grain storage studies. XIX. Influence of mold infestation and temperature on the deterioration of wheat during storage at approximately 12% moisture. *Cereal Chem.,* **33,** 45–52.

Hammond, E. G. (1983). Oat lipids. In: *Lipids in Cereal Technology,* P. J. Barnes (Ed.), Academic Press, London, pp. 331–52.

Huang, A. C. H. and Moreau, R. A. (1978). Lipases in the storage tissues of peanuts and other oilseeds during germination. *Planta,* **141,** 111–16.

Hummel, C., Cuendet, L. S., Christensen, C. M. and Geddes, W. F. (1954). Grain storage studies. XIII. Comparative changes in respiration, viability and chemical composition of mold-free and mold-contaminated wheat on storage. *Cereal Chem.,* **31,** 143–50.

Hunt, S. M. V. and Matheson, N. A. (1959). The relationship between the quality of dehydrated raw beef and the adenosine triphosphatase activity after storage at various moisture contents. *Fd Research,* **24,** 262–70.

Hutchinson, J. B. (1944). The drying of wheat. III. The effect of temperature on germination capacity. *J. Soc. Chem. Ind. Trans.,* **63,** 104–7.

Iglesias, H. A. and Chirife, J. (1982). *Handbook of Food Isotherms: Water Sorption Parameters for Food and Food Components,* Academic Press, New York.

Inagami, T. and Sturtevant, J. M. (1960). The trypsin catalysed hydrolysis of benzoyl-L-arginine ethyl ester. *Biochem. Biophys. Acta,* **38,** 64–79.

Joffee, F. M. and Ball, C. O. (1962). Kinetics and energetics of thermal inactivation and the regeneration rates of a peroxidase system. *J. Food Sci.,* **27,** 587–92.

Jones, D. B. and Gersdorf, C. E. F. (1941). The effect of storage on the protein of wheat, white flour and whole wheat flour. *Cereal Chem.,* **18,** 417–34.

Joslyn, M. A. (1951). The action of enzymes in contentrated solution and in the dried state. *J. Sci. Fd Agric.,* **2** 289–94.

Kertesz, Z. I. (1935*a*). Water relations of enzymes. I. Influence of viscosity on invertase action. *J. Am. Chem. Soc.,* **57,** 345–7.

Kertesz, Z. I. (1935*b*). Water relations of enzymes. II. Water concentration required for enzyme action. *J. Am. Chem. Soc.,* **57,** 1277–9.

Kiermeier, F. and Coduro, E. (1954). Über den diastatischen Starkeabbau in lufttrockenen Substanzen. *Biochem. Z.,* **325,** 280–7.

Kitchen, B. J. (1985). Indigenous milk enzymes. In: *Developments in Dairy Chemistry—3. Lactose and Minor Constituents,* P. F. Fox (Ed.), Applied Science Publishers, London and New York, pp. 239–80.

Kuntz, I. D. Jr and Kaufmann, W. (1974). The hydration of proteins and polypeptides. *Adv. Protein Chem.,* **28,** 265–79.

Laidler, K. J. (1955). Some kinetic and mechanistic aspects of hydrolytic enzyme action. *Discuss. Faraday Soc.,* **20,** 83–96.

Lechert, H. T. (1981). Water binding on starch: NMR-studies on native and gelatinised starch. In: *Water Activity: Influences on Food Quality,* L. B. Rockland and G. F. Stewart (Eds), Academic Press, New York, pp. 223–45.

Linko, Y.-Y. and Linko, P. (1986). Enzymes in baking. In *Chemistry and Physics of Baking,* RSC Special Publication No. 56, J. M. V. Blanshard, P. J. Frazier and T. Galliard (Eds). The Royal Society of Chemistry, London, pp. 105–16.

Lukow, O. M. and Bushuk, W. (1984). Influence of germination on wheat quality. II. Modification of endosperm protein. *Cereal Chem.,* **61,** 340–4.

Mackenzie, A. P. (1975). The physico-chemical environment during the freezing and thawing of biological materials. In: *Water Relations of Foods,* R. B. Duckworth (Ed.), Academic Press, London, pp. 477–503.

Matheson, M. A. (1962). Enzyme activity at low moisture levels and its relation to deterioration in freeze-dried foods. *J. Sci. Fd Agric.,* **13,** 248–54.

Matthews, B. W. (1970). Solvent content of protein crystals. *J. Molec. Biol.,* **33,** 491–7.

Meredith, P. and Pomeranz, Y. (1984). Sprouted grain. *Adv. Cer. Sci. Technol.,* **7,** 239–320.

Milner, M. (1950). Biological processes in stored soybeans. In: *Soybeans and Soybean Products, Vol. 1,* K. S. Markley (Ed.), Academic Press, New York, pp. 483–501.

Milner, M., Christensen, C. M. and Geddes, W. F. (1947). Grain storage

studies. VII. Influence of certain mold inhibitors on respiration of moist wheat. *Cereal Chem.*, **24,** 507–17.
Money, R. W. and Born, R. (1951). Equilibrium humidity of sugar solutions. *J. Sci. Fd Agric.*, **2,** 180–5.
Morris, H. J. and Wood, E. R. (1956). Influence of moisture on the keeping quality of dry beans. *Fd Technol.*, **10,** 225–9.
Morrison, W. R. (1978). Cereal lipids. *Adv. Cer. Sci. Technol.*, **2,** 221–348.
Morrison, W. R. (1983). Acyl lipids in cereals. In: *Lipids in Cereal Technology,* P. J. Barnes (Ed.), Academic Press, London, pp. 11–32.
Morrison, W. R. and Barnes, P. J. (1983). The distribution of wheat acyl lipids and tocols into millstreams. In: *Lipids in Cereal Technology,* P. J. Barnes (Ed.), Academic Press, London, pp. 149–164.
Morrison, W. R., Coventry, A. M. and Barnes, P. J. (1982). The distribution of wheat acyl lipids in flours millstreams. *J. Sci. Fd. Agric.*, **33,** 925–33.
Multon, J. L. and Guilbot, A. (1975). Water activity in relation to the thermal inactivation of enzymic proteins. In: *Water Relations of Foods,* R. B. Duckworth (Ed.), Academic Press, London, pp. 379–96.
Naessens, W., Bresseleers, G. and Tobback, P. (1982). Diffusional behaviour of tripalmitin in a freeze-dried model system at different water activities. *J. Food Sci.*, **47,** 1245–9.
Nelson, J. M. and Schubert, M. P. (1928). Water concentration and the rate of hydrolysis of sucrose by invertase. *J. Am. Chem. Soc.* **50,** 2188–93.
Nicolas, J. and Drapron, R. (1983). Lipoxygenase and some related enzymes in breadmaking. In: *Lipids in Cereal Technology,* P. J. Barnes (Ed.), Academic Press, London, pp. 213–35.
Partmann, W. (1975). The effects of freezing and thawing on food quality. In: *Water Relations of Foods,* R. B. Duckworth (Ed.), Academic Press, London, pp. 505–37.
Potthast, K., Hamm, R. and Acker, L. (1975). Enzymic reactions in low moisture foods. In: *Water Relations of Foods,* R. B. Duckworth (Ed.), Academic Press, London, pp. 365–77.
Potthast, K., Hamm, R. and Acker, L. (1977). Einfluss der Wasseraktivität auf enzymatische Veränderung in gefriergetrockenem Muskelfleisch. IV. Änderung der Aktivität glykolytischer Enzyme während der Lagerung. *Z. Lebensmittelunters. u.-Forsch.*, **165,** 18–20.
Poulsen, K. L. and Lindeløv, F. (1981). Acceleration of chemical reactions due to freezing. In: *Water Activity: Influences on Food Quality,* L. B. Rockland and G. F. Stewart (Eds), Academic Press, New York, pp. 651–78.
Reid, D. S. (1976). Water activity concepts in intermediate moisture foods. In: *Intermediate Moisture Foods,* R. Davies, G. G. Birch and K. J. Parker (Eds), Applied Science Publishers, London, pp. 54–65.
Rockland, L. B. (1957). A new treatment of hygroscopic equilibria application to walnuts (*Juglans regia*) and other foods. *Food Res.*, **22,** 604–28.
Rockland, L. B. and Nishi, S. K. (1980). Influence of water activity on food product quality and stability. *Fd Technol.*, **34,** 42–51.
Roozen, J. P. and Pilnik, W. (1970). Über die Stabilität adsorbierter Enzyme

in wasserarmen Systemen. I. Die Stabilität von Peroxidase bei 25°C. *Lebensmitt.-Wiss. u. Technol.*, **3,** 37–40.

Roozen, J. P. and Pilnik, W. (1971). Über die Stabilität adsorbierter Enzyme in wasserarmen Systemen. II. Der Einfluss von pH-Verschreibungen auf die Stabilität von Peroxidase bei 2°C. *Lebensmitt. -Wiss. u. Technol.*, **4,** 24–7.

Ruchti, J. and McLaren, A. D. (1964). Enzyme reactions in structurally restricted systems. V. Further observations on the kinetics of yeast β-fructofuranosidase activity in viscous media. *Enzymologia,* **27,** 185–98.

Schwimmer, S. (1981). *Source Book of Food Enzymology,* AVI, Westport, Conn.

Scott, W. J. (1957). Water relations of food spoilage organisms. *Adv. Fd Res.*, **7,** 83–127.

Shearer, G. and Warwick, M. J. (1983). The effect of storage on the lipids and breadmaking properties of wheat flour. In: *Lipids in Cereal Technology,* P. J. Barnes (Ed.), Academic Press, London, pp. 253–68.

Sisler, J. P. and Johnson, W. H. (1965). The effect of temperature and moisture on the inactivation of *o*-diphenol oxidase. *Pl. Cell Physiol.*, **6,** 645–51.

Skujins, J. J. and McLaren, A. D. (1967). Enzyme reaction rates at limited water activities. *Science,* **158,** 1569–70.

Soekarto, S. T. and Steinberg, M. P. (1981). Determination of binding energy for the three fractions of bound water. In: *Water Activity: Influences on Food Quality,* L. B. Rockland and G. F. Stewart (Eds), Academic Press, New York, pp. 265–79.

Sorger-Domenigg, H., Cuendet, L. S., Christensen, C. M. and Geddes, W. F. (1955). Grain storage studies. XVII. Effect of mold growth during temporary exposure of wheat to high moisture contents upon the development of germ damage and other indices of deterioration during subsequent storage at approximately 12% moisture. *Cereal Chem.*, **32,** 270–85.

Stead, D. (1984). A fluorimetric method for determination of *Pseudomonas fluorescens* Ar11 lipase in skim milk powder, whey powder and whey protein concentrate. *J. Dairy Res.*, **51,** 623–8.

Stearn, A. E. (1949). Kinetics of biological reactions with special reference to enzymic processes. *Adv. Enzymol.*, **9,** 25–74.

Tappel, A. L. (1966). Effects of low temperatures and freezing on enzymes and enzyme systems. In: *Cryobiology,* H. T. Meryman (Ed.), Academic Press, London and New York, pp. 163–77.

Theimer, R. R. and Rosnitschek, I. (1978). Development and intracellular localisation of lipase activity in rapeseed cotyledons. *Planta,* **139,** 249–56.

Tome, D. Nicolas, J. and Drapron, R. (1978). Influence of water activity on the reaction catalysed by polyphenoloxidase from mushrooms in organic liquid media. *Lebensmitt.-Wiss. u. Technol.*, **11,** 38–41.

Warwick, M. J. and Shearer, G. (1980). The identification and quantification of some non-volatile oxidation products of fatty acids developed during prolonged storage of wheat flour. *J. Sci. Fd Agric.*, **31,** 316–18.

Warwick, M. J., Farrington, W. H. H. and Shearer, G. (1979). Changes in total fatty acids and individual lipid classes on prolonged storage of wheat flour. *J. Sci. Fd Agric.*, **30,** 1131–8.

Weaste, R. C. (Ed.) (1984). *CRC Handbook of Chemistry and Physics,* 65th Edn, CRC Press Inc., Boca Raton, Florida.

Whittaker, J. R. (1958). Proteolytic enzyme content of figs. *Food Res.*, **23,** 371–9.

Wolf, W., Spiess, W. E. L. and Jung, G. (1973). Die Wasserdampfsorptions-isothermen einiger in der Literatur bislang wenig berücksichtigter Lebensmittel. *Lebensmitt.-Wiss. u. Technol.*, **6,** 94–6.

Wolf, W., Spiess, W. E. L., Jung, G., Weisser, H., Bizot, H. and Duckworth, R. B. (1984). The water-vapor sorption isotherms of microcrystalline cellulose (MCC) and of purified potato starch. The results of a collaborative study. *J. Food Eng.*, **3,** 51–73.

Chapter 6

PROTEIN-STABILIZED EMULSIONS AND THEIR PROPERTIES

D. G. DALGLEISH
Hannah Research Institute, Ayr, UK

SYMBOLS

a Radius of a particle
A_H Hamaker constant
d Distance between surfaces of interacting particles
D Dielectric constant of the medium
e Charge on the electron
I Ionic strength of the solution
k Boltzmann's constant
N Avogadro's number
pI Isoelectric pH: that pH at which net charge on the protein is zero
R Centre-to-centre distance of approaching particles
T Temperature

ε_0 Permittivity of free space
ε_r Dielectric constant
κ Debye–Huckel parameter $\kappa = \{8\pi Ne^2I/(1000DkT)\}^{0.5}$
ψ_0 Surface potential of a particle

INTRODUCTION

It is a truism that oil (or fat) and water do not mix: in nature, fat or oil is always found encapsulated either within a cell or, if it is outside the cell, within a membrane composed mainly of phospholipid. For example, the butter fat in milk is contained within the natural phospholipid membrane (the milk fat globule membrane) and is stable towards coagulation (Mulder and Walstra, 1974). Breaking of this membrane by churning leads to coagulation of the fat particles, and ultimately to phase separation and the formation of the water-in-oil emulsion, butter. In many foods, the natural stabilizing materials are often broken down, removed or present in insufficient quantity, and, when emulsions are to be made, they must be replaced by suitable surface-active agents. Even in the presence of the original membranous materials, the reduction in the size of the fat particles when they are subjected to homogenization leads to a considerable increase in the surface area of the fat. Stabilization of this newly formed surface requires binding of surface-active materials. Among these are small amphiphilic molecules of the detergent type, but proteins and certain peptides also exhibit the ability to stabilize the dispersed fat in food emulsions. The aim in the manufacture of such emulsions is to create a suspension of fat particles in an aqueous medium which is stable over extended periods of time, which may be required to be as long as one year for products such as concentrated homogenized milks. The stabilization arises primarily from the binding of surfactant proteins to the oil/water interface (i.e. they cover the surface), but the interactions of the adsorbed proteins are also critical in preventing coagulation of the emulsion droplets.

To successfully adsorb to the oil/water or fat/water interface, a protein must possess certain essential properties. Like all surface-active molecules, the proteins must be amphiphilic, that is, some parts of their structure must be composed mainly of amino acids which possess hydrophobic side-chains (which will bind to or even dissolve in the lipid or non-aqueous side of the interface), while other parts of the molecule must contain a predominance of hydrophilic amino acids which allow the protein to interact favourably with the aqueous phase. In principle, these two regions should be on distinct parts of the protein, but because the polypeptide chain may be flexible, residues from apparently distant parts of the molecule may be brought into close proximity. Because of the very different compositions and structures of proteins, it must be apparent that different proteins should possess different emulsifying powers. This is indeed found to be true. Especially, there is a range of surface activities among proteins which relates to their abilities to adsorb to oil/water interfaces.

Once protein has bound to the oil/water interface during homogenization to form an emulsion, it can stabilize the newly-formed particles by a number of mechanisms. Particles in an emulsion should be considered as colloidal particles, and therefore the general mechanisms of colloid stability can be taken to apply to them (Dickinson and Stainsby, 1982; Hunter, 1987). In one of these, it is assumed that the hydrophilic part of the protein (which is not itself bound to the interface) is sufficiently highly charged as to create significant repulsion between two approaching particles. Coagulation of the emulsified material is prevented simply because of this repulsive energy barrier. An alternative mechanism by which bound macromolecules confer stability on the system is that the bound protein, if it is sufficiently flexible (and not all are), can create a highly hydrated and mobile layer around the surface of the fat globule, and hence sterically stabilize the particle. The particles are stabilized because the highly hydrated layers of bound macromolecules on different particles cannot interpenetrate because of the mobility of the surface protein chains. Either or both of these mechanisms of colloid stability may be effective in stabilizing any particular emulsion. The ideal emulsifying protein would, on these criteria, be disordered rather than globular, and possess extensive charged and hydrophobic areas. There are such proteins which are good surfactants: for example, caseins from milk (Swaisgood, 1982). These proteins and their properties will be considered extensively in this chapter, partly because of the author's interest, but also because they have been more extensively studied than most surface-active proteins.

What the caseins lack, however, is the ability to form a strong surface film, and therefore emulsions stabilized by α_s or β-caseins do not possess the additional stability which can be conferred by the formation of such rigid films. Other proteins do have this ability. It may therefore be important to distinguish between surface activity and emulsifying capacity on the one hand, and emulsion stability on the other. An important corollary of this is that it is not necessarily the most surface-active proteins which give the most stable emulsions. Also, it may not be possible to extrapolate the results of studies of proteins in, for example, spread films, to the stability of emulsions.

This chapter aims to describe the properties of protein-stabilized oil-in-water emulsions from a molecular point of view. Thus, no attempt is made to describe the practical aspects of the production of emulsions either in the laboratory or industrially. Rather, it is the principles behind the systems which are discussed, such as the properties of the protein, the binding of the proteins to oil/water interfaces, and the causes and mechanisms of stability of the emulsions. A further restriction has been applied, in that the emulsions considered will all fall into the category of oil-in-water emulsions, to the deliberate exclusion of other emulsified systems. Moreover, no consideration is given to the non-aqueous material in the emulsion: this, it appears, is not of great importance in defining the properties of the emulsions. Although emulsions are the main topic, much of our understanding of interfacial phenomena has come from model studies of surfaces, and it is not possible to consider emulsions only: much reference will be made to these model studies. It is not intended that the chapter should be an exhaustive review of the extensive literature on emulsions and the interactions of proteins at interfaces. There has consequently been a conscious attempt at selection, although it is hoped that no important aspects of the subject have been ignored.

STRUCTURES OF PROTEINS AND THEIR RELATION TO INTERFACIAL BEHAVIOUR

The properties of oil-in-water emulsions are almost wholly dependent on the properties of the surfactant proteins: there is little evidence that the nature of the lipid phase is important. For an understanding of the emulsions, it is essential to consider a number of properties of proteins. An individual protein molecule consists of a chain of amino acids whose sequence is defined by the genetics of the source animal or plant. There are some 20 common amino acids, which have different structures and properties, differing in the nature of the side-chain attached to the α-carbon of the amino acid. The amino acids are linked together by amide linkages to form the polypeptide backbone of the protein. It is therefore the different side-chains of the amino acids which confer upon a protein much of its individual character. In some cases, these side-chains may be modified post-translationally in the original source organism, which further enhances the differences between one protein and another. Examples of such modifications are the phosphorylation of certain seryl residues in the caseins (Mercier *et al.*, 1972) and in egg phosvitin (Taborsky, 1974) or the glycosylation of milk κ-casein (Fiat *et al.*, 1980). Artificial modifications may be induced during processing such as the modification of lysyl residues by the Maillard reaction (Reynolds, 1965). It is possible to identify three aspects of the properties of side-chains which are likely to be important in emulsions. First is their hydrophobic or hydrophilic character, which is the major factor in determining the interactions between protein and fat and the adsorption to the interface. Second is the charges which are carried by the side-chains: these are more important in defining the stability of the emulsions and the interactions between emulsion droplets. Finally, the side-chains are the major determinants of the native structures of the proteins.

Of the amino acids, some have aliphatic or aromatic side-chains which are obviously hydrophobic, some have positively or negatively charged groups, and some have hydroxyl substituents. Some properties of the individual side-chains are collected in Table 1. Importantly, it can be seen from the table that not all of the hydrophobic side-chains have the same hydrophobicity: a scale of

TABLE 1
Properties of the amino acids

Amino acid	*Hydrophobicity*[a] (*kcal/residue*)	*Charge at pH 7*
Trp	3·00	0
Ile	2·95	0
Tyr	2·85	0
Phe	2·65	0
Pro	2·60	0
Leu	2·40	0
Val	1·70	0
Lys	1·50	+1
Met	1·30	0
Cys/2	1·00	0
Ala	0·75	0
Arg	0·75	+1
Thr	0·45	0
Gly	—	0
Ser	—	0
His	—	+1
Asp	—	−1
Asn	—	0
Glu	—	−1
Gln	—	0
Serine phosphate	—	−2

[a] Values taken from Bigelow (1967): alternative values have been proposed (Cornette *et al.*, 1987).

values can be drawn up (Bigelow, 1967). However, it is not always the case that the behaviour of a protein can be assessed from its overall hydrophobicity (or, indeed its overall charge), as determined by the sum of the contributions from the individual side chains (Keshavarz and Nakai, 1979). What is also important is the distribution of the hydrophobic or charged residues along the polypeptide chain and how the chain itself is folded. Other things being equal, it would be expected that in the folding of the protein many of the hydrophobic side-chains would be found together towards the interior of the protein structure, and that the charged and hydrophilic groups would be found on the exterior. This ideal is seldom fulfilled in practice, since a number of partly conflicting factors determines the structure of a protein, such as the possibility of ionic reactions, the nature of the solvent, the formation of disulphide bonds, and the possibility of the formation of α-helices or β-sheets as well as the primary structure of the protein itself. The tendency for hydrophobic side-chains to cluster together is only one of these, and in many cases not the dominant, force in determining the protein's conformation.

Thus, a protein in solution will almost necessarily possess exposed hydrophobic side-chains in its surface, and it is possible to make some correlations between the 'surface hydrophobicity' of a protein and its emulsifying ability (Kato *et al.*, 1981, 1983; Voutsinas *et al.*, 1983). This measure of hydrophobicity can be estimated experimentally using a probe, *cis*-parinaric acid (Kato and Nakai, 1980; Nakai, 1983; Li-Chan *et al.*, 1984). Although this leads to a better estimate of emulsifying capacity than calculation of the overall hydrophobicity, it is not totally adequate, since other factors come into play when the protein binds to the interface (e.g. possibility of denaturing and unfolding the protein). Also, there is some effect of the material to which the hydrophobic portions of the proteins bind: for example, β-casein is a more surface-active protein than is α_{s1}-casein when they are bound to an oil/water interface (Dickinson *et al.*, 1985*b*), but when the proteins are bound to a 'hydrophobic' chromatography column, the two proteins may bind with similar strengths (Visser *et al.*, 1986).

The folding of the proteins (i.e. their conformation) is largely determined by the interactions between the side-chains and other side-chains, and with the solvent. Thus, hydrophobic residues will tend to be buried and hydrophilic residues will tend to be exposed, when the protein is in an aqueous environment. The protein's conformation may then be stabilized by a number of bonds. Hydrogen bonds between different parts of the polypeptide backbones stabilize the α-helix and β-sheet structures which are possessed by many proteins, and which confer a degree of rigidity in the structure. Not all of the amino acids permit the formation of these regular structures. In particular, proline, in which the amino and carboxyl groups of the peptides form part of a ring, cannot participate in formation of an α-helix, and therefore proteins which contain many proline residues (e.g. β-casein (Swaisgood, 1982)) cannot have extensive secondary structure.

A second important stabilizer of structure in proteins, which restricts the flexibility of the molecule, is the disulphide bond, which is formed by the oxidation of cysteine residues which are brought together by the folding of the protein. The bonds are generally intramolecular in globular proteins, but can also be formed intermolecularly: this last may lead to dimer or higher polymer formation. Disulphide bonds are of especial importance because, in contrast to hydrogen bonds, which are fairly weak, disulphide bonds require chemical action or high temperature treatment for breakage. Conformations of proteins which contain intramolecular disulphides are therefore more stable than those which are dependent only upon hydrogen bonds. Reduction of the bonds generally involves exchange reactions between sulphydryl and disulphide groups. For example, β-mercaptoethanol will react with disulphide groups to form mixed disulphides between the cysteinyl residues and the mercaptoethanol. Free cysteinyl residues in proteins can also catalyse the rearrangement of disulphides during denaturation, as in the case of β-lactoglobulin, which denatures (and polymerizes) via disulphide exchange reactions of this type (De Wit and Klarenbeek, 1981).

The success of a protein as a surfactant will depend significantly on the flexibility of its polypeptide chain, since high flexibility will allow the maximum interaction of hydrophobic side-chains with the non-aqueous surface. Therefore, proteins which possess a large amount of secondary structure and contain disulphide bonds will be less surface active than proteins with less structure and more flexible chains. The two cases of lysozyme and β-casein have been contrasted specifically (Graham and Phillips, 1979*c*), and will be considered in more detail below.

Thus, the primary, secondary and tertiary structures of proteins are important in determining the nature of emulsions which incorporate them. The quaternary structure of the protein, namely the extent and manner in which it aggregates when in an aqueous solution, is at least potentially important, since such aggregation often occurs because of the interaction of hydrophobic portions of the proteins. Many small surfactant molecules exist in solution in the form of micelles, which are particles containing a number of individual amphiphilic molecules, arranged so that their hydrophobic moieties are in the centres of the particles and their hydrophilic moieties are exposed to the solution. It has not been conclusively demonstrated that proteins behave in this way, partly because they are structurally larger and more complex than the simple surfactants, (i.e. their 'head' and 'tail' groups are not so clearly defined as in the more simple molecules). However, β-casein is known to form particles containing of the order of 30 molecules, and these have been considered as micelles in the sense described above (Evans *et al.*, 1979). (Unfortunately, the term 'micelle' is generally applied to caseins in another sense from that used here.) The effect which this aggregation has on the emulsifying capacity of the protein is not established. It is apparent that under normal conditions β-casein is not aggregated when at the interface (Graham and Phillips, 1979*c*), so that the complex which exists in solution must be broken into its individual units before or during the adsorption to the fat/water interface. This may readily happen with the β-casein complex which is thought to be maintained in solution by hydrophobic interactions. Intermolecular bonds of a more specific type, for instance disulphides, would not be broken during emulsion formation, and the protein in this case would bind as polymers or perhaps, if the hydrophobic portions of the monomers were concealed in the polymer, not bind at all (Dickinson *et al.*, 1987). Also, casein micelles and parts of micelles bind to the fat surface in homogenized milk, their original polymeric form being maintained in its stable configuration by calcium phosphate (Holt *et al.*, 1982).

An illustration can be made of all of these factors by considering the protein β-casein. Figure 1 shows the amino acid sequence of the protein (Grosclaude *et al.*, 1972), and identifies the hydrophobic and the charged residues. It is apparent that nearly all of the charge on the protein is carried by a relatively short peptide at the N-terminal end of the protein: the remainder of the protein contains residues which are either hydrophobic or uncharged. It is not possible to crystallize the protein, so that there is no definite information on its conformation. However, the large number of proline residues in the C-terminal

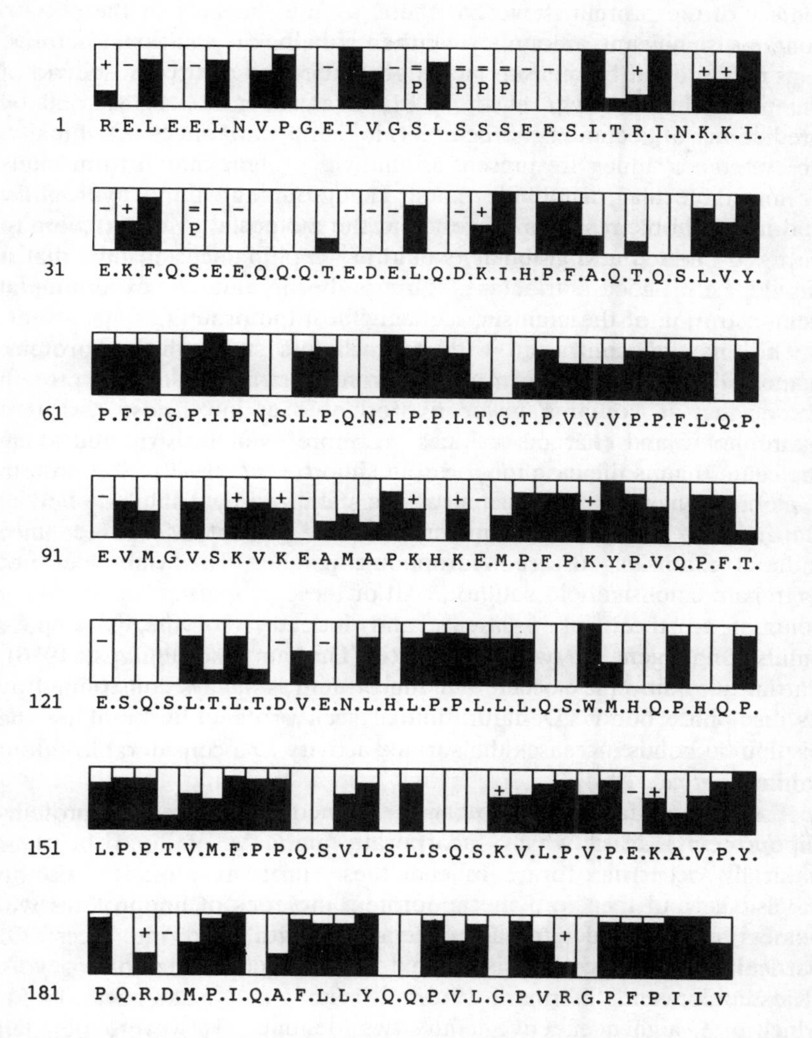

FIG. 1. Sequence of bovine β-casein-A, using single-letter code for amino acids. Charged residues are shown, together with phosphorylated serines, designated with a P in the box above the residue. The heights of the bars above the residues denote the residue hydrophobicity as defined by Bigelow (1967). See also Table 1.

moiety of the protein show that there is little chance that the protein contains significant amounts of either α-helix or β-sheet structures. This is borne out by measurements of the circular dichroism spectra of the molecule (Creamer *et al.*, 1981; Graham *et al.*, 1984), and by predictions of secondary structure (Holt and Sawyer, 1988). Finally, no cysteine residues are present so that the protein cannot form inter- or intramolecular disulphide links. The grouping of the hydrophilic and hydrophobic residues suggest that the molecule, even if it were to be tightly folded and globular, would be amphiphilic, and thus that it should be a good surfactant. This is borne out by experimental demonstration of the high surface activity of the protein.

This may be contrasted with a much less surface-active protein, namely lysozyme. A study of the primary structure of the protein shows that it contains fewer hydrophobic amino acids, and that hydrophobic and charged residues are more evenly distributed along the chain than is the case for β-casein (Imoto *et al.*, 1972). The protein is globular, having a compact structure stabilized by disulphide bonds. Both α-helix and β-sheet conformations are present. Moreover, since it lacks extensive surface hydrophobic patches, lysozyme does not aggregate extensively in solution. All of these properties are in strong contrast to those of β-casein, and indeed lysozyme is a poor emulsifying agent in its native state (Graham and Phillips, 1976). Partial unfolding may occur, but the protein is always constrained by its disulphide bonds. Denaturation of such proteins by breaking the disulphide bonds increases the surface activity to a considerable extent (Shimizu *et al.*, 1984).

A structural factor not so far mentioned is that certain proteins (lipoproteins) exist which incorporate amounts of lipid in their naturally occurring form. Indeed these proteins already exist in emulsions, and therefore the apoprotein moieties of lipoproteins will possess ready-made sites for interaction with lipid interfaces. Of particular importance in this context are the lipoproteins of egg yolk (Kiosseoglou and Sherman, 1983*a*; Mizutani and Nakamura, 1985), which are highly effective emulsifying agents. However, detailed discussion of these entities is difficult since they are rather poorly characterized. For example, the proteins are aggregated in their native state, but it is not known whether they bind to the interface in this form during the creation of an emulsion, or whether the interaction with the interface involves the breakdown of these aggregates. Nor is it known what part the original lipid and phospholipid moieties of the lipoprotein play in emulsion formation.

PROTEIN STRUCTURES AND THE FORMATION OF EMULSIONS

Surfactants, including proteins, bind to fat surfaces because the free energy of the system is minimized when the interface is coated with surfactant, and the surface tension of the interface is decreased. A number of contributions to the overall free energy of the system can be identified. There are ionic interactions between the charged residues of the surfactant protein and ions in the aqueous layer: also it is possible to form hydrogen bonds between the water molecules and the polar groups in the protein. These are phenomena associated with the formation of solutions as the proteins dissolve, and there may be little difference in such contributions to the overall free energy between the free and bound states of the protein. The dominant effects on the optimization of free energy arise from so-called hydrophobic bonding. This phenomenon occurs between the hydrophobic side-chains of the protein molecule and the surface of the non-aqueous material, and has the overall effect of creating an apparent bond between the two. The term 'hydrophobic bond' has been used earlier in this article, and will continue to be used as a convenient shorthand for the apparent interaction of hydrophobic groups.

In fact, hydrophobic bonds are not true bonds, since there is no physical attractive force (apart from van der Waals' interactions when the groups approach each other) between the two adjacent entities. The apparent bonding effect arises because the clustering of hydrophobic groups decreases the free energy of the system, via changes in the entropy (Tanford, 1980). An isolated hydrophobic group in an aqueous environment causes water molecules to be ordered in its vicinity, more than is the case for a polar group. This restriction on the movement of the water molecules results in a considerable drop in entropy. When the hydrophobic entities cluster together, to form volumes from which water is virtually excluded, there is less ordering of the water structure, and the more mobile water will increase the entropy of the system. Since increasing entropy results in a decrease of the free energy, clustering of the hydrophobic entities is favoured.

This gives the appearance of formation of a bond, and hence the loose term 'hydrophobic bonding' has been applied to it. A more suitable term is 'hydrophobic dehydration' since it is from the effect of hydrophobic groups on water that the apparent bond arises.

In the presence of large areas of lipid surface, the increase in free energy which can be gained from hydrophobic effects is sufficiently large that the oil will bind hydrophobic materials very efficiently (i.e. the oil droplets will coagulate in the absence of an emulsifier, or will bind hydrophobic groups on a protein if it is present). Therefore, the most energetically favourable position for an amphiphile to adopt in oil/water mixtures is to be on the oil/water interface. While this may be axiomatic for small surfactant molecules such as detergents which have well defined hydrophobic and polar moieties in their structures, it is not so obvious for polymers in general and proteins in particular. Specifically, since many proteins contain both hydrophobic and hydrophilic regions in close proximity, and are, at least prior to adsorption, structured, they cannot, except in a few cases, be regarded as simple amphiphiles. It is not always possible to predict what part of a protein is likely to be responsible for adsorption. An additional complexity is that it is traditionally imagined that the non-polar moiety of a small surfactant such as a detergent dissolves in the oil, but it may be unlikely for a part of the protein molecule to totally penetrate the interface. This is for two reasons: the hydrophobic groups of the protein are somewhat dissimilar to the triglyceride molecules in the oil (soaps and detergents have long aliphatic groups similar to those of fatty acids, whereas none of the amino acids possesses side-chains of these dimensions), and more importantly, the polypeptide backbone of the protein is not of itself hydrophobic. Thus, although the side-chains of the hydrophobic amino acids might tend to dissolve in the oil, it is probable that the polypeptide chain would not. It may therefore be more correct to envisage the hydrophobic groups of the protein as lying on the surface rather than being dissolved in it. However, there is some evidence that β-casein can completely penetrate a lecithin monolayer (Phillips *et al.*, 1975) and that κ-casein, after reduction of its intermolecular disulphide groups, can penetrate a lipid monolayer (Griffin *et al.*, 1984).

A protein which lacks disulphides in its structure and which also is deficient in secondary structure is likely to be flexible and to behave in some ways as a flexible homopolymer. A number of studies of such materials have been made, and three structures can be envisaged for the polymer in contact with or dissolved in the interface (Vincent and Whittington, 1982). There are continuous stretches of the polymer which are in contact with the interface: these are termed trains. Parts of the polymer not in contact with the interface are termed loops if the chain is attached at both ends of the free portion, and tails if there is one free end. When a homopolymer is adsorbed, the distribution of loops, tails and trains is dependent on thermodynamic and statistical forces, and, because the material is homopolymeric, the distribution of the material can be calculated (Hesselink, 1977; Scheutjens and Fleer, 1980). Proteins, on the other hand, cannot be envisaged as behaving in this essentially random way. Although the polypeptide backbone is chemically similar in all proteins, the affinity of even a totally flexible protein molecule is dependent on the nature and distribution of its side chains. Thus there will almost inevitably be specific parts of any protein which will have no affinity for the oil/water interface. Conversely, there will be parts of the protein where binding to the interface is highly favoured, but they will be specific residues or groups of residues. If the protein molecule possesses sufficient flexibility, then there is a greater chance that a maximum number of its hydrophobic residues will bind to the surface and the more hydrophilic residues will project from the surface as loops or tails, but it is certain that calculation of the proportions of residues bound to the interface will be more complicated than for homopolymers. Also, some success may be obtained in understanding protein adsorption by considering the proteins as polyelectrolytes, but this is in qualitative rather than quantitative terms (Bonekamp *et al.*, 1983). It is obvious that the less flexible proteins with globular structures will be even less susceptible to this type of analysis.

The increased flexibility of a protein favours maximum hydrophobic interactions between side-chains and the interface and will therefore increase its surface activity. One way of increasing this flexibility is to denature the protein. This may be achieved in a number of ways, such as heating or treatment with high concentrations of urea. The former treatment can break disulphide bonds but the latter does not: separate reaction with reducing agents is required to completely urea-denature a protein which contains disulphide bonds. Denaturation will also occur after adsorption of the protein to the interface (Graham and Phillips, 1976). If the free energy is favourable, a protein will change its conformation, even although it is not originally highly flexible. This denaturation or alteration of the protein conformation on the surface may be a lengthy process, requiring in some cases up to some days after the primary binding of the protein to the interface (Castle *et al.*, 1986). The controlling factor in this process is not the overall free energy (this will define only that the denatured state is possible), but the activation energy which is required to denature the protein. Because the latter may be large, the denaturation may be slow. The process is also dependent on the surface concentration of the protein: when it is less, then more denaturation will occur (Graham and Phillips, 1979*a*), to allow maximum spreading on the interface.

This clearly establishes that time may be an important factor when emulsions are stabilized by protein: there may be no such entity as an emulsion whose properties are not time dependent. During the formation of an emulsion in a homogenizing system, the primary reaction must be rapid binding of the protein to the newly-created fat surface: if this does not happen, then the oil droplets will rapidly coalesce. The initial binding of protein must therefore occur within fractions of a second, and can therefore only be a diffusion-controlled process (Walstra and Oortwijn, 1982). Once adsorbed, the protein will then have time to adopt the most favourable conformation, and most simply this involves rearrangement of the structure of a flexible protein as the more surface-active residues displace the initially adsorbed segments: this can be fairly rapid in, for example, β-casein. If the rearrangement involves denaturation and unfolding it will take longer (Castle *et al.*, 1986). This alteration in the conformation of the protein may also cause the desorption of some of the originally bound protein as the structure of the protein monolayer becomes more established and some of the bound protein spreads on the surface (van Dulm and Norde, 1983).

The rearrangements of protein structure described above relate to the secondary and tertiary structures. The significance of the original quaternary structure of the protein is less clearly established, but it also can be subject to change in the presence of an oil/water interface. Changes in the quaternary structure must depend on the forces which maintain it, and the possibility of disrupting them. As has been described for intramolecular bonds, it is likely to be intermolecular disulphide bonds which cause the most permanent structures. Theory suggests that quaternary structure is important: it is known that the adsorption of a polymer molecule is dependent on its molecular weight (Cohen Stuart *et al.*, 1980; Koopel, 1981). Theory and experiment agree that the larger polymers will be preferentially adsorbed, although as before the kinetics of equilibration are very slow (Felter and Ray, 1970; Furusawa *et al.*, 1982). Thus, if homopolymers can be regarded as models for proteins, it would be expected that the molecular weight or the degree of aggregation of the protein should affect its binding to the interface, especially in a competitive situation. Inter-protein comparisons with respect to molecular weight are complicated by compositional and structural differences, but it has been shown that the dimer of a protein (in this case, albumin) adsorbs preferentially to the monomeric form, on the surfaces of lattices (Brooks and Greig, 1981; Lensen *et al.*, 1984). A more important case might be the binding of caseins in emulsions based on milks: in such preparations the caseins exist in many states of aggregation. There is some evidence for preferential adsorption of larger aggregates in homogenized milks (Oortwijn *et al.*, 1977), but this, as described below, may not be specifically caused by thermodynamic factors.

FACTORS IN EMULSION STABILITY

We have already seen how the formation of the interfaces of emulsions is controlled by the nature of the interactions between the fat, the protein and the aqueous phase. However, the fact that protein binds to the interface does not in itself guarantee that the emulsions will be stable: they may well coalesce or coagulate over a long time (and many food emulsions are designed to be stored for appreciable periods). The mechanisms which contribute to emulsion stability are therefore of considerable importance. Since emulsions are colloidal particles, their behaviour is governed by general aspects of colloid and emulsion stability as well as specific factors relating to the presence of protein at the interface. There are two main mechanisms which are widely used in considering the stability of colloids, one of which is based on the interactions between charged molecules, and the other of which depends on steric considerations.

The first of these is the mechanism described by Derjaguin and Landau (1941) and Verwey and Overbeek (1948), and which is named after them as the DLVO mechanism. In this theory, the forces which act between the diffuse double layers of ions surrounding charged particles in aqueous solution are determined. When a charged particle is in solution, it attracts counterions from the solution towards the charged surface, to form a layer of the counterions. This in turn causes the formation of a second layer whose charge is of the same sign as the original charges on the particle. This is the electrical double layer. Since all proteins carry some net charge, it is certain that the adsorption of proteins to a lipid surface will lead to the formation of double layers not only around the individual protein molecules but effectively around the whole emulsion droplet. As two such particles approach, a repulsive force is established between the particles, which may be considered in electrostatic terms, but also understood as an osmotic pressure effect. The excess concentration of counterions in the space between the double layers produces a local osmotic pressure difference between the interacting layers and the bulk solution (Langmuir, 1938).

It has been demonstrated (Verwey and Overbeek, 1948) that it is possible to calculate reasonable solutions for this energy of interaction by considering either particles which are large and have thin double layers (where the product of the reciprocal of the Debye length and the radius of the particle is much greater than 1, i.e. $\kappa a \gg 1$), or which are small and have large double layers ($\kappa a \ll 1$). In the case of protein-stabilized emulsions, it is clear that the first of these two cases will be applicable, and so the repulsive potential between the particles is given by

$$E_R = 2\pi\varepsilon_r\varepsilon_0 a\psi_0^2 \ln\{1 + \exp(-\kappa d)\} \tag{1}$$

The repulsive force between two particles is therefore dependent upon their initial surface potentials, their distances apart, and the concentration and charge of the small electrolyte ions present in the solution, which determine the parameter κ. This then gives the repulsive potential between the interacting colloid particles as a function of distance.

The attractive forces in the DLVO model arise from van der Waals' dispersion forces between the particles. They may be calculated to be given by the formulation (Hamaker, 1937)

$$E_A = -(A_H/6)[\{2a^2/(R^2 - 4a^2)\} + (2a^2/R^2) + \ln\{1 - (4a^2/R^2)\}] \tag{2}$$

A_H is the Hamaker constant, which depends upon the density and the polarizability of the material. In principle, this constant can be calculated, but in practice considerable uncertainty attaches to its estimation, especially as the structures of the particles become more complex (Payens, 1978).

The interaction energy between the colloidal particles can then be

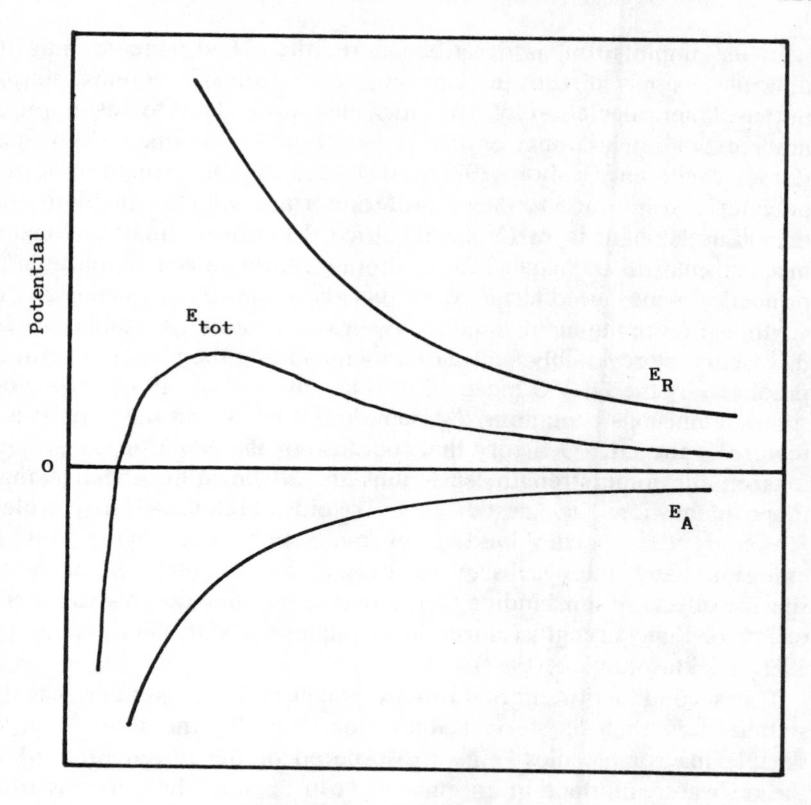

FIG. 2. Schematic diagram of the DLVO mechanism. The forms of the attractive (E_A) and repulsive (E_R) potentials are given, as well as the resultant potential E_{tot}, given by the sum of the two.

calculated as a function of their separation, by adding the attractive and repulsive potentials. Schematic results, as in Fig. 2, show that the forms of the functions for E_A and E_R combine to give a maximum in the repulsive potential so that the particles are prevented from coalescing. From eqns (1) and (2) it is clear that the precise form of the potential depends effectively on the Hamaker constant, the original charge, and the Debye length. The last of these is readily altered by changing the ionic strength of the medium, and it is a matter of observation, as well as theory, that increasing ionic strength destabilizes some colloidal suspensions.

In its simple form as described here, the DLVO theory may be difficult to apply directly to emulsions stabilized by proteins, simply because the calculation of the attractive dispersion forces requires modification to account for the presence of the stabilizing polymer (Israelachvili and Tabor, 1972), and because the structure of the polymer as it is bound to the oil/water interface will alter the nature of the diffuse double layer (Vincent, 1973). Moreover, other stabilizing mechanisms will be active. Thus, although the theory is applicable in principle, it may be difficult to predict the behaviour of emulsions in anything more than a qualitative way. Emulsions stabilized by α_s-caseins are reversibly flocculated by increasing the ionic strength, as predicted by the DLVO mechanism (Dickinson *et al.*, 1987): however, similar emulsions containing β-casein do not behave in this way. It is a feature of the DLVO theory that, because of the effect of multivalent ions on the ionic strength, such ions should be more effective than monovalent ions in destablizing colloids (Schulze–Hardy rule). However, the specific binding of multivalent ions by caseins in emulsions and their subsequent coagulation depend as much on specific effects of ion binding to the protein as they do on the general effects of ionic strength (Horne and Dalgleish, 1980; Dalgleish *et al.*, 1985; Dickinson *et al.*, 1987).

The second major mechanism by which colloidal systems can be stabilized is that of steric stabilization. Simply, the adsorption of flexible macromolecules (e.g. a disordered or denatured protein) to the oil/water interface in an emulsion can prevent the particles from approaching closely enough for the attractive van der Waals' interactions to be sufficiently powerful to permit coagulation. The repulsion effectively arises from an entropic contribution to the free energy, rather than being a true repulsive potential. This is for two reasons. Either the approach of two interacting particles will tend to compress their surface layers of adsorbed macromolecules, or some interpenetration of the molecules in the layers occurs. In the first case, the compression diminishes the volume available to the macromolecule, so that it suffers a loss of configurational entropy (i.e. it is constrained to be effectively less flexible) Meier, 1967; Hesselink, 1969), and in the second case, the entropy of mixing of the polymer chains is unfavourable to close approach (Fischer, 1958). On both counts, therefore, there is a tendency for the adsorbed macromolecule to promote stability of the colloid. Calculations are available for the two different mechanisms, and it is clear that they are both feasible.

However, whether the interactions between two particles can be described by the sum of the two effects is not certain (Gaylord, 1982).

A further stabilizing factor associated with the presence of adsorbed hydrated macromolecular material at the interface is that the interpenetration of the layers may disrupt the binding of water to the macromolecules. This will contribute unfavourably to the overall free energy of the aggregation. Especially this would be true for proteins, which are often highly hydrated. For successful steric stabilization, there will be an optimal structure for the stabilizing protein: best stabilization will be achieved by large flexible (disordered rather than globular) proteins. While this is in general accord with experience that the more flexible proteins appear to be better emulsifiers, it is not the only reason, as has been described above. It is simply another factor which is improved by flexibility.

A well-documented example of steric stabilization of particles by proteins is to be found not in emulsions but in the case of the casein micelles in milk. These particles consist of an interior formed from mainly hydrophobic proteins linked to calcium phosphate, and surrounded by a layer of the protein κ-casein, which is highly amphiphilic (Holt and Dalgleish, 1986). It has been suggested that the hydrophilic moiety of this protein projects into the solution and prevents coagulation by steric means (Holt, 1975; Walstra, 1979). A good deal of evidence is now available to confirm this hypothesis: the loss of the hydrophilic portion of the protein (by enzymatic means) destabilizes the micelles, and also reduces their radii and their voluminosity and hydration (Walstra *et al.*, 1981; Horne, 1984), all of which is consistent with steric stabilization of the original particles by the macropeptide of κ-casein.

There is perhaps less evidence for proteins acting in this way in emulsions: however, it is known that polymers behave in this way, and it is highly probable that at least some of the components of stabilization of emulsions by flexible proteins may have their causes in this mechanism. For example, the β-casein-stabilized emulsions referred to above (Dickinson *et al.*, 1987) may be partly sterically stabilized.† Protein conformations, at least in solution, depend upon the ionic strength, and the most extended conformation will occur at the lowest ionic strength. Thus, addition of salt to increase the ionic

† Although the footnote in Dickinson and Stainsby (1982), p. 75, may be relevant here!

strength may alter charge-dependent interactions of emulsion particles by altering the structures of the sterically stabilizing layer as well as by minimizing inter-particle repulsive forces.

The two mechanisms for the stabilizing of emulsions by bound macromolecules which have just been described act essentially by preventing the particles from approaching sufficiently close for attractive forces to be effective in inducing coagulation. A final mechanism is that large-scale coagulation can be prevented by the rigidity of the interfacial adsorbed layer and this surface rheology has been considered in some cases to be the dominant factor in emulsion stability (Biswas and Haydon, 1963*a,b*; Sherman, 1973; Vernon-Carter and Sherman, 1981), although alternative arguments have been given (Phillips, 1981). In such case, repulsive free energy would be too weak to prevent collision of the particles, but the interfacial layers could not be moved aside to allow the interfaces of the particles to touch and fuse. This would happen if the adsorbed protein were mobile within the time-scale of an encounter between two particles (or were readily desorbed). Experiments suggest that there are a range of proteins of different cohesive tendencies, giving a range of emulsion stabilities. This cohesive tendency is not necessarily related to the surface activity of the protein. For instance, β-casein is highly surface active but, being a flexible molecule, gives a poorly cohesive surface film and relatively unstable emulsions (Kinsella, 1982). On the other hand, proteins such as bovine serum albumin and gelatin give more structured interfacial films and greater emulsion stability (Dickinson *et al.*, 1985*a*). Although this rigidity may not develop for some hours or indeed days after the initial protein adsorption, it may ultimately prevent coagulation by in effect preserving a total protective layer around the particle surface, which is more difficult to break as time goes on (Castle *et al.*, 1986).

The tendency of proteins to form cohesive films depends upon their initial structures and their possibilities for molecular interaction. β-Casein has a strong tendency to self-associate in solution (Evans *et al.*, 1979; Swaisgood, 1982), mainly caused by the association of the hydrophobic C-terminal moieties of the protein. When β-casein is adsorbed, however, the hydrophobic parts are bound to the interface, and are no longer available for self-association, so that there is little cohesive tendency between the other parts of the molecules, which carry the bulk of the molecular charge. Since the β-casein unfolds completely at the interface, there is a likelihood of large adjustments in its conformation as the surface is compressed (Graham and Phillips, 1979*c*). Lysozyme, on the other hand, unfolds to a very limited extent (Graham and Phillips, 1979*c*), and, since the protein does not naturally self-associate strongly, will again form a poorly cohesive film. But, in presence of negatively-charged proteins the positively-charged lysozyme may interact by charge-attraction to form a cohesive mixed film. This has been demonstrated for foams (Garibaldi *et al.*, 1968; Sauter and Montoure, 1972), and it is likely that a similar mechanism occurs in emulsions (Poole *et al.*, 1984). Bovine serum albumin provides a stronger more cohesive film: it contains flexible hydrophobic domains, but also has limited possibility for structural change since it also contains a number of intermolecular disulphide bonds. The protein also contains domains of positive and negative charge. The combination of these properties apparently allows interactions between the molecules in the interfacial film to render it strongly cohesive (MacRitchie, 1978). Reduction experiments confirm that the conformations stabilized by intramolecular disulphide bonds are essential to the formation of a cohesive film (Waniska *et al.*, 1981).

It is apparent that, if the stability of emulsions is controlled, at least in part, by the viscoelastic properties of the surface films, then the effect of aging must be taken into account. For it is demonstrable that the surface viscosity of certain proteins increases for some days after their adsorption to the interface. This can only be the result of slow denaturation of the polypeptide, and therefore will not be apparent in films of β-casein but in other proteins where structural changes are slower (Kiosseoglou and Sherman, 1983*b*; Dickinson *et al.*, 1985*a*). This method of stabilizing emulsions appears to assume that the inter-protein forces of two colliding emulsion droplets were weak, so that a DLVO mechanism could not be applied. It is not completely accepted that surface rheology and coalescence of emulsions are linked, but a number of studies have established a correlation between the two. It may be more important to consider the energy of desorption as an essential factor in this form of destabilization, rather than the strength of the interfacial film *per se* (Halling, 1981).

In all of the foregoing discussion, it has been assumed that the instability of an emulsion will arise from attractive dispersion forces between the adsorbed proteins, or by hydrophobic interactions. This is not necessarily the case, since specific interactions can also contribute to the instability of the emulsion, either by simply reducing the charge or in other more complex ways. An example of this is the emulsion stabilized by α_{s1}-casein. This protein is capable of being precipitated in its native state by calcium ions, but not by sodium chloride, and the specific effect of calcium arises from specific binding of the ion to the phosphoserine groups of the casein, thereby reducing its overall charge (Horne and Dalgleish, 1980). The emulsion is, unlike the protein, susceptible to being destabilized by increasing the ionic strength of the medium with sodium chloride (Dickinson *et al.*, 1987). However, addition of calcium ions also destabilizes the emulsion: this takes place at such a low ionic strength that it cannot simply be the effect of shielding of the charges, and therefore arises from the specific neutralization of the negative charges on the casein by the calcium, or even by calcium bridging between the caseins.

A final mechanism for instability of polymer-stabilized emulsions is that of bridging flocculation (Hunter, 1987, p. 489). This phenomenon has been demonstrated more for other polymers than for proteins, and effectively arises from the ability of a single macromolecule to adsorb to two non-polar surfaces. It is characteristic of low polymer concentrations, and it is possible that most proteins are too small to act as effective bridging agents. The formation of meat emulsions which contain fat within a gel network of protein can be regarded as a form of bridging (Schut, 1976). Also, emulsions made using skim milk at low concentrations show evidence of bridging, with casein micelles rather than individual proteins as the agents linking between the particles (Ogden *et al.*, 1976; Oortwijn *et al.*, 1977).

THE BINDING OF PROTEINS AND EMULSION FORMATION

As has been seen, the proteins bind to the interface through hydrophobic regions of their polypeptide chains. Originally, the binding behaviour was thought to be via a Langmuir type of isotherm, but this is now rather more in doubt (Lyklema, 1984). In at least some cases, the individual binding isotherms for a particular protein/surface combination show evidence that there is very strong, almost stoichiometric, binding at low bulk concentrations of protein and low surface coverage (e.g. β-casein on polystyrene surfaces (Dickinson *et al.*, 1983) or at an oil/water interface (Graham and Phillips, 1979*b*)). The amount of bound protein increases fast, as the bulk concentration is increased, until saturation coverage is achieved. This process appears to be irreversible, and some doubts have been expressed as to whether the protein adsorbed on the surface and free in bulk solution are in equilibrium. One reason why the adsorption cannot be regarded as a classical Langmuir process is that a flexible protein molecule, as has been described, will attach to the interface via a number of points along the polypeptide. In such circumstances, adsorption and desorption of the protein are not one-stage processes, but depend on the flexibility of the protein and effectively the interaction of a number of different sites. The probability of spontaneous desorption is therefore small.

In a true equilibrium, the concentrations of protein which are bound to the interface and which are free in the bulk solution are linked. Hence, when the concentration of protein in the bulk phase is reduced by dilution, then protein should desorb from the interface to compensate for this change: this is of course the essence of a binding isotherm. High affinity binding isotherms such as are typical for many disordered proteins, make it necessary to dilute the protein in the bulk phase to extremely low values before desorption effects are likely to be observed. Even then, because of the number of points of attachment, and the low probability of them all being desorbed at the one time, desorption will be slow. Thus, binding may appear to be irreversible when it is in fact kinetically slow. In practical terms, it may take days or longer for the systems to approach equilibrium. Since emulsions are formed very rapidly if they are to be formed at all, it is probable that true equilibrium is not established at the time of their formation: this will be discussed later. The kinetics of desorption appear to depend on the surface pressure (MacRitchie, 1985): the higher the pressure, the more rapid the desorption, and the larger the protein, the slower the desorption.

The situation is rendered more complex by the fact that the proteins can change conformation and spread on the surface after the initial adsorption. Thus, a further tendency towards irreversibility becomes apparent, since the free energy change for the initial binding event is only part of the overall change. Therefore, even the adsorption of globular proteins, which show more apparently Langmuir-type binding isotherms, can be seen to be a multi-stage process. Depending on the amount of protein available, therefore, a number of interfacial events may be possible when an emulsion is formed. When the concentration of protein is low, the protein will be forced to spread to stabilize the maximum possible interfacial surface area. This however may take time, and is in some cases unlikely to occur in the time scale of the passage of a material through a homogenizer (Pearce and Kinsella, 1978), and only a poor emulsion will be formed. It has been shown that the complete adjustment of protein on the interface, to form a steady-state system, may take some hours (Graham and Phillips, 1979*a*).

As the concentration of protein in the system is increased, there may be less necessity for the protein to spread and to denature extensively after binding to the interface, since there will, at least in the time scale of homogenization, be sufficient protein to cover a larger surface. The sizes of the emulsion particles may be smaller (Pearce and Kinsella, 1978), with the formation of more interface, but this is not true for all systems (Oortwijn and Walstra, 1979). In any case, there comes a limit where further increases in the concentration of the protein will cause no further decrease in the particle size of the emulsion, because of mechanical limitations of the homogenizer. Above this, the protein load may increase without the particle size changing, although the protein load will generally stabilize at this point. Protein loads will of course depend to some extent on the particular protein which is used: this is evident from spread film results (Graham and Phillips, 1979*b*) and from actual emulsions (Tsai *et al.*, 1972; Satterlee and Free, 1973; Pearce and Kinsella, 1978; Tornberg, 1978*a*; Oortwijn and Walstra, 1979) but the effects are not strikingly large. The conditions of homogenization are also important (Tornberg and Hermansson, 1977; Tornberg, 1978*a,b*; Tornberg and Lundh, 1978). Some desorption can occur after the first emulsion-forming step (van Dulm and Norde, 1983). The protein load for monolayer coverage of an interface appears to be generally of the order of a few milligrams per square metre of interface. β-Casein, for example, gives a surface loading of the order of 2–3 $mg\,m^{-2}$ (Graham and Phillips, 1979*b*; Dickinson *et al.*, 1983), and lysozyme a value of about 2 $mg\,m^{-2}$ (Graham and Phillips, 1979*b*), although this latter is complicated by the fact that the adsorption isotherm does not possess a plateau. Whey proteins (Oortwijn and Walstra, 1979) gave values of 0·5–2·5 $mg\,m^{-2}$, and caseins a value of about 2 $mg\,m^{-2}$ (Dalgleish and Robson, 1985) or about 3 $mg\,m^{-2}$ (Dickinson *et al.*, 1988).

At even higher concentrations of protein, it is possible to form multilayers. For stabilization of an emulsion, only monolayers are essential, and it has been demonstrated that an equilibrium may be established between adsorbed protein and protein which is free in the bulk solution. In monolayers, the protein layer bound to the interface is one molecule thick. Multilayer formation involves the coverage of the interface with a layer of protein which is more than one molecule in thickness. It is possible to conceive of a separate equilibrium between protein in the bulk phase and protein which is bound, not to the interface itself, but to the adsorbed interfacial protein. The formation of multilayers will depend largely on the nature of the protein: some proteins (e.g. β-casein) have a strong tendency to aggregate, and, when there is excess protein in solution above that required for stabilization of the emulsion, multilayers are formed, as evidenced by increased binding (Graham and Phillips, 1979*b*). The formation of multilayers can only occur when the bulk concentrations of protein are high, and the monolayer coverage can be simply achieved in most cases by dilution of the emulsion, since the tendency of the protein to aggregate is less strong than its tendency to adsorb to the interface. The equilibrium leading to formation of multilayers will be more easily disrupted than will be the binding to the interface itself.

In some cases the protein may adsorb to the interface as an oligomeric or polymeric unit, because of the nature of the system. It is known that sodium caseinate (which contains a mixture of the four major casein proteins) has a strong tendency to self-aggregation, and also that it is highly surface active. We have no direct evidence as to the exact composition of the interface at the moment of homogenization, but it is probable that the protein adsorbs as an aggregated mixture (Robson and Dalgleish, 1987*b*). Subsequently the mixture of caseins on the interface rearranges, since, when emulsions made from oil and sodium caseinate are stored, the concentration of β-casein in solution is found to decrease while the α_{s1}-casein concentration in solution increases. This process occurs over a period of several hours, and the indications are that this represents a displacement of one casein by another either on the interface itself or in a multilayer of aggregated protein.

A particularly specialized, but none the less important, form of multilayer binding is the stabilization of oil–water interfaces by large protein particles, specifically in homogenized milk. In this material, some of the stabilizing agencies which are adsorbed at the interface are semi-intact casein micelles (Oortwijn *et al.*, 1977; Oortwijn and Walstra, 1979; Oortwijn and Walstra, 1982; Walstra and Oortwijn, 1982), which are particles containing caseins bound together with calcium phosphate, and stabilized by a surface excess concentration of the protein κ-casein (Schmidt, 1982). Electron microscopy of these particles has shown that there are apparently intact casein micelles at the fat–water interface, although there are large gaps on the interface, between these bound micelles,which must presumably be filled with much smaller units of protein, invisible to the electron microscope. Also, smaller casein micellar fragments are present, resulting perhaps from 'spreading' of the micelles (Oortwijn *et al.*, 1977). The casein micelles which are apparently bound to the interface as intact entities, must, however either be the major fragments of original micelles which have been disrupted by passage through the homogenization valve, or else particles which have undergone conformational change, since the surfaces of intact native casein micelles are acknowledged to be fairly hydrophobic and probably sterically stabilized (see above), and would therefore not interact with the fat/water interface. The protein loads of the fat globules in homogenized milk are very much higher than those quoted above for the monolayer coverage of the fat surface. Values from ~17 to ~40 mg m^{-2} have been found for particles of different sizes and densities (Dalgleish and Robson, 1985), the latter being the limit for close packing of casein micelles on the surface (Oortwijn and Walstra, 1979). The range of values has been shown to extend down to about 10 mg m^{-2} and demonstrates the different extents of coverage of the droplet surfaces by casein micelles and smaller units. Whey proteins do not appear to be very important when casein is present, but it is obvious that homogenized milk contains a range of particles with widely differing protein/fat ratios (Oortwijn and Walstra, 1979; Dalgleish and Robson, 1985).

It would not be unexpected to find that the properties of these different particles, in which the protein load is distributed in different ways, differed, but there is little information regarding the properties of the different emulsions. A limited amount is known to suggest that the coverage of the fat droplets by casein micelles does affect their coagulation properties (Robson and Dalgleish, 1987*a*), but further research is necessary to investigate this in detail.

EXCHANGE OF PROTEINS AT THE INTERFACE

It has been described above that, although the adsorption of proteins at an oil/water interface may appear to be irreversible, more detailed investigation shows this is unlikely to be true. In such a case, therefore, it is possible, at least conceptually, to define individual affinity constants for the adsorption of the different surface-active proteins at the interface. In many real food emulsions, the emulsion contains not a single type of protein but several, and this will have an effect on the composition of the interfacial layer. When the emulsion is formed in the homogenizer, proteins from the mixed solution will bind to the interface in what is essentially a kinetically controlled process (Pearce and Kinsella, 1978). This is because, if an emulsion is to be successfully formed, the surfaces of the oil droplets must be rapidly coated with surfactant to prevent coalescence. It is therefore more important that the surface be coated with any available protein than that selective adsorption should occur. If the initial adsorption is slow, then no emulsion can be formed, since proteins do not appear to be able to act as detergents and to disperse the fat from large particles: the sizes of the particles of an emulsion are determined largely by the conditions in, and the design of, the homogenizer valve (Tsai *et al.*, 1972; Satterlee and Free, 1973; Tornberg and Hermansson, 1977; Pearce and Kinsella, 1978; Tornberg, 1978*a,b*; Tornberg and Lundh, 1978; Oortwijn and Walstra, 1979).

In the emulsion just after formation the initial composition of the interfacial protein will reflect mainly the number concentrations of the protein particles (rather than molecules, since the protein may be in an aggregated state) which were originally in the aqueous solution, and to a smaller extent the diffusion coefficients of the molecules. This has been demonstrated to be the case in emulsions made using homopolymers: the smaller polymer units, being more numerous, bind first, so that the freshly-formed emulsion has a surface layer which is composed largely of the small units (Felter and Ray, 1970; Bain *et al.*, 1982; Furusawa *et al.*, 1982). However, thermodynamics favours the adsorption of the larger molecules, and the composition of the interfacial layer alters, over lengthy periods of time (Bain *et al.*, 1982), as smaller molecules on the interface exchange with larger ones from solution. The conformation of a protein adsorbed to an interface can change with time: we can now see that the composition of the interfacial layer may also be subject to change in mixed protein systems. Since the exchange is slow (i.e. time scales are at least of the order of minutes and are often much longer), it is to be expected that the properties of an emulsion (e.g. its stability) will be time-dependent, because of the alteration in the composition and structure of the interfacial material.

The polymer molecules just described differ in their affinity towards the interface because of their different sizes. With proteins, other possibilities arise, because of their unique primary structures, and hence unique adsorption properties. Once the initial emulsion has been successfully formed, optimization of the surface layer will occur by interchange of proteins between free and adsorbed states, according to their surface activities. Even identical proteins appear to exchange in this way (Brash and Samak, 1978), although these studies involve radiolabelled proteins which may have slightly different properties from the unlabelled ones (Lensen *et al.*, 1984). Thus, if there are appreciable amounts of protein in the bulk phase, thermodynamic considerations will favour the displacement of a protein of lower surface activity by a higher-affinity protein. If the bulk phase contains only low concentrations of proteins, then they will virtually all adsorb to the interfaces as the emulsion is formed and no exchange will be possible because of the need to maintain surface coverage. The exchange of proteins between interface and solution has been observed in a number of cases where the emulsions have been made with mixed proteins, in particular combinations of individual caseins, or casein/gelatin mixtures.

The exchange reactions involving caseins have been studied in systems of varying degrees of complexity. The most simple experiments involved the formation of an emulsion from oil and a solution containing only purified α_{s1}-casein. When the oil of the emulsion phase was removed from the initial aqueous phase and resuspended in a solution of β-casein, then it was apparent that the concentration of β-casein in the solution decreased and the concentration of α_{s1}-casein increased over a period of 24 h (Dickinson *et al.*, 1987). In the initial experiments, the β-casein did not completely replace the α_{s1}-casein, but later studies have shown that stoichiometric exchange is possible (Dickinson *et al.*, 1988). Conversely, it was also demonstrated that α_{s1}-casein displaced β-casein only with difficulty from the emulsion surface (Dickinson *et al.*, 1987). β-Casein is of course more surface active, with a higher binding constant than α_{s1}-casein. Therefore this displacement of the latter by the former is in accord with thermodynamic principles. In a more complex emulsion where whole sodium caseinate (containing α_{s1}-, α_{s2}-, β- and κ-caseins) was used as the emulsifier, it was shown that the composition of the interfacial protein changed with time as the emulsion was aged after formation, with the interface being progressively enriched, over a period of hours, in

β-casein at the expense of the other caseins (Robson and Dalgleish, 1987*b*). This demonstrates also the non-selectivity which occurs in the adsorption during homogenization: all of the casein types were initially adsorbed.

This time-dependent exchange reaction involving β-casein was, not surprisingly, enhanced if the aqueous phase was enriched with β-casein once the emulsion had been formed, although, in contrast to the experiments involving individual purified caseins, it was not found possible to displace all of the α_{s1}-casein from the oil/water interface when the emulsion was initially formed using sodium caseinate. Also in contrast to the more simple mixtures was the observation that, when the aqeuous phase of the sodium caseinate emulsion was enriched with α_{s1}-casein, there was no obvious displacement of β-casein, but simply more α_{s1}-casein was adsorbed to the interfaces. This may reflect the fact that β-casein does not form cohesive surface films, so that the interface containing β-casein is perhaps readily rearranged to admit more of the other proteins when the latter are in large excess. Alternatively, the added α_{s1}-casein may be involved in multilayer formation.

It has also been demonstrated that the adsorbed protein in homogenized milk contains *para*-κ-casein but no κ-casein (McPherson *et al.*, 1984). This in principle may arise from proteolysis of adsorbed κ-casein, but other results suggest that *para*-κ-casein is preferentially adsorbed (Itoh and Nakanishi, 1974). Such an interpretation was also suggested by studies of κ-casein stabilized emulsions (Dickinson *et al.*, 1987). Treatment of these with rennet caused changes in the surface only if κ-casein was present in the solution, and it was concluded that κ-casein was split by rennet only in solution and not when adsorbed. The *para*-κ-casein produced by the proteolysis in solution could then displace the adsorbed κ-casein.

It has been established for some time that caseins are preferentially adsorbed at a planar air/water interface from mixed solutions of gelatin and casein (Musselwhite, 1966). Measurements of surface properties have demonstrated that at short times after formation of the interface the adsorption of the protein is diffusion controlled and that, for example, the surface pressure depends only on the total concentration of protein and not its composition. Within a few minutes, the adsorbed proteins start to rearrange and to exchange with those in solution, and finally at long times it is apparent that the caseins have completely replaced gelatin at the interface (individual casein exchange reactions were not studied) (Dickinson *et al.*, 1985*b*; Castle *et al.*, 1986). In emulsions, rather than spread films, the same process seems to apply, with casein being able to totally displace the gelatin (Chesworth *et al.*, 1985). However, this displacement of gelatin by casein is not so simple as the displacement of one casein by another, since the surface rheology of the gelatin at the interface is important in the exchange (Dickinson *et al.*, 1985*a*). Gelatin forms stronger films at the interface than do caseins, and also the strength of the films is time-dependent. Thus, the replacement of gelatin by casein in aged emulsions is slower and less complete than in freshly prepared systems (Castle *et al.*, 1986). The composition of the interfacial layer will reflect the processing history of the sample.

More complex systems still exist in homogenized milks. In these systems, the caseins are present, but in the form of highly-aggregated protein particles (the casein micelles): also present are the whey proteins β-lactoglobulin, α-lactalbumin and serum albumin. These latter are all less surface active than are caseins. It is possible that, in the production of homogenized milk, the whey proteins are initially adsorbed because of (number) concentration effects, as is apparently the case in homogenized cream (Darling and Butcher, 1978; Darling, 1982). However, after only a short time, it is likely that only the caseins (as fragments of casein micelles) remain on the surfaces of the fat globules (Oortwijn and Walstra, 1979, 1982). A similar displacement of protein by caseins is found when soy protein isolate is competed with casein and displaced (Aoki *et al.*, 1984). The whey proteins themselves, although less surface active than caseins, are nevertheless more surface active than some other proteins, for example soy protein isolate (Waniska *et al.*, 1981; Tornberg *et al.*, 1982). They also give very viscous adsorbed films, especially β-lactoglobulin, which has the capacity to form disulphide linked polymers on the interface, especially if binding of the protein to the interface is accompanied by denaturation of the protein. The whey proteins also compete against one another in their natural mixtures (Yamauchi *et al.*, 1980; Shimizu *et al.*, 1981). The relative adsorptions of the different proteins are pH-dependent, with α-lactalbumin being more strongly adsorbed at low pH and β-lactoglobulin at high pH.

THE EFFECT OF pH AND OTHER FACTORS ON PROTEIN-BASED EMULSIONS

One of the most important aspects of proteins is that their behaviour is pH-dependent. This is a result of the numbers of differently charged residues which they contain, and which have different p*K* values. Regrettably, this aspect of emulsion studies, although widely recognized, has received less than the attention which it deserves. Particularly, each protein has an isoelectric point, where the contributions from positive and negative charges cancel out to give the molecule no net charge. At this pH, proteins tend to coagulate, and therefore it is to be expected that surface rheological parameters of proteins close to their isoelectric point will be maximal (Halling, 1981). It is also apparent that, at the pI, charge-based contributions to repulsion will be minimal, and also that steric stabilization will also be minimized because the proteins will be in their least expanded state. Nevertheless, it has been found that emulsion stability in some systems is greatest at the pI: this has been demonstrated for a number of proteins such as gelatin (Nielsen *et al.*, 1958), bovine serum albumin (Biswas and Haydon, 1962), pepsin (Biswas and Haydon, 1962) and soluble muscle protein (Swift and Sulzbacher, 1963). The explanation for this probably lies in the greater surface coverage at the pI, with the compacted protein structures. This, together with the tendency of the protein to coagulate at the pI gives cohesive films, with their enhanced stabilizing action. Some proteins have been shown to give less stable emulsions at the pI: low concentrations (0·004%) of either bovine serum albumin or of lysozyme are unstable, but this may be in part because of low surface coverage. Likewise, there is no simple explanation for the behaviour of milk fat/whey protein emulsions which are highly unstable at pH 4·5–5, close to the isoelectric points (De Wit *et al.*, 1976). It is probable that the stability of emulsions at the isoelectric point is dependent on protein concentration and the volume and surface area of the oil phase as well as simply the pH.

At other pH values, proteins may show a distinct dependence on pH. Bovine serum albumin shows increasing emulsifying activity as the pH is increased between 4 and 9, and then decreases sharply as the protein changes conformation (Aoki *et al.*, 1974; Pearce and Kinsella, 1978; Waniska *et al.*, 1981). In the range of pH 3–8, β-lactoglobulin does not, according to one study (Waniska *et al.*, 1981), change in its emulsifying capacity, although it too undergoes conformational transitions. However, another investigation (Shimizu *et al.*, 1984) shows a strong dependence of the emulsifying power with pH, increasing from pH 3–9. It has already been pointed out that the whey proteins show specific adsorption which is pH-dependent (Yamauchi *et al.*, 1980; Shimizu *et al.*, 1981), and it is probable that part of the change seen in the properties of β-lactoglobulin arises from this cause.

The hydrophobicities of the different whey proteins vary with pH, but all of the proteins behave similarly, in that their surfaces become less hydrophobic as the pH is increased (Shimizu *et al.*, 1981).

Proteins are also susceptible to changes in the ionic composition of the solution, either because of general effects of ionic strength or, more rarely, because of specific effects of particular ions. Increasing the ionic strength diminishes the charge-based interactions between proteins and will produce the same type of effects as changing the pH towards the isoelectric point. A number of demonstrations of this is available (Mita *et al.*, 1974; McWatters and Holmes, 1979). Perhaps the most obvious specific effect of ions once again involves caseins. Emulsions stabilized by individual caseins are destabilized by the presence of relatively low concentrations of calcium ions (Dickinson *et al.*, 1987), and the effect is even more pronounced when sodium caseinate emulsions are suspended in alcoholic solutions to produce the so-called cream liqueurs. An important source of instability in these is the presence of small quantities of calcium ions, and the removal of these by complexing with citrate has a great effect upon the stability (Banks *et al.*, 1981).

MODIFIED PROTEINS IN EMULSIONS

Since the behaviour of an individual protein at an interface is defined largely by the chemical nature of its side chains, which determine not only the conformation of the protein but its tendency to bind to an oil/water interface, it is almost axiomatic that the emulsifying properties of proteins will be modified by either deliberate or inadvertent chemical modification. The latter may occur as a side-effect of processing. By no means all changes are beneficial: for example the glycosylation of β-lactoglobulin reduces its rate of adsorption and, importantly for this protein, alters the rate of rearrangement at the oil/water interface in spread films (Kinsella, 1982). However, the modification appeared to improve emulsion formation and stability (Waniska *et al.*, 1981). The effect of glycosylation is to render the protein more hydrophilic, so that the enhancement of emulsion formation may arise from the increased stabilization of the aqueous part of the interface, and the improved stability from the increased hydration and steric stabilization of the interface. The effects, however, were relatively small. On the other hand the increased solubility that can be provided via modification can lead to enhanced emulsifying capacity. The detailed effect of glycosylation depends on the nature and number of the substituents introduced.

Succinylation of proteins causes the replacement of positively charged lysyl side chains by negatively charged carboxyl groups. Some plant proteins show considerable improvements in their emulsifying capacity as a result of the modification: part of this appears to be simply caused by improved solubility of the proteins (Franzen and Kinsella, 1976*b*). Similarly, soy proteins show an increasing emulsifying activity as a result of progressively increasing succinylation (Franzen and Kinsella, 1976*a*), as do fish proteins (Groninger, 1973). In addition to increasing emulsifying activity, the modification enhances the stability of the emulsion, presumably as a result of the increased repulsion between emulsion droplets. Bovine serum albumin when succinylated undergoes a molecular expansion, and increases in net negative charge (Kinsella, 1982). Similar effects are caused by succinylation of heat-denatured mixtures of whey proteins (Thompson and Reyes, 1980). Parallel with this is an increase in emulsifying activity at pH values above 5. It has been suggested that by this modification it is possible to form a protein carrying too much negative charge, so that a cohesive interfacial film cannot be formed because of repulsions between the individual protein molecules (Kinsella, 1982): a similar phenomenon has been identified in modified soy glycinins (Kim and Kinsella, 1986), so that an optimal level of modification can be defined. It is demonstrable, however, that it is not simply the charge which is responsible: methylation of lysines in bovine serum albumin, although it increased the net negative charge by removing the positive charge on the lysyl side chains, did not alter the emulsifying activity to a significant extent. Similarly, acetylation does not affect the emulsifying properties of the proteins to nearly the same extent as does succinylation (Franzen and Kinsella, 1976*a,b*).

It has been pointed out that disulphide bonds are important in determining the emulsifying capacity of a protein since they are important in maintaining the conformation of the protein, and may affect intermolecular complex formation. Bovine serum albumin shows reduced emulsifying capacity when all disulphides are reduced (Waniska *et al.*, 1981), and it is probable that the diminished emulsifying power of denatured β-lactoglobulin (Kato *et al.*, 1983) arises at least in part from an alteration in the disulphide bonding, although reduction of the bonds by mercaptoethanol enhances the emulsification ability at low pH (Shimizu *et al.*, 1984). κ-Casein has been shown to be a better emulsifier when the intermolecular disulphides are reduced (Dickinson *et al.*, 1987), and the same protein penetrates a lipid monolayer more readily when the disulphides are similarly broken (Griffin *et al.*, 1984). Formation of intermolecular disulphides between κ-casein and β-lactoglobulin by heating improves the functional properties of milk powder (Morr, 1975), although only limited information is available on possible interfacial effects of such mixtures or complexes (Murray, 1987). It is clearly not possible to predict the effect of disulphide breakage, since both advantageous and disadvantageous cases are known. The number of such bonds in the molecule may be a critical factor, since a protein with many disulphides as soy glycinin is a poor emulsifier until these bonds are at least partially reduced (German *et al.*, 1985).

A more drastic means of modification compared with those just described is the treatment of the proteins with proteases: in some cases this leads to an increase in the emulsifying activity, although the stability of the emulsion may be impaired because the cohesiveness of the surface film is generally reduced by the shortness of the peptides. Whey protein can be hydrolysed by pronase, prolase or pepsin and gives decreased emulsifying capacity (although foaming capacity is improved) (Kuehler and Stine, 1974). More specific proteolysis by trypsin or chymotrypsin forms peptides which are less surface active than the original whey proteins, but more active than the products of non-specific proteolysis (Jost and Monti, 1982). A number of other proteases can be used to solubilize the heat-denatured whey protein (Monti and Jost, 1978). This last effect is in effect improving the emulsifying capacity by increasing solubility: this is also the case for soy protein isolates (Puski, 1975), where emulsification capacity is increased, although stabilities are decreased (Monti and Jost, 1978). A related study on tryptic digests of glycinin (Kamata *et al.*, 1984) shows essentially the same, although the extent of digestion is important (Ochiai *et al.*, 1982), with some early digests being more surface active than the original protein.

Specific proteolysis of α_{s1}-casein to remove the N-terminal 23 residues results in a protein which has reduced adsorption to the oil/water interface but the removal of the C-terminal 54 residues does not affect the behaviour (Shimizu *et al.*, 1983). In α_{s1}-casein the N-terminal region is hydrophobic, and presumably is responsible for the binding of the intact protein.

A further type of enzymatic modification is the plastein type of reaction, where the protein is 'reformed' following proteolysis. A reaction of this type has been used to prepare proteinaceous surfactants incorporating the leucine *n*-dodecyl ester, which have considerable surfactant ability (Shimada *et al.*, 1982; Watanabe *et al.*, 1982; Arai *et al.*, 1984). These preparations are derived from gelatin and from succinylated fish protein, and give emulsions which compare favourably with those prepared using the Tween-80 detergent. Suggestions have been made that the proteinaceous emulsifiers are capable of binding a great deal of water: stabilization of the emulsions may therefore be by a steric mechanism rather than charge stabilization, since the addition of high concentrations of salt did not cause coagulation of the emulsion. A different polymerization reaction of proteins, namely transglutamination, has been shown to be effective in polymerizing α_{s1}- and κ-caseins and soybean globulins (Motoki *et al.*, 1984). The modified casein is a less effective emulsifier than the native protein, although the modified proteins bind considerable amounts of water.

A final type of enzymatic modification does not affect the protein part of the system: this is the treatment of egg yolk by phospholipase to provide a better emulsifier than the native material (Dutilh and Groger, 1981). The fermentation of the egg lecithin affects the whole of the emulsification system.

CONCLUSION

It will have become obvious from the foregoing descriptions that the properties of proteins in emulsions are not easy to predict (Kinsella and Whitehead, 1987). It has been pointed out (Dickinson, 1986) that the detailed aspects of the stabilization of emulsions by proteins have not been sufficiently precisely defined, and that the theories of polyelectrolyte adsorption and emulsion stabilization are ill-equipped to deal with such complex structures as proteins. To some extent, therefore, the study of emulsions stabilized by proteins depends heavily on qualitative estimates, rather than any exact correspondence with particular theoretical models. Especially, the effects of pH and ionic strength in introducing structural changes in the proteins themselves, require to be more extensively studied. Two points require to be made: first, it should have become apparent that a preponderance of the experiments to study proteins on interfaces do not involve the formation of emulsions, but rely upon spread film and other techniques. Leading from this, it must be acknowledged that the means of looking at proteins on interfaces, especially in emulsions rather than spread films, are severely limited. For example, electron microscopy may be effective in studying the particles in such materials as homogenized milk (Oortwijn *et al.*, 1977), but can offer little information on the conformation of individual proteins at the interface. Thus, the development of new techniques which can be applied to the molecules on the surfaces of emulsions is to be awaited.

It has also become apparent that the structure and the behaviour of an emulsion are not fixed at the moment of homogenization. We have seen that proteins may exchange on the interface, and that this may take time. In addition, the spreading and denaturation of the protein after the preliminary adsorption are time-dependent phenomena. Further information on the time-scales of such processes and the factors influencing them is required before a picture can be constructed of the behaviour of complex emulsions. The description of competitive phenomena in an earlier section concerned protein exchange reactions only: while these are obviously important, there are also significant reactions where proteins are displaced by non-protein surfactant molecules. The equilibrium and dynamic properties of such systems are poorly understood except in the most qualitative way, and extensive research effort if required to define these effects in totality.

REFERENCES

Aoki, K., Murata, M. and Hiramatus, K. (1974). *Analyt. Biochem.*, **59,** 146.
Aoki, H., Shirase, Y., Kato, J. and Watanabe, Y. (1984). *J. Food Sci.*, **49,** 212.
Arai, S., Watanabe, M. and Fujii, N. (1984). *Agric. Biol. Chem.*, **48,** 1861.
Bain, D. R., Cafe, M. L., Robb, I. D. and Williams, P. A. (1982). *J. Coll. Int. Sci.*, **88,** 467.
Banks, W., Muir, D. D. and Wilson, A. G. (1981). *J. Food Technol.*, **16,** 587.
Bigelow, C. C. (1967). *J. Theor. Biol.*, **16,** 187.
Biswas, B. and Haydon, D. A. (1962). *Kolloid Z.*, **185,** 31.
Biswas, B. and Haydon, D. A. (1963*a*). *Proc. Roy. Soc. London,* **A271,** 296.
Biswas, B. and Haydon, D. A. (1963*b*). *Proc. Roy. Soc. London,* **A271,** 317.
Bonekamp, B. C., Van Der Schee, H. A. and Lyklema, J. (1983). *Croat. Chem. Acta,* **56,** 695.
Brash, J. L. and Samak, Q. M. (1978). *J. Coll. Int. Sci.*, **65,** 495.

Brooks, D. E. and Greig, R. G. (1981). *J. Coll. Int. Sci.*, **83,** 661.
Castle, J., Dickinson, E., Murray, A., Murray, B. S. and Stainsby, G. (1986). In: *Gums and Stabilisers for the Food Industry*. G. O. Phillips, D. J. Wedlock and P. A. Williams (Eds). Elsevier Applied Science, vol. 3, p. 409.
Chesworth, S. M., Dickinson, E. and Stainsby, G. (1985). *Lebensm.-Wiss. Technol.*, **18,** 230.
Cohen Stuart, M. A., Scheutjens, J. M. H. M. and Fleer, G. J. (1980). *J. Polym. Sci., Polym. Phys. Ed.*, **18,** 559.
Cornette, J. L., Cease, K. B., Margalit, H., Spouge, J. L., Berzofsky, J. A. and DeLisi, C. (1987). *J. Mol. Biol.*, **195,** 659.
Creamer, L. K., Richardson, T. and Parry, D. A. D. (1981). *Arch. Biochem. Biophys.*, **211,** 689.
Dalgleish, D. G. and Robson, E. W. (1985). *J. Dairy Res.*, **52,** 539.
Dalgleish, D. G., Dickinson, E. and Whyman, R. H. (1985). *J. Coll. Int. Sci.*, **108,** 174.
Darling, D. F. (1982). *J. Dairy Res.*, **49,** 695.
Darling, D. F. and Butcher, D. W. (1978). *J. Dairy Res.*, **45,** 197.
De Wit, J. N. and Klarenbeek, G. (1981). *J. Dairy Res.*, **48,** 293.
De Wit, J. N., Klarenbeek, G. and Swinkels, G. A. M. (1976). *Zuivelzicht*, **68,** 442.
Derjaguin, B. V. and Landau, L. (1941). *Acta Physichochim. URSS*, **14,** 633.
Dickinson, E. (1986). *Food Hydrocolloids*, **1,** 3.
Dickinson, E. and Stainsby, G. (1982). *Colloids in Food*, Applied Science Publishers, London.
Dickinson, E., Robson, E. W. and Stainsby, G. (1983). *J. Chem. Soc. Farad. Trans. 1*, **79,** 2937.
Dickinson, E., Murray, B. S. and Stainsby, G. (1985*a*). *J. Coll. Int. Sci.*, **106,** 259.
Dickinson, E., Pogson, D. J., Robson, E. W. and Stainsby, G. (1985*b*). *Colloids Surf.*, **14,** 135.
Dickinson, E., Whyman, R. H. and Dalgleish, D. G. (1987). In: *Food Emulsions and Foams*, E. Dickinson (Ed.), RSC Special Publication No. 58, p. 40.
Dickinson, E., Rolfe, S. and Dalgleish, D. G. (1988). *Food Hydrocolloids*, **2,** 397.
Dutilh, C. E. and Groger, W. (1981). *J. Sci. Food Agr.*, **32,** 451.
Evans, M. T. A., Phillips, M. C. and Jones, M. N. (1979). *Biopolymers*, **18,** 1123.
Felter, R. E. and Ray, L. N. (1970). *J. Coll. Int. Sci.*, **32,** 349.
Fiat, A.-M., Jollès, J., Aubert, J.-P., Loucheux-Lefebvre, M.-H. and Jollès, P. (1980). *Eur. J. Biochem.*, **111,** 333.
Fischer, E. W. (1958). *Kolloid-Z.*, **160,** 120.
Franzen, K. L. and Kinsella, J. E. (1976*a*). *J. Agric. Food Chem.*, **24,** 788.
Franzen, K. L. and Kinsella, J. E. (1976*b*). *J. Agric. Food Chem.*, **24,** 914.
Furusawa, K., Yamashita, K. and Konno, K. (1982). *J. Coll. Int. Sci.*, **86,** 35.
Garibaldi, J. A., Donovan, J. W., Davis, J. G. and Cimino, S. L. (1968). *J. Food Sci.*, **33,** 514.

Gaylord, R. J. (1982). *J. Coll. Int. Sci.*, **87,** 577.
German, J. B., O'Neill, T. E. and Kinsella, J. E. (1985). *J. Amer. Oil Chem. Soc.*, **62,** 1358.
Graham, D. E. and Phillips, M. C. (1976). In: *Theory and Practice of Emulsion Technology,* A. L. Smith (Ed.), Academic Press, London, p. 75.
Graham, D. E. and Phillips, M. C. (1979*a*). *J. Coll. Int. Sci.*, **70,** 403.
Graham, D. E. and Phillips, M. C. (1979*b*). *J. Coll. Int. Sci.*, **70,** 415.
Graham, D. E. and Phillips, M. C. (1979*c*). *J. Coll. Int. Sci.*, **70,** 427.
Graham, E. R. B., Malcolm, G. N. and McKenzie, H. A. (1984). *Int. J. Biol. Macromol.*, **6,** 155.
Griffin, M. C. A., Infante, R. B. and Klein, R. A. (1984). *Chem. Phys. Lipids,* **36,** 91.
Groninger, H. S. (1973). *J. Agric. Food Chem.*, **21,** 978.
Grosclaude, F., Mahé, M.-F., Mercier, J.-C. and Ribadeau Dumas, B. (1972). *Eur. J. Biochem.*, **26,** 328.
Halling, P. J. (1981). *CRC Crit. Rev. Food Sci. Nutr.*, **15,** 155.
Hamaker, H. C. (1937). *Physica,* **4,** 1058.
Hesselink, F. Th. (1969). *J. Phys. Chem.*, **73,** 3488.
Hesselink, F. Th. (1977). *J. Coll. Int. Sci.*, **60,** 448.
Holt, C. (1975). *Proc. Int. Conf. Coll. Surf. Sci.*, E. Wolfram (Ed.), Akademiai Kiado, Budapest, p. 641.
Holt, C. and Dalgleish, D. G. (1986). *J. Coll. Int. Sci.*, **114,** 513.
Holt, C. and Sawyer, L. (1988). *Protein Eng.*, **2,** 251.
Holt, C., Hasnain, S. S. and Hukins, D. W. L. (1982). *Biochim. Biophys. Acta,* **719,** 299.
Horne, D. S. (1984). *Biopolymers,* **23,** 989.
Horne, D. S. and Dalgleish, D. G. (1980). *Int. J. Biol. Macromol.*, **2,** 154.
Hunter, R. J. (1987). *Foundations of Colloid Science,* Oxford University Press.
Imoto, T., Johnson, L. N., North, A. C. T., Phillips, D. C. and Rupley, J. A. (1972). In: *The Enzymes,* P. D. Boyer (Ed.), 3rd Edition, Vol. 7, Academic Press, p. 666.
Israelachvili, J. N. and Tabor, D. (1972). *Proc. Roy. Soc. London,* **A311,** 19.
Itoh, T. and Nakanishi, T. (1974). *J. Agric. Chem. Soc. Japan,* **48,** 239.
Jost, R. and Monti, J. C. (1982). *Lait,* **62,** 521.
Kamata, Y., Ochiai, K. and Yamauchi, F. (1984). *Agric. Biol. Chem.*, **48,** 1147.
Kato, A. and Nakai, S. (1980). *Biochim. Biophys. Acta,* **624,** 13.
Kato, A., Tsutsui, N., Matsudomi, N., Kobayashi, K. and Nakai, S. (1981). *Agric. Biol. Chem.*, **45,** 2755.
Kato, A., Osako, Y., Matsudomi, K. and Kobayashi, K. (1983). *Agric. Biol. Chem.*, **47,** 33.
Keshavarz, E. and Nakai, S. (1979). *Biochim. Biophys. Acta,* **576,** 269.
Kim, S. H. and Kinsella, J. E. (1986). *J. Agric. Food Chem.*, **34,** 623.
Kinsella, J. E. (1982). *ACS Symp. Ser.*, **206,** 301.
Kinsella, J. E. and Whitehead, D. M. (1987). *ACS Symp. Ser.*, **343,** 629.
Kiosseoglou, V. D. and Sherman, P. (1983*a*). *Coll. Polym. Sci.*, **261,** 502.
Kiosseoglou, V. D. and Sherman, P. (1983*b*). *Coll. Polym. Sci.*, **261,** 520.
Koopel, L. K. (1981). *J. Coll. Int. Sci.*, **83,** 116.

Kuehler, C. A. and Stine, C. M. (1974). *J. Food Sci.*, **39,** 379.
Langmuir, I. (1938). *J. Chem. Phys.*, **6,** 893.
Lensen, H. G. W., Bargeman, D., Bergveld, P., Smolders, C. A. and Feijen, J. (1984). *J. Coll. Int. Sci.*, **99,** 1.
Li-Chan, E., Nakai, S. and Wood, D. F. (1984). *J. Food Sci.*, **49,** 345.
Lyklema, J. (1984). *Colloids Surf.*, **10,** 33.
Meier, D. J. (1967). *J. Phys. Chem.*, **71,** 1861.
Mercier, J.-C., Grosclaude, F. and Ribadeau Dumas, B. (1972). *Milchwiss.*, **27,** 402.
Mita, T., Iguchi, E., Yamada, K., Matsumoto, S. and Yonezawa, D. (1974). *J. Texture Stud.*, **5,** 89.
Mizutani, R. and Nakamura, R. (1985). *Lebensm.-Wiss. Technol.*, **18,** 60.
Monti, J. C. and Jost, R. (1978). *J. Dairy Sci.*, **61,** 1233.
Morr, C. (1975). *J. Dairy Sci.*, **58,** 977.
Motoki, M., Nio, N. and Takinami, K. (1984). *Agric. Biol. Chem.*, **48,** 1257.
Mulder, H. and Walstra, P. (1974). *The Milk Fat Globule*; Farnham Royal, Commonwealth Agricultural Bureau.
Murray, E. K. (1987). In *Food Emulsions and Foams,* E. Dickinson (Ed.), RSC Special Publications No. 58, p. 170.
Musselwhite, P. R. (1966). *J. Coll. Int. Sci.*, **21,** 99.
MacRitchie, F. (1978). *Adv. Prot. Chem.*, **32,** 283.
MacRitchie, F. (1985). *J. Coll. Int. Sci.*, **105,** 119.
McPherson, A. V., Dash, M. C. and Kitchen, B. J. (1984). *J. Dairy Res.*, **51,** 289.
McWatters, K. H. and Holmes, M. R. (1979). *J. Food Sci.*, **44,** 770.
Nakai, S. (1983). *J. Agric. Food Chem.*, **31,** 676.
Nielsen, G. E., Wall, A. and Adams, G. (1958). *J. Coll. Int. Sci.*, **13,** 441.
Ochiai, K., Kamata, Y. and Shibasaki, K. (1982). *Agric. Biol. Chem.*, **46,** 91.
Ogden, L. V., Walstra, P. and Morris, H. A. (1976). *J. Dairy Sci.*, **59,** 1727.
Oortwijn, H. and Walstra, P. (1979). *Neth. Milk Dairy J.*, **33,** 134.
Oortwijn, H. and Walstra, P. (1982). *Neth. Milk Dairy J.*, **36,** 279.
Oortwijn, H., Walstra, P. and Mulder, H. (1977). *Neth Milk Dairy J.*, **31,** 134.
Payens, T. A. J. (1978). *Farad. Disc. Chem. Soc.*, **65,** 164.
Pearce, K. N. and Kinsella, J. E. (1978). *J. Agric. Food Chem.*, **26,** 716.
Phillips, M. C. (1981). *Food Technol.*, **35,** 50.
Phillips, M. C., Evans, M. T. A. and Hauser, H. (1975). *ACS Adv. Chem. Ser.*, **144,** 217.
Poole, S., West, S. I. & Walters, C. L. (1984). *J. Sci. Food Agr.*, **35,** 701.
Puski, G. (1975). *Cereal Chem.*, **52,** 655.
Reynolds, T. M. (1965). *Adv. Food Res.*, **14,** 167.
Robson, E. W. and Dalgleish, D. G. (1987*a*). In: *Food Emulsions and Foams,* E. Dickinson (Ed.), RSC Special Publications No. 58, p. 64.
Robson, E. W. and Dalgleish, D. G. (1987*b*). *J. Food Sci.*, **52,** 1694.
Satterlee, L. D. and Free, B. (1973). *J. Food Sci.*, **38,** 306.
Sauter, E. A. and Montoure, J. E. (1972). *J. Food Sci.*, **37,** 918.
Scheutjens, J. M. H. M. and Fleer, G. J. (1980). *J. Phys. Chem.*, **84,** 178.
Schmidt, D. G. (1982). In: *Developments in Dairy Chemistry—1,* P. F. Fox (Ed.), Applied Science Publishers, Ch. 2.

Schut, J. (1976). In: *Food Emulsions,* S. Friberg (Ed.), Marcel Dekker, 1976, Ch. 8.
Sherman, P. (1973). *J. Coll. Int. Sci.,* **45,** 427.
Shimada, A., Yazawa, E. and Arai, S. (1982). *Agric. Biol. Chem.,* **46,** 173.
Shimizu, M., Kamiya, T. and Yamauchi, K. (1981). *Agric. Biol. Chem.,* **45,** 2491.
Shimizu, M., Takahashi, T., Kaminogawa, S. and Yamauchi, K. (1983). *J. Agric. Food Chem.,* **31,** 1214.
Shimizu, M., Saito, M. and Yamauchi, K. (1984). *Agric. Biol. Chem.,* **49,** 189.
Swaisgood, H. E. (1982). In: *Developments in Dairy Chemistry—1,* P. F. Fox (Ed.), Applied Science Publishers, 1982, Ch. 1.
Swift, C. E. and Sulzbacher, W. L. (1963). *Food Technol.,* **17,** 224.
Taborsky, G. (1974). *Adv. Prot. Chem.,* **28,** 1.
Tanford, C. (1980). *The Hydrophobic Effect. Formation of Micelles and Biological Membranes,* John Wiley, London.
Thompson, L. U. and Reyes, E. S. (1980). *J. Dairy Sci.,* **63,** 715.
Tornberg, E. (1978*a*). *J. Sci. Food Agr.,* **29,** 867.
Tornberg, E. (1978*b*). *J. Food Sci.,* **43,** 1559.
Tornberg, E. and Hermansson, A.-M. (1977). *J. Food Sci.,* **42,** 468.
Tornberg, E. and Lundh, G. (1978). *J. Food Sci.,* **43,** 1553.
Tornberg, E., Granfeldt, Y. and Håkansson, C. (1982). *J. Sci. Food Agr.,* **33,** 904.
Tsai, R., Cassens, R. G. and Briskey, E. J. (1972). *J. Food Sci.,* **37,** 286.
Van Dulm, P. and Norde, W. (1983). *J. Coll. Int. Sci.,* **91,** 248.
Vernon-Carter, E. J. and Sherman, P. (1981). *J. Dispersion Sci. Technol.,* **2,** 399.
Verwey, E. J. W. and Overbeek, J. Th. G. (1948). Theory of the Stability of Lyophobic Colloids, Elsevier, North-Holland.
Vincent, B. (1973). *J. Coll. Int. Sci.,* **42,** 270.
Vincent, B. and Whittington, S. (1982). In: *Surface and Colloid Science,* E. Matjevic (Ed.), John Wiley, 1982, Vol. 12, Ch. 1.
Visser, S., Slangen, K. J. and Rollema, H. S. (1986). *Milchwiss.,* **41,** 559.
Voutsinas, P. L., Cheung, E. and Nakai, S. (1983). *J. Food Sci.,* **48,** 26.
Walstra, P. (1979). *J. Dairy Res.,* **46,** 317.
Walstra, P. and Oortwijn, H. (1982). *Neth. Milk Dairy J.,* **36,** 103.
Walstra, P., Bloomfield, V. A., Wei, J. G. and Jenness, R. (1981). *Biochim. Biophys. Acta,* **669,** 258.
Waniska, R., Shetty, J. & Kinsella, J. E. (1981). *J. Agric. Food Chem.,* **29,** 826.
Watanabe, M., Fujii, N. and Arai, S. (1982). *Agric. Biol. Chem.,* **46,** 1587.
Yamauchi, K., Shimizu, M. and Kamiya, K. (1980). *J. Food Sci.,* **45,** 1237.

Chapter 7

GEL STRUCTURE AND FOOD BIOPOLYMERS

A. E. BELL
Department of Food Science and Technology,
University of Reading, UK

INTRODUCTION

Food materials are multicomponent, multiphase systems exhibiting textural properties which are due to the interactions of substructures, usually in an aqueous environment. Substructure size, size distribution and deformability all contribute to the functional behavior of the material. Such networks and structures are usually made up from carbohydrates, proteins and lipids and subsequently modified by the presence of other smaller molecules, such as simple sugars, salts, etc.

Much has been written on the nature of the interactions of specific polysaccharides, lipids and proteins, especially in dilute solution, where measurements of viscosity may provide a relatively direct and simple indication of the flexibility and shape of the hydrated polymer. However, food materials tend to exist as 'higher order' structures such as gels, emulsions and foams and it is the gelation of food biopolymers I shall concentrate on within this chapter as well as giving some consideration to the rheological, and hence textural, properties of such systems.

DEFINITION OF A GEL STRUCTURE

All food biopolymers when added to solution have thickening properties, but not all form gels. This leaves us with the problem of defining gelation. A gel structure has been described as the association or cross-linking of long polymer chains to form a continuous three-dimensional network which immobilises a liquid phase (Glicksman, 1982). However, these 'associations and cross-links' may or may not be permanent in nature and are usually heavily dependent on external influences (temperature, shear, etc). Such a structure then, may be maintained by the making and breaking of its 'bonds' and gives rise to a dynamic system with an 'effective' or 'averaged' association of the components (Ferry, 1980). It therefore follows that any assessment of a 'gels' structure is dependent on the time scale of the measurement that we use to determine it.

CHARACTERISATION OF FOOD GELS

The behaviour of a gel can be explained in terms of its elastic and viscous behaviour. This is usually modelled using 'springs', 'dashpots' and 'sliders', representing a true elastic solid, a Newtonian liquid and a limiting frictional force respectively (Muller, 1973). These three elements can be arranged in series or parallel to describe the deformation–time responses of any model system, however what is often not clearly appreciated is that the resultant model only describes the behaviour of the system at one rate of applied force (e.g. shear rate) and at one set deformation (strain) and makes no allowances for any structural changes caused by the testing procedures.

Many food structures are strain amplitude dependent, i.e. their structure is lost when they are moved too far in testing; therefore, the first step in characterising such materials must be to ascertain the limits to non-destructive testing. For example, mayonnaise or salad dressings stabilised by electrostatic or steric means have 'critical' strains of the order of 0·5% and 5% respectively, and it is important that testing be carried out below these values. Once limits have been established, it is possible to characterise the structure itself by changing the frequency at a strain lower than the 'critical' strain. It is therefore essential for the test machine to be capable of varying these two parameters over a wide range in order to build up as complete a 'picture' as possible of the material under test. Classical and fundamental rheological measurements of this type were usually found to take so long as to be impractical for food materials, but, with the advent of more advanced microcomputer interfacing techniques, this is no longer the case and there is more and more interest in such measurements and the information they afford.

Periodic sinusoidal experiments provide information corresponding to short time scale actions, the applied stress being altered at a given frequency (cycles s^{-1} or, ω, radians s^{-1}). If the behaviour of a viscoelastic material is linear, the strain will also vary sinusoidally with the stress, but will be out of phase with it. This behaviour is intermediate between the plots for an ideally elastic material and a true Newtonian liquid where the stress is in phase ($\delta = 0°$) and 90° out of phase respectively with the strain (Fig. 1).

For a viscoelastic system it is possible to resolve the stress in terms of an in phase component ($\sin \omega t$) and an out of phase term ($\cos \omega t$)

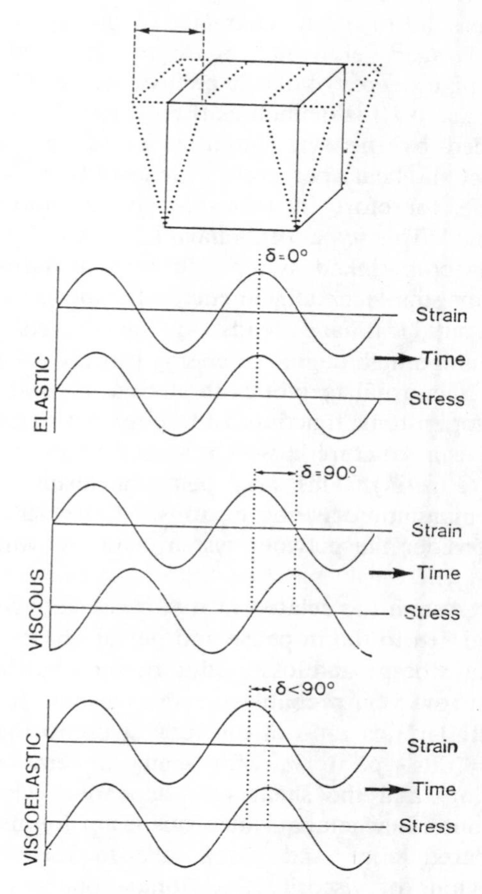

FIG. 1. Dynamic rheological behaviour.

related by the phase angle (δ) between the sinusoidal functions. Just as a modulus is defined as the stress/strain ratio in any constant deformation experiment, then, for a dynamic sinusoidal experiment it follows that two moduli can be defined; stress in phase/strain and stress out of phase/strain. These are the storage modulus (G') and loss modulus (G'') respectively.

The storage modulus (G') is a measure of the energy stored in the material and recovered from it per cycle. On a molecular basis, the magnitude of G' is dependent upon what rearrangements can take place within the period of oscillation, (Ferry, 1980), and is taken as an indication of the solid or elastic character of the material under test. For example, an agar gel, which is essentially permanently cross-linked, shows a high degree of elastic behaviour, i.e. G' is high.

The loss modulus (G'') is defined as the stress 90° out of phase with the strain divided by the strain and is a measure of the energy dissipated or lost (as heat) per cycle of sinusoidal deformation. The loss modulus G'', therefore, is taken as an indication of liquid or viscous behaviour. This type of behaviour is usually exhibited by non-permanently cross-linked systems such as hyaluronate solutions which interact by simple entanglement of the polysaccharide chains (Morris *et al.*, 1980*b*), and leads to high levels of molecular rearrangement, and a high degree of energy loss (G'', the loss modulus predominates). It is usual to plot both storage moduli (G') and loss moduli (G'') as logarithmic functions of frequency (ω) and we can now redefine our gel as a structure in which the elastic nature of the system predominates ($G' > G''$). This may be as a result of 'permanent' cross-linking or many 'time averaged' transient associations and it may be helpful to consider the polymer system as a 'network' rather than an idealised 'gel'.

A viscoelastic system has related to it two viscosity components (η' and η'') again related to the in phase and out of phase stresses. These are related to the storage and loss moduli by the equations $\eta' = G''/\omega$ and $\eta'' = G'/\omega$, however, probably the most useful parameter is the dynamic viscosity (η^*), a ratio of the total stress to the frequency of oscillation $\eta^* = (G'^2 + G''^2)^{1/2}/\omega$; for many systems the frequency dependence of η^* and the shear rate dependent viscosity, η, are closely superimposable when equivalent values of frequency and shear rate are compared (Cox and Merz, 1958). Shear or frequency independent regions of viscosity (Newtonian plateaux) indicate the ease of rearrangement in any system under study, and give some idea as to the degree of interaction present.

Another parameter which is often useful in indicating the physical behaviour of a system, but which conveys no physical magnitude and is dimensionless in the loss tangent (tan δ). It is the ratio of the energy lost to the energy stored for each cycle of the deformation, i.e. $\tan \delta = G''/G'$. The logarithmic plot of the loss tangent (tan δ) gives rise to several characteristic levels. For example, for a dilute solution, tan is high, as G'' is a function of both the solvent and the solute, while G' is representative of only the solute, a relatively minor component. However, for a highly cross-linked system e.g. agar gel, G' the storage modulus becomes the major component and G''/G' falls markedly. Unfortunately it is often difficult to apply a theoretical interpretation to loss tangent data alone and for this reason it is usually only used as a general indicator of viscoelastic function, and then only in conjunction with the storage moduli, loss moduli and dynamic viscosity data (Ferry, 1980). A detailed review of the structural and mechanical properties of various biopolymer gels is given by Clark and Ross-Murphy (1987).

When we eat or drink we are all acutely aware of the textures involved and that such qualities are major determinants in our assessment of food acceptability. When properly executed, sensory panels give very good data, but they are both time consuming, expensive, and require considerable statistical analysis to verify the results. In contrast instrumental methods are usually more reproducible, simpler, and less expensive to conduct. Attempts to correlate basic physical measurements and sensory characteristics have, in general, met with little success, but, since a food material's behaviour is described by a three-dimensional 'plane' (Fig. 2(a)), yet most measurements have been of the single point type (fixed strain, frequency, or both), this is not surprising. By analogy, this type of approach is like attempting to assess the colour of an object by considering only one component wavelength of its emitted light. Thus, a single rheological measurement on two very different samples may, with, for example, an Instron at fixed cross head speed and distance of travel, demonstrate weaker, stronger or identical consistency, depending on the conditions chosen (Fig. 2(b)).

Dynamic mechanical testing techniques have unique advantages for characterising food systems because the strain amplitude and frequency can be independently controlled. In this way, the structure build-up and breakdown or recovery characteristics can be measured in real time and continuously, and can show differences between the behaviours of apparently similar materials when the appropriate test

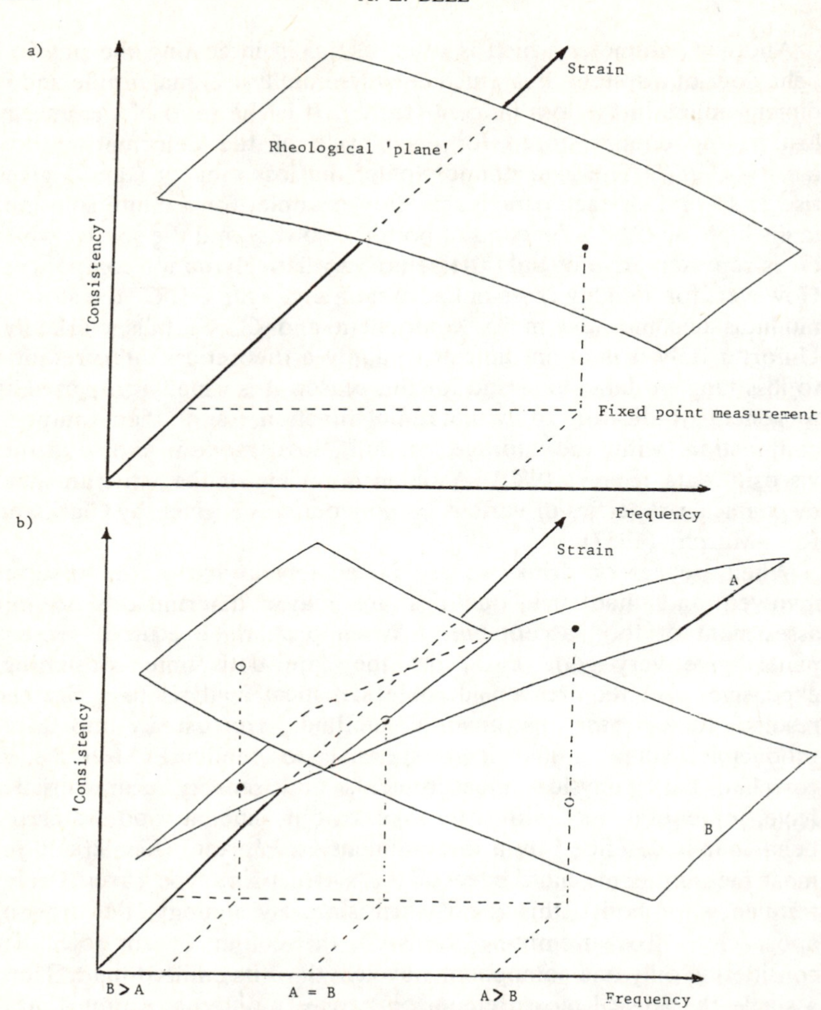

FIG. 2. 3-D nature of rheological behaviour of gels.

conditions are used. Single point measurements (viscosity, penetration, cutting force, etc.) can give no real indication of the structure or interactions present, and may, depending on the test conditions used, give totally misleading information.

A wide variety of such empirical test equipment is available making use of this type of fixed point measurement and comprehensive reviews of their use do exist (Bourne, 1982). It may be possible to take such data and use it for a quality control/product assessment, but only after the system under test has been fully characterised. Such problems form the basis of establishing the relationships between 'mechanical properties' and 'eating characteristics' of foodstuffs.

FOOD BIOPOLYMERS

Gelling substances used in the food industry are derived from various sources, however they fall into two main subgroups, 'disordered' and 'ordered'. Disordered biopolymers are those in which the individual polymers interact to give rise to some sort of network, for example, gelatin and various polysaccharides. Ordered biopolymers are those which already have some 'unit' structure which then subsequently interact to produce 'higher order' systems (myosin, collagen, etc.).

Gelatin Gels

Industrially gelatin is derived from collagen by the acid or alkaline hydrolysis of a number of raw materials. Commercially produced gelatin is a complex mixture of differing molar masses and structures (α, β and γ gelatins), the final composition being dependent on the source material (hide, bone, etc.) and the extraction method used (Veis, 1964). The amino acid content and sequences are again heterogeneous, however there is always a considerable proportion of proline, hydroxyproline and glycine. The pyrolidine content is of major significance as it promotes the helical nature in the polypeptide chains needed to provide the triple helix structure of collagen (Parry and Creamer, 1979) while the glycine promotes interchain hydrogen bonding (Bornstein and Traub, 1979). Extensive studies of both the inter- and intra-molecular bonding of collagen have been undertaken and detailed explanations of structure given; we shall consider here only the interactions of gelatin itself in network formation.

Gelatin itself may be regarded as an attempt by the individual protein chains to regain their original collagen structure (triple helix). When used in normal food processing concentrations (for example 1–2% w/v) individual chains, are sufficiently separated spacially, to allow only limited areas of reassociation.

It is generally believed that the junctions formed are stabilised by hydrogen bonding similar to these found in the native collagen and that their formation is centred around the proline-rich regions of the gelatin chains (Ledward and Stainsby, 1966; Glanville and Kuhn, 1979). Some water molecules are also believed to be bound during gelation and retained in the subsequently formed structure (Naryshkina *et al.*, 1982). These gelatin molecules whether purified or mixed together dissolve in aqueous solution above about 40°C and on cooling undergo a sol/gel transition to form transparent elastic gels. These have been assumed to be formed by the formation of junction zones of individual triple helix although some electron microscopic evidence suggests that these junctions may be several triple helix structures aggregated together (Lewis, 1981). This again reaffirming the 're-formed' collagen-like structure for gelatin gels rather than a random association of individual chains.

Gelatin gel strength is highly temperature dependent, when matured at high temperatures only a few triple helix-type junctions will be formed producing a relatively disordered weak gel. On further cooling this system becomes more ordered and the gel strength increases. This suggests the growth and increasing stability of the junctions due to steric restraints within the network (Nijenhuis, 1981).

It is well established that the gel strength of a gelatin system increases with time when held at a fixed temperature. If this is considered in the light of the potential ease of rearrangement of the hydrogen bonding which stabilises the triple helix and the 'making and breaking' of the network structure, then an increase in the proportion of energetically more favoured conformations with time can be postulated. This would explain the above phenomena, provided the holding temperature was not so high as to prevent initial nucleation of the system (Ledward, 1968). The random nature of gelatin aggregation on cooling produces regions of interaction which are not all inter-chain and the final cross-link density is not only dependent on concentration but also on the ratio of intra- to inter-junctions (Bobrova *et al.*, 1972). This ratio is dependent on the absolute concentration and also on the weight averaged molecular size of the gelatin moiety (Saunders and Ward, 1958).

The outcome of all of the above is that a gelatin gel at a fixed concentration and pH may have very different characteristics depending on its thermal and processing history and there is no simple correlation between the chemical composition, rigidity, viscosity or melting point between the wide range of commercial samples available. (An excellent overview is given by Ledward, 1986).

There are still several aspects of gelatin which need further explanation even with regard to its own interactions with itself and when we also consider the complexities introduced by its addition to a food system containing other macromolecular components (polysaccharides, polypeptides, etc.) the range of possible associations becomes almost infinite. However further research has been slow due to the relative decrease in importance of gelatin as a food ingredient due to the labile nature of the peptide linkages. Thermal processing of gelatin-based foods leads to a considerable loss of gel-forming character, and despite being macromolecules which are cheap, water soluble and commercially acceptable as food additives they are being replaced in many instances by polysaccharides as thickening and gelling agents. Considerable work is being undertaken to chemically modify gelatin structure and it may be that future work will produce a more functional derivative of more general use to the food industry as a whole.

Marine Polysaccharides

The use of marine polysaccharides already yields large profits not only in the food industry but also in other areas such as marine fouling and drag reduction, enhanced oil recovery, the animal feed industry and pharmaceuticals. In the context of marine polysaccharides the most familiar candidate for discussion would appear to be agar, however examination of current food industry literature suggests that the carrageenans would be the most interesting representative to describe initially in terms of their properties and characteristics.

Carrageenan Gels

Carrageenan is a generic term referring to the group of water soluble sulphated galactan extracts from certain types of red seaweed or macroalgae (Rhodophyceae). They are usually unbranched polysaccharides, based on an alternating saccharide structure (Rees, 1969) and have a general structure $(AB)_n$. The repeating structure is derived from a 1–3 linked β-D-galactose residue (A) and a 1–4 linked 3,6-anhydro-α-D-galactose (B) both of which may be sulphated and substituted to give the various types found in nature (Fig. 3). The name carrageenan derives from Carrageen, County Waterford, on the coast of Ireland where a trade in 'Irish moss' has flourished since the early 1800s, since which time, polysaccharide-bearing seaweeds have been discovered all over the world.

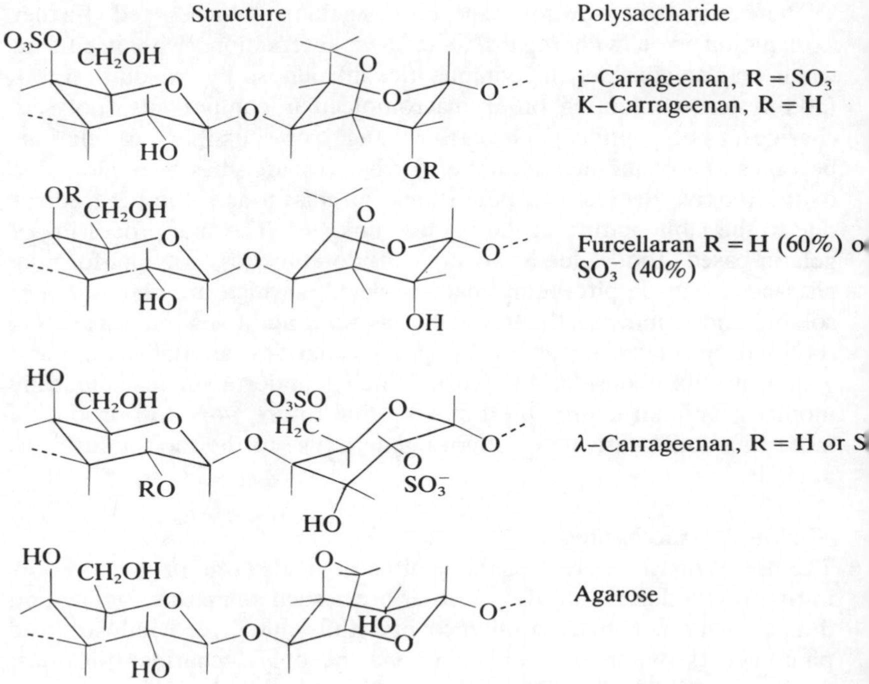

Structure	Polysaccharide
	i–Carrageenan, $R = SO_3^-$ K–Carrageenan, $R = H$
	Furcellaran $R = H$ (60%) o SO_3^- (40%)
	λ–Carrageenan, $R = H$ or S
	Agarose

FIG. 3. The agarose carrageenan polysaccharides.

The carrageenans are analogous in many respects to the charged animal proteoglycans such as chondroitin sulphate and heparin (Fig. 4) and are produced in a wide range of molar masses (10^4–10^6 daltons). They are used in a wide range of food products including ice cream, chocolate milk, milk puddings, desserts, pet foods and low calorie jellies and as stabilisers/emulsifiers in many food emulsions. In general most of these products use carrageenans with molar masses of about 10^5–10^6 daltons and are usually a combination of kappa and iota types.

There are six possible idealised structures for carrageenan. However since mu, nu and lambda can be converted to kappa, iota, and theta respectively by alkaline treatment or enzyme modification, to close the 'oxygen bridge' in the 'B' residue, they may be regarded essentially as three pairs of structural analogues. A ratio of one or more of these pairs is found in each type of plant source used for the commercial production of carrageenan.

Regular repeating unit	*Polysaccharide*
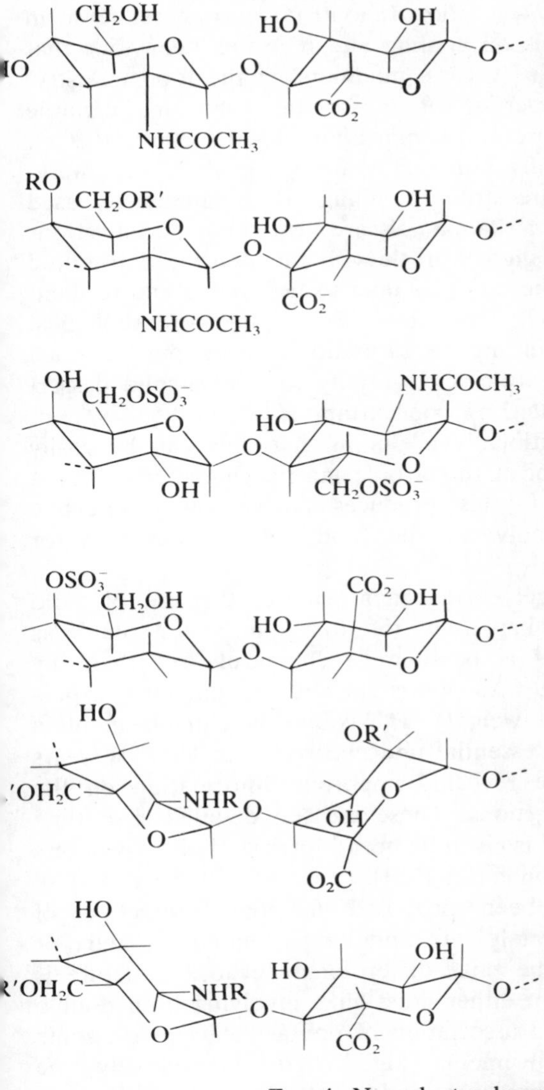	Hyaluronate
	Chondroitin 4-sulphate $R = SO_3^-$ $R' = H^+$ Chondroitin 6-sulphate $R = H^+$, $R' = SO_3^-$
	Keratan sulphate
	Dermatan sulphate
	Heparin $R = SO_3^-$ (major) or $COCH_3$ (minor) $R' = SO_3^-$ (major) or H (minor)
	Heparan sulphate $R = COCH_3$ or SO_3^- $R' = H$ (major) or SO_3^- (minor)

FIG. 4. Non-plant polysaccharides.

Carrageenans, depending on type and concentration, form ion dependent gels or viscous solutions. Heated solutions of carrageenans, even at relatively low concentration (about 1% w/v) tend to form rigid, thermoreversible gels on cooling and it is this behaviour that makes them most useful as 'texture modifiers' in the food industry. The actual gelation properties of the carrageenan are variable depending on their structure and composition (Morris *et al.*, 1980*a*), varying from the essentially non-gelling forms (lambda and theta) through to those which are strongly gelling and commercially used (kappa and iota). In general kappa gels are firm gels (concentrations as low as 0·5% w/v) being slightly brittle with a tendency to loose fluid by syneresis. In contrast the iota gels tend to be flexible and resilient (even at about 0·3% w/v). This allows for a range of rheological behaviour on mixing, permitting the formation of a gel matrix which can be used for anything from a clear jelly to a semi-solid dessert which will suspend fruit, etc. Low concentrations (about 0·5% w/v) of iota carrageenan are additionally useful in that they can be easily thinned by shear producing a thixotropic matrix characterised by a relatively low yield point, this produces an extremely effective suspension medium not only for the food industry but also for pharmaceutical usage.

It is also possible to gel carrageenan without the use of 'cold casting' from hot solution, kappa and iota forms are water soluble as their sodium salts, but it is possible to chemically set them by replacing the sodium ions, with potassium and calcium ions respectively (about 3–5% salt by weight). The counter-ion involvement in carrageenan systems is an essential one and over the last few years several investigations have revealed important information on the mechanisms of such interactions. These are too detailed to outline here but excellent coverage is given by both Morris and his co-workers (Morris *et al.*, 1980*c*; Norton *et al.*, 1983).

Various models have been proposed for the interaction of carrageenan molecules (mainly iota and kappa) based on helix or double helix formation. The three-dimensional network structure is considered to be built from either cross-links involving both double helix formation and limited association of double helices with some possible counter-ion involvement (Fig. 5) or alternatively the development and subsequent association of single helices again with some ion involvement. The structures of such carrageenan gels have been studied by various techniques, fibre diffraction (Anderson *et*

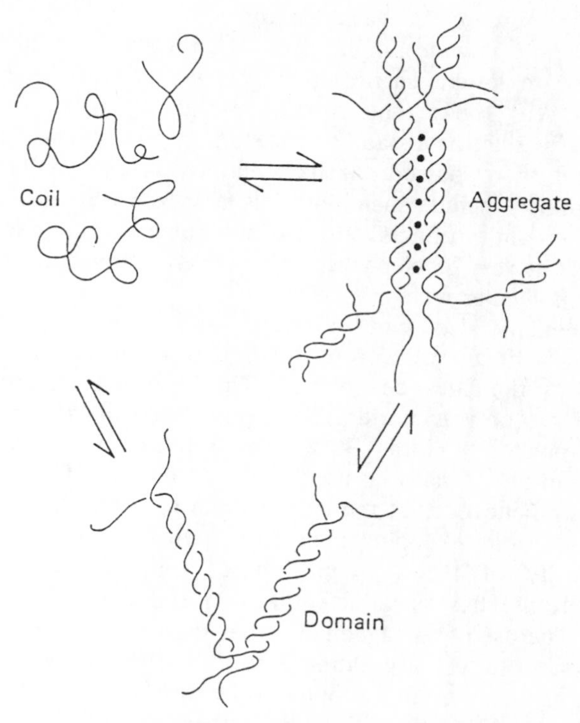

FIG. 5. Interactions possible for iota-carrageenan.

al., 1969), optical rotation (Rees *et al.*, 1970), thermodynamics (Reid *et al.*, 1974), nuclear magnetic resonance (Bryce *et al.*, 1974) and fast reaction kinetics (Norton *et al.*, 1983) and a detailed overview of the arguments for each proposed model is given by Clark and Ross-Murphy (1987).

Carrageenans have a number of interesting and useful properties including their interactions with proteins and their freeze–thaw characteristics. The degree of interaction with proteins will depend upon several factors, the degree of denaturation of the protein, the type of carrageenan, the degree of sulphation, the molecular size of the chain and the pH of the system. If the protein is above its isoelectric point increased viscosity or gelation may occur, however at or below this point coprecipitation is likely. Both situations may be of use in food systems and/or processing.

Again because of the varied nature of carrageenans their freeze–thaw behaviour is wide ranging from the iota type which is essentially unaffected, to kappa carrageenan gels which fracture and undergo syneresis (Witt 1984). The lambda non-gelling form of carrageenan finds extensive use as a stabilising system in products which need to be frozen and then thawed, as its solubility is essentially unchanged. These products include such materials as frozen whiteners, natural and synthetic cream products and various sauces. The natural charges present act as very efficient dispersants even at low concentrations and with low molecular weight species.

Carrageenans also find extensive use in various milk systems interacting with the milk proteins, the nature of the interaction being governed by the carrageenan used. This in turn is determined by the source of the polysaccharide. For example *Chondrus crispus* extracts (65% kappa, 5% iota, 30% lambda) are the major type of carrageenan-based material used in the formation of milk products. It is generally believed that the carrageenan reacts with the protein coat of the fat droplets keeping them in suspension, as well as increasing the viscosity of the aqueous phase, until ultimately at gelling concentrations these products take on the consistency of an 'egg custard'. The use of carrageenan rather than a traditional starch-based gelling system to 'set' a pudding ensures a structure which is generally lighter, creamier, and more 'open' in texture, and gives a great deal of latitude in the design of milk-based dessert products (Moirano, 1969).

In conclusion carrageenan finds many uses in the food industry only a few of which have been mentioned. Other uses including medical, industrial and pharmaceutical abound and are too numerous to mention here. However it must be said that the carrageenans are without doubt some of nature's most versatile and useful polysaccharides.

Alginate Gels

Alginates are composed of two distinct monosaccharide units, 1–4 linked β-D-mannuronic acid and 1–4 linked α-L-glucuronic acid (M + G units respectively). They are extracted from the intercellular material of various brown algae (Smidsrod, 1974) and the extracellular mucilage of some bacteria (Sandford, 1979). The proportion of M to G is dependent on the source of the alginic acid. They may be present either as block copolymers, homopolymers or mixed blocks containing amorphous sequences of M and G (Haug *et al.,* 1967). A structural description of this material in terms of M and G is difficult and as such they are usually classified only in terms of their gross M/G ratio.

Alginates do not form thermoreversible gels and gel structure or network formation is brought about by divalent cations, for example, calcium, setting up the classical 'egg box' structure (Grant *et al.*, 1973). These gels have varied rheological behaviours depending on the type of alginate used (M/G ratio), the divalent cation used (Ca^{2+}, Mg^{2+}, etc.) and mode of introduction of the ion (diffusion or *in situ* 'salting' by changing the pH), but in general fairly stiff, if somewhat brittle, gels are produced. Even in dilute solution and at high ionic strength alginate molecules have been shown to be highly expanded 'coils' suggesting their rheological behaviour to be that of essential inflexible 'rods' interacting together (Smidsrod and Haug, 1968). Gel formation occurs when the divalent ion (Ca^{2+}) binds preferentially to the glucuronic acid residues in the alginate, this process is cooperative and shows a 'zipping' effect if more than twenty glucuronate moieties are found in a single block (Kohn and Larson, 1972). These cross-links form ordered junction zones with divalent cations occupying the sites within these structures. Combinations of either several chains interacting or simple dimers have been postulated giving the proposed 'egg box' structure previously mentioned (Fig. 6). Gels may be produced at concentrations as low as 0·1% w/v but are more usually prepared at about 1% w/v (McDowell, 1955).

Structural modifications of alginates and their effects on gelation have not been extensively studied. Acetylation is known to inhibit divalent ion-induced gelation (Schweiger, 1964) and hinder association of the glucuronic acid residues (Rees 1969).

In general, alginates are used as thickeners, suspending agents and gels where the heating of a thermoset gelling agent would cause the 'degradation' of other heat labile ingredients.

Agar Gels

Agar is the collective name given to a complex mixture of polysaccharides which may be extracted from certain red seaweeds (*Rhodophyceae*). The crude agar may be fractionated into two components agaropectin and agarose (Araki, 1956) of which agarose is the primary gelling component. Agarose has a repeating disaccharide subunit based on the 1–3 linked β-D-galactose residue and a 1–4 linked 3,6 anhydro-α-L-galactose moiety. Various substitutions of the two units give rise to its most common derivative, agarose

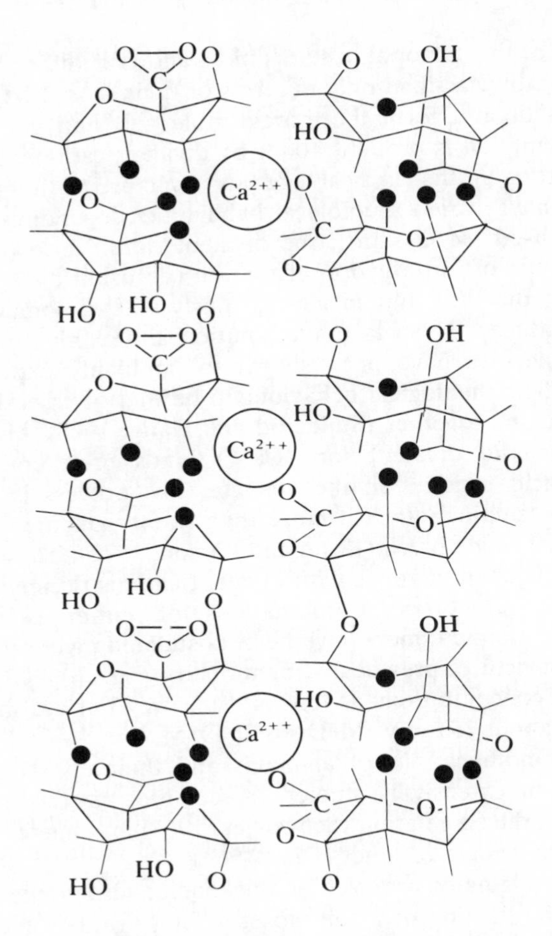

FIG. 6. Alginate chelated by calcium ions.

sulphate. Molar masses of agarose polymers have been estimated at about 10^5 daltons (Letherby and Young, 1981).

Commercial preparations of agar contain various other chemical species apart from the idealised disaccharide repeating unit. This includes various amounts of the 6-methyl ether and 6-sulphate derivatives of the β-D-galactose and substitution of the 3,6-anhydro-α-L-galactose by galactose or galactose-6-sulphate (Dea *et al.*, 1972).

These latter modifications introduce some structural 'kinking' and prevents 'perfect' ordering during the gelation process.

Agarose undergoes thermoreversible gelation when hot solutions are cooled (below about 40°C) giving rise to stiff, brittle, turbid gels showing both syneresis and thermal hysteresis (e.g. a gel formed at about 40°C will 'melt' at 80–90°C). The 6-methyl ether derivative can be made to gel at concentrations as low as 0·1% w/v, however sulphation of agarose increases the gelling concentration considerably (Dea and Morrison, 1975). Agar or agarose gels have two major usages, food additives (Gilmer, 1972) and microbiological media. In the latter use it has been found to be impervious to most organisms, but there are some that can produce lytic enzymes and metabolise agar (Fenchel and Jorgensen, 1977). Agar, although not readily digestible in the human, has been used in such food products as fruit pie fillings, dessert toppings and meringues. However due to its relatively high cost as a gelling agent (about four times the cost by weight as carboxymethyl cellulose or twice that of carrageenan) it has been replaced in many food applications, by other, cheaper, colloidal agents.

Gelation of agarose has been observed using light scattering (Letherby and Young, 1981), fluorescence (Hayasmi *et al.*, 1978), nuclear magnetic resonance (Ablett *et al.*, 1978) and X-ray diffraction (Arnott *et al.*, 1974). X-ray evidence suggests that a double helix structure may be involved in agarose gel networks (left handed as opposed to the right handed helices of iota carageenan). However evidence from the other techniques mentioned suggests that the structure provides more than a double helix structure and that considerable aggregation may occur to produce the suitable 'junction zones'. While this proposed model goes some way to explain the characteristic hysteresis evident in the optical rotation plots (as a function of temperature) found for agarose gels (Dea *et al.*, 1972), this double helix junction zone hypothesis has been challenged (Foord, 1980) and it seems that for agarose at least, that there is still some controversy about the exact nature of the linkages and molecular arrangements involved in network formation.

The rheological properties of agarose gels have been studied under conditions of both small and large deformation. Excellent coverage of this aspect is given by Clark and Ross-Murphy (1985), and McEnvoy *et al.*, (1985).

Plant-Based Polysaccharides

Pectin Gels

Pectins are polysaccharides derived from a wide range of plant cell wall material (especially fruits—Rees, 1969). Commercially they are important as thickening and gelling agents in the production of many types of jam and preserve. The actual structures of the pectins are complex and will vary, not only with the plant source, but also with the method of extraction.

Native pectins are branched polysaccharide structures but when extracted are mainly linear polymers of α 1–4 linked D-galacturonic acid containing L-rhamnose residues. These are partially esterified to varying degrees to form the methyl esters and this forms the basis of the major classification used for pectins in general i.e. high or low methoxy pectins (Aspinall, 1970).

Pectins can be made to produce two types of gels: (i) those containing a high proportion of methoxy groups which form at low pH and high sugar concentrations (for example jams) and (ii) pectin gels prepared using pectin having a lower proportion of methoxy groups, which can be formed without sugar over a wide pH range, but which usually have a requirement for divalent cations to produce a network structure. It should however, be noted that both high and low methoxy pectins can be made to gel without ions being involved at all (Morris *et al.*, 1980*a*) and that gels made at low pH are sometimes thermoreversible 'melting' at fairly low temperatures (about 40°C).

Pectins containing blocks of galacturonic acid residues mirror the structure of those of glucuronic acid found in alginate (the configuration at one carbon being different) and as such mimics the gelation behaviour of alginate (at high pH and with divalent cations). This in turn would suggest a similar 'egg box' gelation mechanism. This model for cross-linking has received support from a number of analytical techniques such as equilibrium dialysis and circular dichroism (Gidley *et al.*, 1979), however evidence from fibre diffraction studies would suggest the formation of a three-fold helix structure for the galacturonic acid sequences (Walkinshaw and Arnott, 1981). The discrepancies in the physical evidence may most likely be explained in terms of the source of the material and/or the differences in gelation methods with special regard to the hydration states of the systems under study (Gidley *et al.*, 1979; Powell *et al.*, 1982).

Structural modifications of pectins have significant effects upon their gelling abilities. As well as the methoxy distribution acetylation produced serious modifications in behaviour, for example one acetyl grouping per eight residues markedly reduces the gelation ability (Pippin *et al.*, 1950). This effect is more noticable at lower ester levels where calcium binding is inhibited (Kohn, 1975). Amidation also occurs in pectin systems (Lockwood, 1976) possibly affecting the regular uronic acid structure.

Rigorous rheological studies on pectin gels have been limited, however compression test data are available and give information on shear moduli for high methoxy pectin gels (Okenfull, 1984). Previous testing has mainly been confined to empirical techniques for the development of grading methods (Matz, 1962; Black and Smit, 1972). The problems of rheological characterisation is difficult because of the dependence of the gel properties on divalent cation level, pH, composition and temperature. In general such research would benefit greatly from the fractionation of the various subcomponents to determine the contributions to structural properties made by each.

Starch Gels

Starch is a complex polysaccharide and is the major carbohydrate reserve of higher plants (e.g. potato, rice, corn, tapioca, etc.) and is located in tubers, roots, swollen stems and seeds. It is usually produced, as a raw material, in its granular form. These granules are usually roughly spherical in shape with a diameter of some 2–100 microns depending on source, containing predominantly starch polymers along with other components such as lipids and proteins (Banks and Muir, 1980). These polymeric starches are packed in a complex semi-crystalline arrangement which renders them essentially water insoluble (Banks and Greenwood, 1975).

Starch itself consists of two main types of molecule, amylopectin and amylose, which are branched and linear (respectively) polymers of glucose. These polymers may be dissolved intact from starch granules using such solvents as dimethyl sulphoxide and the resultant solutions fractionated if required. Amylose is an α 1–4 linked polymer of D-glucose while amylopectin has both α 1–4 and α 1–6 linkages between the D-glucose residues to give a branched structure. As a general rule amylose usually accounts for some 20–25% of the polymeric starch material, however some exceptional starches occur which have unusually low (e.g. sorgum, about 1%) or high (e.g. amylomaize, 50–60%) amylose levels. Both amylose and amylopectin form stable aqueous solutions and the solution properties (viscosity, etc.) have been extensively studied (Banks and Greenwood, 1975). Both have broad distributions of molar mass with values of hundreds of millions of daltons and several million daltons being reported for amylopectin (Zimm and Thurmond, 1952) and amylose (Foster, 1965) respectively. They exhibit dilute solution behaviour consistent with the model for simple molecular entanglement (Clark and Ross-Murphy, 1987).

Starch granules show a limited degree of swelling in cold water and this has been shown to be an exothermic process (Schierbaum *et al.*, 1962). However when heated in water above a specific temperature, known as the gelatinisation temperature, the tightly packed starch granules will swell, suddenly and irreversibly, and the constituent amylose will be released into solution. The end result is a system of porous amylopectin 'islands' in a hot amylose solution which on cooling produce thick, turbid, viscoelastic pastes and gels (if the inital starch concentration was high enough) which are thermo-irreversible in nature. On quenching a gelatinised starch suspension (e.g. 90°C down to 30°C) the linear amylose chains which have been released undergo a rapid aggregation as indicated by the sudden increase in turbidity and shear modulus (Miles *et al.*, 1984). However the presence of any crystalline character (as indicated by wide angle X-ray diffraction) takes some time to develop and may not be completed after several hours (Morris, 1979). So as the hot solution is cooled rapid network formation occurs through the formation of junction zones, probably by association of amylose double helices, trapping the 'islands' of amylopectin in its structure. Subsequent aggregation and phase separation of the helical regions would then lead to the formation of crystalline areas (Ring and Stainsby, 1982). This behaviour is analogous to classical rheological theory for a matrix (amylose) reinforced with a deformable filler (amylopectin granules) and the physical behaviour of the composite is subsequently very dependent on the matrix–filler interaction and previous processing history. Excellent work on the somewhat complex rheological properties of starch and its constituents are reported by both Morris and co-workers (Morris, 1979) and Ellis and Ring (1985).

Other Plant Polysaccharide Gels

Two other groups of polysaccharide derived from plants should be briefly mentioned. These are cellulose and galactomannans.

Cellulose. This is the most common plant polysaccharide and is based on a linear polymer of β-D-1–4 linked glucose residues, it is almost totally insoluble in water and subsequently is only used as a gelling or thickening agent in the form of one of its many derivatives. The gelling characteristic of cellose derivatives is complex, being a balance of their hydrophilic and hydrophobic tendencies under any given set of processing conditions. This determines whether the polymer interacts in a limited manner to form a gel or produces a precipitate. An excellent discussion of this behaviour has been given by Rees (1969).

Galactomannans. These are usually derived from plant seed material (storage material in the endosperm) and are generally based on β 1–4 backbone of D-mannopyranosyl sugars with single α-D-galactopyranosyl side groups linked α 1–6. These have been shown to be randomly distributed along the sugar 'backbone' (McCleary *et al.*, 1985).

Galactomannan solutions have been shown to be viscoelastic (Robinson *et al.*, 1982), however in practice they do not form strong gels in isolation. Consequently they are used in the food industry as thickening agents and usually only produce gels when combined with other polysaccharides such as xanthan and agarose (Dea and Morrison, 1975).

Non-plant Polysaccharide Gels

Other gelling agents do exist, the animal polysaccharides such as the proteoglycans—keratan sulphate, hyaluronic acid, etc. (Morris *et al.*, 1980*a*) (Fig. 4), and the glycoproteins such as those found lining mammalian respiratory, reproductive and intestinal tracts (Allen, 1983). These however find only limited and indirect use as gelling agents in the food industry. Probably the major non-plant polysaccharides of scientific and commercial interest are the microbial polysaccharides such as those produced by the organism *Xanthomonas campestris* to enable it to become attached to its plant cell host (a form of blight—Morris, *et al.*, 1977). It has a linear sugar backbone (β 1–4 linked glucose) carrying complex side chain structures (trisaccharide chain every second residue).

Gelation of xanthan (Norton *et al.*, 1984) under various conditions (changes in concentration, electrolyte and temperature) may produce quite strong gels. However it should be noted that the junction zones formed by xanthan generally tend to be less stable than those previously mentioned for the plant polysaccharides. These essentially temporary linkages (Morris *et al.*, 1983) lead to the formation of weak pseudo-plastic materials rather than true viscoelastic gels and as a consequence tend to be used extensively in the food industry more as thickening agents.

CONCLUSIONS

Polysaccharide gelation is a complex business depending not only on the structure of the polymer but also on its processing history and other co-factors and I have considered here only some of the most basic systems. If we then proceed to consider in any detail mixed systems (mixed polysaccharides and polysaccharide protein/lipid, etc.) the whole picture becomes extremely difficult. Synergistic interactions between differing food biopolymers has only just started to become a serious area of study and it is this information we must access if we are ever to be able to explain the behaviour of complex multicomponent systems such as real foods. Biopolymer gels still produce considerable controversies with regard to their behaviour, and mixed systems even more so, however this research is of great practical importance in assessing the properties of macromolecular food substances and has relevance in many other scientific areas (molecular biology and physical chemistry and increasingly various aspects of biotechnology).

The general trend of the food industry away from its traditional 'batch' processing toward more 'continuous' production is heavily dependent on the physical properties of food materials. Multipoint rheological measurements (e.g. dynamic mechanical testing) have unique advantages for characterising food systems and have the potential to be used in continuous production for feedback control in processing. 'Structural' products ('bubble' and 'flake' chocolate; multilayer, multistructure frozen dessert products; extruded corn snacks; etc.) are becoming increasingly popular with consumers and are only possible by exerting considerable control over the rheological properties of the product at various stages of manufacture. The 'added value' aspect of these products would seem to justify additional efforts being made in the understanding of this very important area of food research.

REFERENCES

Ablett, S., Lillford, P. J., Rayhdai, S. M. and Derbyshire, W. (1978). *J. Coll. Interface. Sci.*, **67,** 355.
Allen, A. (1983), *Trends Biochem. Sci.*, **8,** 169.
Allen, A., Bell, A. E., Mantle, M. and Pearson, J. P. (1982). *Adv. Exp. Med. Biol.*, **144,** 97.
Anderson, N. S., Campbell, J. W., Harding, M. M., Rees, D. A. and Samuel, J. W. B. (1969). *J. Mol. Biol.*, **45,** 85.
Araki, C. (1956). *Bull. Chem. Soc. Japan,* **29,** 543.
Arnott, S., Fulmer, A., Scott, W. E., Dea, I. C., Moorhouse, R. and Rees, D. A. (1974), *J. Mol. Biol.*, **90,** 269.
Aspinall, G. O. (1970). In: *Polysaccharides,* Pergamon Press, Oxford.
Banks, W. and Greenwood, C. T. (1975). In: *Starch and its Components,* Edinburgh University Press, Edinburgh.
Banks, W. and Muir, D. D. (1980). In: *Carbohydrate Structure and Function,* J. Preiss (Ed.), Academic Press, New York.
Black, S. A. and Smit, G. J. B. (1972). *J. Food Sci.*, **37,** 726.
Bobrova, L. E., Izmailova, V. N. and Rebinder, P. A. (1972). *Kolloid Zh.*, **34,** 6.
Bornstein, P. and Traub, W. (1979). In: *The Proteins IV*. 3rd edn, H. Neurath, R. L. Hill and C. L. Border (Eds), Academic Press, London.
Bourne, M. C. (1982).In:*Food Texture and Viscosity,* Academic Press, New York.
Bryce, T. A., McKinnon, A. A., Morris, E. R., Rees, D. A. and Thom, D. A. (1974). *Farad. Discuss. Chem. Soc.*, **57,** 221.
Clark, A. H. and Ross-Murphy, S. B. (1985). *Brit. Polymer J.*, **17,** 164.
Clark, A. H. and Ross-Murphy, S. B. (1987). Structural and mechanical properties of biopolymer gels. In: *Advances in Polymer Science 83*, pp. 60–184., Springer-Verlag, Berlin, Heidelberg.
Cox, W. P. and Merz, E. H. (1958). *J. Polymer Sci.*, **28,** 619.
Dea, I. C. M., McKinnon, A. A., and Rees, D. A. (1972). *J. Mol. Biol.*, **68,** 153.
Dea, I. C. M. and Morrison, A. (1975). *Adv. Carbohydrate Chem. Biochem.*, **31,** 241.
Ellis, H. S. and Ring, S. G. (1985). *Carb. Polym.*, **5,** 201.
Fenchel, T. M. and Jorgensen, B. B. (1977). *Adv. Microb. Ecol.*, **1,** 1.
Ferry, J. D. (1980). In: *Viscosity Properties of Polymers,* 3rd edn, John Wiley, New York.
Foord, S. A. (1980). PhD Thesis, University of Bristol, UK.
Foster, J. F. (1965). In: *Starch Chemistry and Technology 2*, R. L. Whistler and E. F. Paschall (Eds), Academic Press, New York and London.
Gidley, M. J., Morris, E. R., Murry, E. J., Powell, D. A. and Rees, D. A. (1979). *J. Chem. Soc. Chem. Comm.*, 990.
Gilmer, R. W. (1972). *Science,* **176,** 1239.
Glanville, R. W. and Kuhn, K. (1979). In: *The Fibrous Proteins*: *Scientific Industrial and Medical Aspects—1,* D. A. D. Parry and K. Creamer (Eds), Academic Press, London.

Glicksman, M. (Ed.) (1982). *Food Hydrocolloids Vol 1,* CRC Press, Boca Raton, Florida.
Grant, G. T., Morris, E. R., Rees, D. A., Smith, P. J. C. and Thom, D. (1973). *FEBS. Letts.,* **32,** 195.
Haug, A., Larsen, B. and Smidsrod, O. (1967). *Acta. Chem. Scand.,* **23,** 691.
Hayasmi, A., Kinoshita, K., Kuwano, M. and Nose, A. (1978). *Polym. J.,* **10,** 5.
Kohn, R. (1975). *Pure Appl. Chem.,* **42,** 371.
Kohn, R. and Larsen, B. (1972). *Acta. Chem. Scand.,* **26,** 2455.
Ledward, D. A. (1968). Physicochemical study of some chemically modified gelatins. PhD Thesis, University of Leeds. UK.
Ledward, D. A. (1986). In: *Functional Properties of Food Macromolecules.* J. R. Mitchell and D. A. Ledward (Eds), Elsevier Press, London.
Ledward, D. A. and Stainsby, G. (1966). GGRA Res. Panel Paper No66/1.
Letherby, M. R. and Young, D. A. (1981). *J. Chem. Soc. Faraday Trans. 1,* **77,** 1953.
Lewis, D. F. (1981). *Scanning Electron Microscopy III,* Chicago SEM. Inc, p. 391.
Lockwood, B. (1976). MPhil Thesis, University of Leeds, UK.
Matz, S. A. (1962). In: *Food Texture.* AVI Publishing Co., Westport Connecticut.
McCleary, B. V., Clark, A. M., Dea, I. C. M. and Rees, D. A. (1985). *Carb. Res.,* **139,** 237.
McDowell, R. H. (1955). In: *Properties of Alginates,* Alginate Industries Ltd Publication, London.
McEnvoy, H., Ross Murphy, S. B. and Clark, A. H. (1985). *Polymer,* **26,** 1483 and 1493.
Miles, M. J., Morris, V. J. and Ring, S. G. (1984). *Carbohydr. Polym.,* **4,** 73.
Moirano, A. L. (1969). US Patent 3,443,968.
Morris, E. R. (1979). *NATO Ad. Study Inst. Ser. Ser. C.,* 379.
Morris, E. R., Rees, D. A., Young, G. and Darke, A. (1977). *J. Mol. Biol.,* **110,** 1.
Morris, E. R., Gidley, M. J., Murry, E. J., Powell, D. A. and Rees, D. A. (1980*a*). *Int. J. Biol. Macromol.,* **2,** 327.
Morris, E. R., Rees, D. A. and Welsh, E. J. (1980*b*) *J. Mol. Biol.,* **138,** 383.
Morris, E. R., Rees, D. A. and Robinson, G. (1980*c*). *J. Mol. Biol.,* **138,** 349.
Morris, V. J., Franklin, D. and I'Anson, K. (1983). *Carbohydr. Res.,* **121,** 13.
Muller, H. G. (1973). In: *An Introduction to Food Rheology,* Heineman, London.
Naryshkina, E. P., Volkov, V. Ya., Dolinnyi, A. I. and Izmailova, V. N. (1982). *Kolloid. Zh.,* **44,** 322.
Nijenhuis, K. (1981). *Colloid Polymer Sci.,* **259,** 1017.
Norton, I. T., Morris, E. R. and Rees, D. A. (1983). *J. Chem. Soc. Farad., Trans., 1,* **79,** 2475.
Norton, I. T., Goodall, D. M., Frungou, S. A., Morris, E. R. and Rees, D. A. (1984). *J. Mol. Biol.,* **175,** 371.
Okenfull, D. (1984). *J. Food Sci.,* **49,** 1103.
Parry, D. A. D. and Creamer, L. K. (1979). In: *Fibrous Proteins Scientific*

Industrial and Medical Aspects, Vol. 1, D. A. D. Parry and K. Creamer (Eds), Academic Press, London.
Pippen, E. L., McCready, R. M. and Owens, H. S. (1950). *J. Amer. Chem. Soc.,* **67,** 2122.
Powell, D. A., Morris, E. R., Gidley, M. J. and Rees, D. A. (1982). *J. Mol. Biol.,* **155,** 517.
Rees, D. A. (1969). *Ad. Carbohydr. Chem. Biochem.,* **24,** 267.
Rees, D. A., Scott, W. E. and Williamson, F. B. (1970). *Nature (London),* **227,** 390.
Reid, D. S., Bryce, T. A., Clark, A. H. and Rees, D. A. (1974). *Farad. Discuss. Chem. Soc.,* **57,** 230.
Ring, S. G. and Stainsby, G. (1982). *Prog. Fd. Nutrition Sci.,* **6,** 323.
Robinson, G., Ross Murphy, S. B. and Morris, E. R. (1982). *Carbohydr. Res.,* **107,** 17.
Sandford, P. (1979). *Adv. Carbohydr. Chem. Biochem.,* **36,** 265.
Saunders, P. R. and Ward, A. G. (1958). In: *Recent Advances in Gelatin and Glue Research.,* G. Stainsby (Ed.), Pergamon Press, London, p. 197.
Schierbaum. F., Taufel, K. and Ullman, M. (1962). *Starke,* **14,** 161.
Schweiger, R. G. (1964). *Koll. Zh.,* **196,** 47.
Smidsrod, O. (1974). *Farad. Discuss. Chem. Soc.,* **57,** 263.
Smidsrod, O. and Haug, A. (1968). *Acta. Chem. Scand.,* **23,** 691.
Veis, A. (1964). In: *The Macromolecular Chemistry of Gelatin,* Academic Press, London.
Walkinshaw, M. D. and Arnott, S. (1981). *J. Mol. Biol.,* **153,** 1055 and 1075.
Witt, H. J. (1984). In: *Biotechnology of Marine Polysaccharides,* R. R. Colwell, E. R. Pariser and A. J. Sinskey (Eds), Hemisphere, London.
Zimm, B. M. and Thurmond, C. D. (1952). *J. Am. Chem. Soc.,* **74,** 1111.

Chapter 8

THE MEAT ASPECTS OF WATER AND FOOD QUALITY

K. O. Honikel
Bundesanstalt für Fleischforschung, Kulmbach, FRG

ABBREVIATIONS

ATP	Adenosine-triphosphate
DFD	Dark–Firm–Dry (meat)
DNA	Deoxyribonucleic acid
IP	Isoelectric point
pH_1	pH measured 45 min *post mortem*
PSE	Pale–Soft–Exudative (meat)
RNA	Ribonucleic acid
SR	Sarcoplasmic reticulum
WHC	Water-holding capacity

INTRODUCTION

Lean meat contains about 75% water. Proteins are the second main constituent part with about 22%; thus a water–protein ratio of 3·4:1 exists in meat. A considerable proportion of the water interacts with the proteins and their highly ordered structures, the myofibrils and substructures. The water which interacts is inhibited in its molecular movement, and is called immobilized water. A further limitation of water movement is caused by the lipid bilayers of membrane structures which allow only a slow movement of water across the bilayers. In the extracellular space the water is immobilized by capillary forces. Therefore, water of unlimited movement is rarely found in meat. In contrast, however, tightly bound water in the sense of chemical binding (like crystal water in salt) is a very small proportion with about 0·1% of tissue water (Fennema, 1977). Another 5–15% of the water shows a rather restricted mobility (Hamm, 1975). This water may be called interfacial water, and it is located at the surface of proteins; water–solute and water–water interactions are involved. Their binding energies are larger than those in normal water (Hamm, 1986).

Chemical and physical changes alter the labile state of immobilization, and these changes happen during the formation of meat. We call meat the final state of a sequence of biochemical and physical events which take place in cross-striated muscles of slaughter animals after slaughter. The processes may take one day or last two weeks. During the post-mortem processes the changes of the water structure and water movements into the extracellular space and between cellular substructures are of paramount importance for the quality of meat especially with regard to tenderness, juiciness and water retention (Hamm, 1960, 1985).

However, the attractive forces with which the water is bound and immobilized by muscle proteins do not change only during the post-mortem processes, but also during processing of meat, manufacturing of meat products and preparation of meat for eating. Processing involves chilling, storage, freezing, thawing, drying and mincing; manufacturing furthermore includes salting (curing), and comminuting. Almost all the procedures change the water retention of meat—the water-holding capacity (WHC). Thus knowledge of factors of influence on the WHC of meat handled under various conditions is important for the quality of meat and also of considerable economic interest. Moreover, investigations on the causes of the changes of WHC of meat teach us about alterations in muscle proteins, especially the myofibrillar ones, which play the most important role not only in the function of the muscle but also in its WHC. Changes in WHC are very sensitive indicators of changes in the structure of myofibrillar proteins (Hamm, 1960; Honikel *et al.*, 1986).

Owing to limited space, only selected aspects of water in meat, and its importance in quality aspects, can be discussed in this chapter. Several important factors related to WHC that will not be considered are the influence of animal-specific factors such as species, sex, age, muscle type and treatment of animals before slaughter. Furthermore, the influence of collagen and non-meat proteins on WHC cannot be taken into consideration.

METHODS FOR THE MEASUREMENT OF WATER-HOLDING CAPACITY

General Remarks

WHC is the ability of meat—or more generally of meat systems—to hold all or part of its own and/or added water. This ability depends on the way of handling and the state of the system. As the state of meat and its treatment differ considerably the meaning of WHC varies to a large extent. Therefore, the methods applied and the state of meat at the time of measurement must be exactly defined in order to obtain comparable results. In spite of all the variations of methods used there are three main ways of treatment which can be divided into three different basic methods of measuring WHC.

Methods of Measurement

Applying no External Force

To this group belongs the measurement of evaporation and weight loss, free drip, bag drip (Penny, 1977; Honikel *et al.*, 1986), cube drip and related methods (Howard and Lawrie, 1956), whereby the meat is left to itself under different environmental conditions. These methods are very sensitive but time consuming (one to several days) and are often sped up by the following methods.

Applying External Mechanical Force

By using positive or negative pressure the WHC of meat can be detected within a few minutes or an hour. To this group belong: centrifugation methods (Wierbicki and Deatherage, 1958; Honikel and Hamm, 1987), filter paper press method (Grau and Hamm, 1957) and suction loss methods (Fischer *et al.*, 1976). With these methods the amount of water released is far higher than with methods without external force as the pressure applied enforces the release of water from the intra- and extra-cellular space of the muscle structure. In drip loss measurements only extracellular water exudates from the meat (Offer, 1984). Therefore, a factor must be known to evaluate the actual drip loss of the meat. The matter becomes even more complicated as the state of meat changes during the time of conditioning and ageing which also influences the WHC. Therefore methods applying mechanical force reveal only the tendency of how the meat may behave in the following days but the absolute values are not directly comparable with the drip loss measurements.

Applying Thermal Force

As meat is consumed usually after heating the WHC of meat on cooking is of interest. The cooking loss is measured in a wide variety of methods (Bendall and Restall, 1983). During heating, the meat proteins denature and the cellular structures are disrupted which have a strong influence on the WHC of meat. Extra- and intracellular water will be released by the meat sample on cooking. The influence of the way of the heating and the final temperature are great on the WHC. These influencing parameters are, however, not fully recognized. Also the relationship between the above mentioned methods and cooking loss are unknown in detail.

These different methods of measurement of WHC in research and practice are due to the different interests of people who handle meat. It becomes evident that WHC also means different things to different people.

Factors which Influence the Measurement of WHC Applying the Different Methods

The WHC of all methods used depends on the pH of the meat which changes after death by the formation of lactic acid. Furthermore the WHC depends on the muscle type and species of animal due to their varying composition and structure. Above this:

The evaporation loss depends on the surface cover of the muscle tissue such as adipose tissue or wrapping material, the size of the sample (big or small), the length of the measuring period, the temperature of meat and chilling room and the air speed and air humidity within the chiller.

The drip loss depends on the size of sample, the shape of sample, the treatment during the conditioning period, the temperature of chilling, and the length of chilling period. Drip loss in meat samples can be measured under defined conditions; an optimum procedure has been recommended (Honikel, 1987*a*). The sum of evaporation and drip loss is the weight loss, which the individual is interested in.

Centrifugation method of uncooked meat; besides the factors influencing weight loss, centrifugation loss depends additionally on the speed of centrifugation ($\times g$), the time of centrifugation, and the plasticity of the meat influenced by state of meat and additives.

Filter paper press method (*Grau and Hamm, 1957*); this method is the most easy to handle and fast and thus is widely used. But there is little known about its relationship with drip loss. The results depend on: the pressure applied, the time of pressure applied, and the plasticity of the meat, i.e. prerigor state, grinding and salting.

Cooking loss depends on: the shape and size of sample, the temperature profile during cooking, the final temperature of cooking and the environment during cooking (in water, salt solution, air, wrapping).

A standardized procedure for cooking concerning shape, size, and environment is possible and has been published (Honikel, 1987*a*).

THE COMPOSITION AND STRUCTURE OF A MUSCLE

The constituent parts of lean meat besides water and proteins are lipids and lipoids of cellular and subcellular membranes amounting to 1–2%, inorganic salts (1%), and small amounts of low molecular weight substances such as amino acids and high molecular weight compounds like DNA and RNA. In a muscle of living animal, glycogen (0·7–1%) is also present.

The muscle is composed of fibre bundles which are surrounded by a collagen network, the perimysium. Bundles are an aggregation of fibres, the latter represent the muscle cells. The fibres are between 20 and 100 μm in diameter and vary in length from a few millimetres up to several centimetres. The cells again are surrounded by a connective tissue sheath, the endomysium, and the cell membrane, the sarcolemma, which separates the intracellular water from the extracellular fluid. The transport of water through this cellular membrane is rather slow (Honikel *et al.*, 1986). Within the cell about a thousand myofibrils are arranged in the direction parallel to the long axis of the cell. The myofibrils have a diameter of 1–2 μm and run the whole length of the muscle fibre. The fibrils are composed of thick and thin filaments which are well arranged in repeating units, the so-called sarcomeres (Fig. 1).

The filaments are constructed of various myofibrillar proteins. The thick filament consists mainly of myosin, a protein with a molecular weight of about 520 000 daltons and amounting to about 45% of the myofibrillar proteins (Robson and Huiatt, 1983). The thin filament is composed of actin (20%), troponin and tropomyosin (5% each).

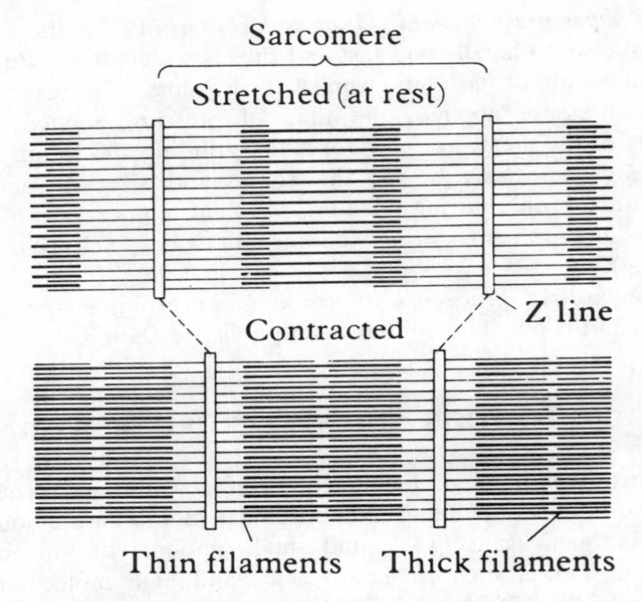

FIG. 1. Scheme of sarcomere structure at rest and contracted.

Numerous other proteins are involved in the sarcomere structure (Robson and Huiatt, 1983). One protein with 10% of the total myofibrillar weight, I want to mention additionally, is connectin (sometimes called titin). It is a high molecular weight protein (one million daltons) of a longitudinal structure running parallel to the thick and thin filaments which connectin seems to arrange in their ordered state (Robson and Huiatt, 1983). Contraction takes place when thick or thin filaments slide into each other thus shortening the length of a sarcomere which is 2–3 μm long in the resting state of a muscle (Fig. 1). One sarcomere can shorten about 0·7 μm.

The filaments as well as other subcellular structures such as sarcoplasmic reticulum, mitochondria, lysosomes, etc. are imbedded in the fluid of the sarcoplasm. This plasma contains dissolved proteins, salts and other low molecular weight compounds. It is supposed that about 20% of the water of the cell is in the sarcoplasm, the main part, however, is located in the myofibrillar space between and within the filaments (Bendall and Restall, 1983).

THE FUNCTION OF THE MUSCLE

Muscles serve the purpose of movement. Muscles contract induced by a nerve stimulus which is transmitted into the cell. Within the muscle cell the contraction of myofilaments is induced by Ca^{2+} ions which are released from the sarcoplasmic reticulum (SR). In the relaxed state the Ca^{2+} concentration around myofilaments is around 10^{-7} M, contraction starts if the Ca^{2+} concentration increases to 10^{-6} M. Contraction uses ATP as fuel. After the nerve stimulus has ended, Ca^{2+} ions are pumped back into the SR also using ATP. Also in muscles after death, before the onset of rigor mortis, at a time when ATP is still present in sufficient concentration the myofibrils can contract if Ca^{2+} ions are released from the SR. This happens if muscles are chilled rapidly or very slowly (Fig. 2 and section on chilling, page 287). In both cases Ca^{2+} ions will be released from the SR as the ATP-driven Ca^{2+} pump does not function properly any more (Honikel *et al.*, 1983).

As mentioned above on contraction the basic units of the fibrils, the sarcomeres, shorten by sliding of the thick and thin filaments into each other. As this occurs the myofibrillar space decreases and a part of the water in the fibrils must become translocated in the sarcoplasmic space. As usually a contraction lasts only a short time in a live animal, the exchange of water from one substructure to another is rather short and reversible. But after a continuous shortening has occurred as may happen *post mortem,* the displacement of water from the myofilaments increases permanently the amount of sarcoplasmic water. This water is then no longer immobilized in and between the filaments. The effect of displacement is further enhanced after death as the pH falls from 7·0 to a value around 5·5, causing a shrinkage of myofilaments.

CHANGES *POST MORTEM* AND WATER-HOLDING CAPACITY

Biochemical Changes *Post Mortem*

In a muscle of a live animal the energy of the reduced organic compounds, the carbohydrates and fatty acids are converted with oxygen into carbon dioxide and water, producing energy which is stored in ATP. Carbon dioxide is transported off by the blood stream. With the death of an animal the blood stream stops and with it the

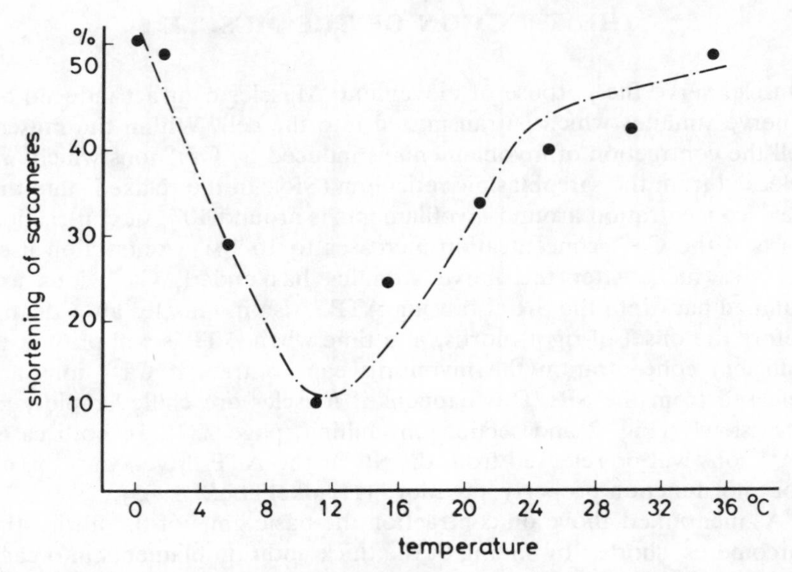

FIG. 2. Influence of temperature on sarcomere shortening in the prerigor state; original sarcomere length (0% shortening was 1·9 μm (M. longissimus dorsi of pork).

supply of energy-rich compounds and oxygen; also the removal of metabolites finishes. But the cell which contains glycogen in a concentration of 0·7–1·0% of its weight, can still convert energy anaerobically and produce ATP. The metabolism ends now with lactic acid. This acid amounts finally to concentrations of about 0·1 M, equal to about 9 g/kg. Due to lactic acid production the pH falls normally from 7·0 to a value between 5·3 and 5·8, depending on the type of muscle and animal. This process may take only one hour in extreme cases. In pigs with a normal rate of glycogenolysis the ultimate pH values are reached within 6–12 h *post mortem*; in beef this process lasts 18–40 h (Honikel and Kim, 1985).

With the exhaustion of glycogen, the concentration of ATP also falls, and finally, just before the ultimate pH is reached, the state of rigor mortis occurs, in which the thick and thin filaments are interacting permanently forming actomyosin. Within the next days or weeks the stiff (firm and tough) meat must age. The ageing is carried out by lysosomal and other proteases. As meat must get chilled soon after death the proteolytic processes occur slowly. Depending on

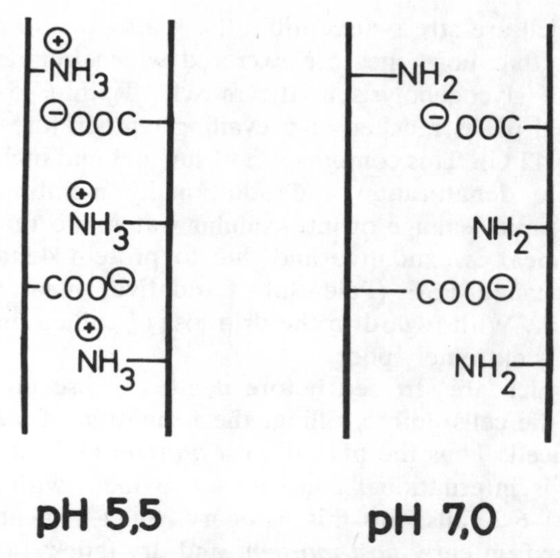

FIG. 3. Scheme of influence on pH value of swelling or unswelling of proteins due to changes in charges of amino side chains of proteins.

temperature of storage, pH of meat, muscle type (amount and crosslinking of connective tissue) and animal species (amount and activity of proteases) ageing takes 1–2 days in chicken, 3–6 days in pork and 10–20 days in beef. During the ageing process the well ordered structure within and across the myofibrils is broken lowering the firmness of the meat.

As said above the post-mortem pH fall to 5·5 causes a shrinkage of myofilaments. This is due to the fact that with falling pH the myofibrillar proteins approach their isoelectric point (IP) which is at pH 5·3 (Grau *et al.*, 1953). At the IP the protein–protein interaction is high, as the number of negative and positive charged amino acid side chains are equal and attraction forces are at a maximum. The attraction decreases the space within and between the myofilaments. Less water can be immobilized in the reduced interfilamental space (Fig. 3). Expressed in colloid chemical terms the swollen gel of myofilaments in a muscle of a live animal turns into a less swollen gel *post mortem*, leading to a reduced immobilization of water.

Abnormal Biochemical Changes *Post Mortem*

In a number of countries of the world the breeds of slaughter animals show due to the handling of the animal ante-, intra- and post-mortem abnormal changes in the muscles.

In pigs which are stress-susceptible the killing of the animal causes such a stress that hormones are excreted which initiate an extreme stimulation of glycogenolysis in the muscle. Within 45–60 min after death the final pH is reached at prevailing temperatures of 35°C and higher (up to 42°C). This combination of low pH and high temperature causes protein denaturation and additionally membrane disorders, leading to a rapid leakage of intracellular water into the extracellular space. This meat is exudative and due to protein denaturation also pale and is called PSE (Pale–Soft–Exudative) meat which occurs mainly in pork. With regard to the drip loss of chilled meat the WHC of PSE pork is extremely poor.

Animals which are stressed before death can use up the glycogen store within the cells before killing; the formation of lactic acid after death is reduced. Thus the pH fall *post mortem* ends at higher values than 5·5. It is international consent that a meat with a final value above pH 6·0–6·2 falls into this category. Such a meat is darker in colour, rather firm early *post mortem*, and dry (sticky) on the surface and is called DFD (Dark–Firm–Dry) meat. It occurs in pigs and cattle.

Structural Changes *Post Mortem*

As mentioned earlier muscles in the prerigor state can contract if in the presence of ATP, sufficient for contraction, the temperature is lowered too fast or slowly (Fig. 2). Contraction enhances drip loss. Therefore biochemical changes and chilling conditions must fit each other (Honikel, 1987*b*). In order to keep shortening to a minimum the temperature at pH values above 6·0 must be between 18°C and body temperature, below pH 6·0, but before the onset of rigor mortis (pH 5·7–5·9) the temperature must be around 12–18°C, after the onset of rigor, the temperature can drop below 10°C. Observing these values contraction or shortening *post mortem* can be avoided (Honikel *et al.*, 1983). The onset of rigor mortis with or without shortening takes place in any case.

After the onset of rigor mortis the ageing process by proteases as said above must take place in order to tenderize meat. During the days of ageing the cellular membranes get leaky and water from the intracellular space is moving into the extracellular fluid. Whereas the proteolytic action of proteases does not change the WHC of meat, the disintegration of membranes affects it.

Summary

Water is immobilized in the muscle in the myofibres and is restricted in movement by subcellular and cellular membranes. In the muscles *post mortem* the pH falls under normal conditions from pH 7 to about 5·5. Approaching the IP (pH 5·3) of the muscle the proteins shrink and the interfilamental space is lowered, translocating water into the sarcoplasm. This 'free' sarcoplasmic water penetrates slowly into the extracellular space and finally appears at the muscle surface.

A fast pH fall occurring in PSE-prone muscles at prevailing temperatures above 35°C causes a partial protein denaturation and membrane leakage. Fluid is moving easily into the extracellular space. Fast or very slow chilling causes shortening of myofibrils decreasing the interfilamental space further, above the action of pH fall.

Ageing itself does not change the immobilization of water in the myofibres, but with progressing time membrane structures disintegrate and water leaves the muscle cell more easily, enhancing the drip loss.

PROCESSING OF MEAT AND WATER-HOLDING CAPACITY

Chilling

Due to the danger of spoilage meat must be chilled soon after slaughter. From a hygienic as well as an economic point of view rapid chilling is preferable as with a faster fall of temperature less microorganisms grow and less water evaporates from the meat. From a consumer's point of view who wants a tender meat an ageing at elevated temperatures seems preferable. All these opinions have introduced a variety of chilling procedures.

We are looking at meat from the point of view of people who are interested in WHC. As mentioned above (Fig. 2) on rapid or very slow chilling contraction is taking place, leading to the increase of 'free' water in the muscle cell. This 'free' water appears with delay as water at the meat surface evaporating or dripping to the floor. Figure 4 shows how the chilling temperature influences the drip loss. One day *post mortem* there is no big difference between the drip loss of meat kept at 0°C or at 15°C (difference about 1%). The difference increases day by day. At seven days *post mortem* meat kept all the time at 0°C exhibited 8·3% drip loss, meat kept the first day at 15°C, from day 2–7 at 0°C showed 3·3% drip loss (difference 5%). At 35°C in the meat

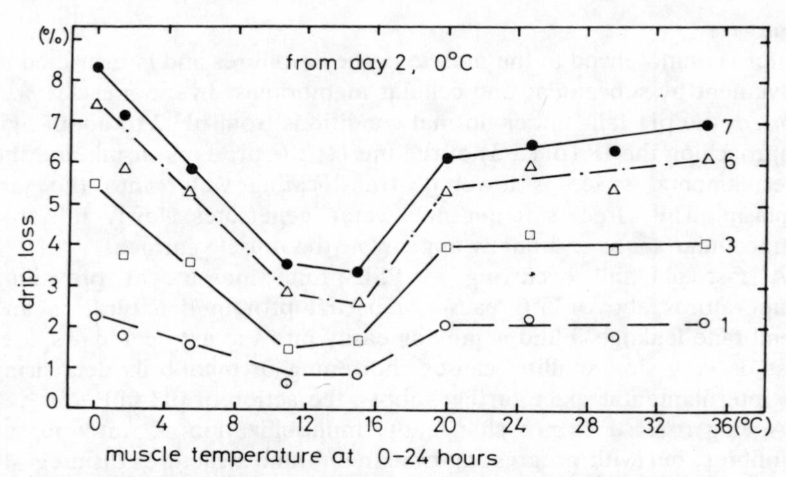

Fig. 4. Influence of temperature *post mortem* (0–24 hours) on drip loss of pork M. longissimus dorsi on 1–7 days *post mortem* (indicated on the right of the graph).

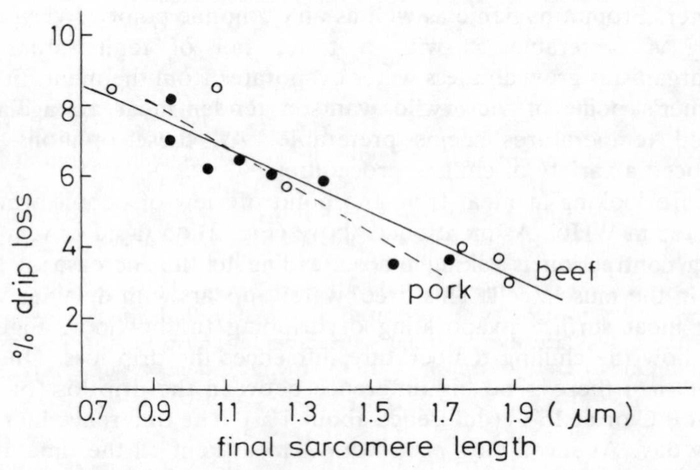

Fig. 5. Relationship between drip loss after seven days and final sarcomere length in pork M. longissimus dorsi and beef M. sternomandibularis.

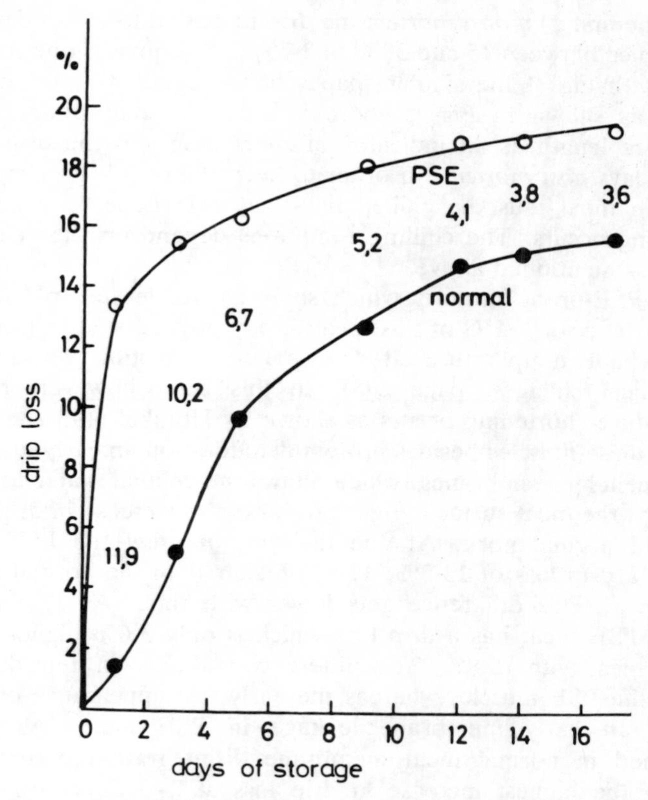

FIG. 6. Drip loss during storage of slices of 'normal' and 'PSE' M. longissimus dorsi with a pH_1 value of 5·8. Slices of M. longissimus dorsi were obtained within 45 min *post mortem* with a pH of 5·8. Two slices were stored at 38°C between 45 min and 2 h, at 35°C between 2 and 3 h, and 33°C between 3 and 4 h *post mortem*. From 4 h up to 17 days *post mortem* the temperature of storage was 0°C. These conditions are called 'PSE'. Two further slices of the same muscle were stored at 20°C between 45 min and 4 h *post mortem*. Then the temperature was 0°C up to 17 days *post mortem*. These conditions are called 'normal'. Drip loss was measured at the days indicated. The figures between the curves are the differences in percentage drip loss at the days of measurement. Data of drip loss are the mean of the two slices used.

within the first 24 h *post mortem* the drip increased to 7% at day 7; i.e. a difference between 15 and 35°C of 3·7% in drip loss. If one compares Fig. 2 with Fig. 4 the similar shapes of the curves become obvious. Indeed as shown in Fig. 5 there is a linear relationship between sarcomere length as an indicator of shortening and the drip loss at several days *post mortem*. In order to keep the drip loss of meat to a minimum meat must be chilled in a way that none or a minimum shortening occurs. The chilling conditions depend on the velocity of pH fall as mentioned above.

With PSE-prone muscles which show an accelerated pH fall *post mortem* the poor WHC of this meat also expressed as drip loss is due to the high temperature at low pH (see section on abnormal biochemical changes, page 285). In this case, however, no high temperature shortening occurs as shown by Honikel (1987*b*). But we have in these muscles besides protein denaturation an early disintegration of cellular membranes, which allows the cellular water to appear rapidly at the meat surface. Figure 6 shows the increase in drip loss of PSE and normal pork. At one day *post mortem* the PSE muscle exhibits a drip loss of 13·5%, 11·9% higher than the normal muscles with 1·6%. This difference gets lower with time. At 17 days *post mortem* PSE meat has a drip loss which is only 3·6% higher than in normal meat with 15·8%. This difference is due to protein denaturation of the PSE muscle, whereas the early fast appearance of drip is due to an early membrane leakage in PSE meat. As already mentioned in normal meat membrane disintegration occurs slowly showing the highest increase in drip loss at 2–6 days *post mortem* (Fig. 4).

TABLE 1
Influence of chilling rate *post mortem* on drip loss in slices (*c.* 75 g) of PSE-prone ($pH_1 = 5{\cdot}45$) M. longissimus dorsi of pork

Min post mortem *to 34°C*	*% Drip after days* post mortem				
	0·3	*1*	*2*	*5*	*8*
45	1·1	2·7	5·0	9·1	10·8
51	4·8	6·9	8·3	10·3	11·7
78	7·1	10·0	11·8	13·1	14·5
138	9·9	10·8	12·2	13·6	14·4

The rapid pH fall in PSE-prone muscles cannot be stopped, but the chilling rates can be enhanced. The increase in chilling rate has a very pronounced effect on drip loss as Table 1 shows. The time to reach 34°C in the muscles expresses the velocity of chilling. At 0·3 days *post mortem* the drip loss varies from 1·1% with the fastest chilling rate to 9·9% in those samples where 34°C is reached 138 min *post mortem*. At day one differences from 2·7–10·8% are observed. With progressing time the difference gets smaller, being 3·6% at eight days *post mortem*.

PSE pork means that the pale meat loses water very early and fast (Figs 6 and 7). It is extremely exudative in the first 5–6 days *post mortem*, in the period of time when pork is sold and consumed. Rapid chilling (34°C must be reached within 50–55 min *post mortem* (Fig. 7)) improves the WHC of PSE-prone muscles, i.e. it reduces drip loss.

In conclusion muscles with a normal rate of glycolysis must not be chilled too fast and below 10–15°C before the onset of rigor mortis, PSE-prone muscles must be chilled rapidly very early in order to reduce the high drip loss. Unfortunately in a carcass both types of muscles occur and the chilling conditions must be adjusted to the need of different muscle types. Chilling within 30–45 min *post mortem* in a

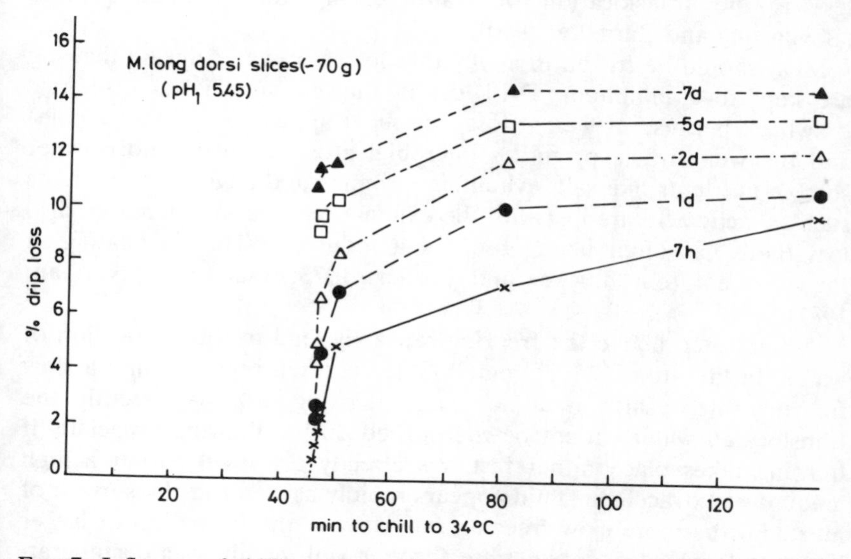

FIG. 7. Influence of chilling velocity (expressed as min *post mortem* to chill to 34°C) on drip loss of PSE-prone pork muscles between 7 h and 7 days *post mortem*.

freezing tunnel of about −25°C for a period of time followed by further chilling in cold room at 3–5°C will help to solve the problem.

Freezing and Thawing

The water in meat starts to form ice crystals at about −1°C; at −5°C about 80% of the freezable water is frozen. At −22°C 90% of the tissue water has formed ice crystals. Further decrease of temperature shows little effect as a small part of water does not freeze (Riedel, 1961; Hamm, 1985). Freezing of meat under commercial conditions is done usually in pieces with rather slow freezing rates. The formation of ice crystals begins in the extracellular fluid, increasing the concentration of soluted substances, e.g. salt ions. This draws water osmotically from within the still unfrozen intracellular space. Slow freezing of muscular tissue results in large ice crystals, which are located entirely in the extracellular areas. The meat fibres in this slow frozen product have a shrunken appearance. By rapid freezing numerous small ice crystals are formed located within and outside the cells, and the faster the transition from −1°C and −7°C occurs, the less is the translocation of water during the freezing process (Bevilacqua and Zaritzky, 1980).

Meat should be frozen in a way that losses of water during thawing are kept to a minimum. Drip loss in thawed meat is an economic drawback because of weight loss; the meat appears unpleasant (pale) and the wet surface promotes microbial growth. Furthermore water soluble nutrients like salts, vitamins, proteins and flavour components such as lactic acid are lost with the exudate. Besides an increased drip loss there has often been observed a reduced WHC on heating of thawed meat (e.g. Locker and Daines, 1973; Tsai and Ockerman, 1981).

As indicated above the freezing rate may lead to a translocation of water. In literature (see Hamm, 1986) it is often reported that a slow freezing rate results in an increased thawing drip. Apparently the translocated water cannot be reabsorbed during thawing especially if thawing takes place rather fast. As already discussed earlier a high amount of extracellular fluid appears rapidly as water at the surface of meat. Furthermore slow freezing rates favour the formation of larger ice crystals which may penetrate through cell membranes disintegrating the structures. It has also been postulated that in slowly frozen muscle myofibrillar proteins are denatured due to an increase in salt concentration in the unfrozen fluid during freezing. There seems, however, to exist little experimental evidence to prove the latter hypothesis (Hofmann and Hamm, 1978; Anon and Calvelo, 1980) and also in our own experiments we found that low freezing did not change the protein solubility considerably.

Differences in the effect of freezing and thawing on the WHC of meat can be explained with the pH of the meat, muscle and species specific characteristics and the velocity of freezing and thawing. Bevilacqua *et al.* (1979), Anon and Calvelo (1980) and Hamm (1986) reported about this. These authors found a rather complicated relationship between the freezing time (−1 to −7°C) and the amount of thaw drip. This relationship is governed by a sequence of events of changes in location and size of ice crystals, intra- and extracellular ice formation, translocation of water, and damage of fibres.

We carried out experiments by freezing postrigor meat with two different freezing rates, a short storage time at −20°C, followed by thawing at different rates. Figure 8 shows the results. Freezing increases water loss in every case. Fast freezing and fast thawing showed by far the lowest thaw loss for seven days, followed at three days post thawing by slow freezing and fast thawing, but exhibited the highest thaw loss in the first day after thawing. Fast freezing and slow thawing was third. The worst combination was slow freezing and slow thawing. The results can be explained as shown in the scheme of Fig. 9. On fast freezing small ice crystals are formed rapidly and no water is

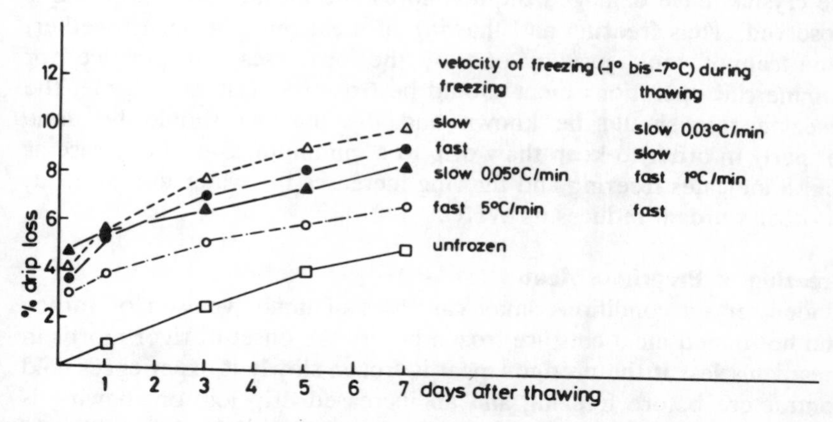

FIG. 8. Drip loss of thawed meat. Postrigor M. sternomandibularis of beef was frozen fast and slow, stored for two days at −20°C and thawed fast and slow. The drip loss was measured for seven days and compared with unfrozen meat.

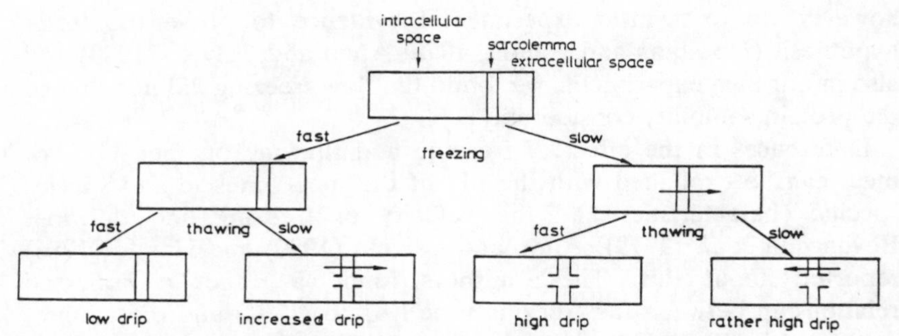

FIG. 9. Scheme for the shift of water between the muscle cell and the extracellular space (for details see the text).

translocated. The intra- and extracellular spaces stay unchanged. By slow freezing translocation of water occurs, the extracellular fluid increases. Fast thawing of fast frozen meat shows little effect and the thaw loss is low. Slow thawing of fast frozen tissue does not cause any considerable translocation but ice crystals grow during the slow thawing which increases by membrane leakage the amount of water in the extracellular space. On fast thawing of slowly frozen meat, the reabsorption of fluid into the intracellular space cannot take place and a high thaw loss will occur immediately. Slow thawing of slowly frozen meat allows a partial reabsorption of water by the myofibres. But large ice crystals have damaged the myofibres and an increased drip loss is observed. Thus freezing and thawing of meat must be seen together. One cannot look just at one of the processes. In practice for commercial operations meat should be frozen as fast as possible, the freezing rate should be known and the thawing should be done properly in order to keep thaw drip to a minimum. But in any case as Fig. 8 indicates freezing and thawing increases the water loss of meat, in other words it reduces its WHC.

Freezing of Pregrigor Meat

Under certain conditions small carcasses of lamb, venison or turkey and hot-boned meat cuts are frozen before the onset of rigor mortis in these muscles. If the prerigor meat is frozen slowly it experiences cold contracture before freezing and an increased drip loss on thawing is observed. Rapid freezing means that the temperature range of cold contracture (+10°C to the freezing point) is passed quickly before extensive cold shortening can occur. In the frozen meat, however, glycogen and ATP are still present. Thawing this meat causes a thaw contracture (Marsh and Thompson, 1958; Fischer and Honikel, 1980; Honikel and Fischer, 1980). Thaw contracture takes place in red and white prerigor frozen muscle. It is prevented only if the level of ATP concentration is as low as 0·1 mM (Davey and Gilbert, 1976). Thaw contracture causes like cold shortening not only a toughening of meat but also a strong increase of drip or thaw loss (Fischer and Honikel, 1980). The sarcomere shortening in thaw contracture is extremely extensive. On cold contracture the sarcomeres can contract by 60% usually between 30–45%, reducing the original sarcomere length from 2 μm to maximally 0·8–1·4 μm. In thaw contracture sarcomere shortening by 70% to a length of 0·6 μm is rather common. It takes place immediately after the thawing of the ice crystal matrix and lasts only a few minutes. Within 2–3 h 15–20% of the weight of the meat is lost as thaw drip. This fast contracture is induced by Ca^{2+} ions as cold shortening is initiated. Calcium ions are translocated during freezing from the sarcoplasmic reticulum into the myofibrillar space.

If prerigor meat is frozen fast without shortening there exists, however, a way to prevent thaw contracture. Glycogen and ATP must be metabolized before the ice matrix melts. The post-mortem metabolism of glycogen and ATP is interrupted below −18°C (Fischer and Honikel, 1980). But increasing the temperature slowly to the melting point increases the glycogen and ATP turnover. Thus fast frozen prerigor meat must be thawed slowly in order to keep its WHC high; the temperature increase from −20°C to −1°C must take about 12 h (Honikel and Fischer, 1980). For the similar purpose prerigor meat can be stored at −12°C for about 20 days (Davey and Gilbert, 1976) or at −3°C for about two days (Bendall, 1974). Due to these problems prerigor freezing of meat cannot be recommended except if some precautions are taken and the facts are known to the persons who thaw the meat.

Heating

During heating the most drastic changes occur in meat, such as shrinkage and hardening of tissue and release of cooking juice. These changes are caused by structural changes of myofibrillar proteins and of membrane structures. The structure changes can be observed with differential scanning colorimetry (Wright *et al.*, 1977) and are shown in Fig. 10.

The denaturation of myosin (peak A) starts at about 40°C ending at

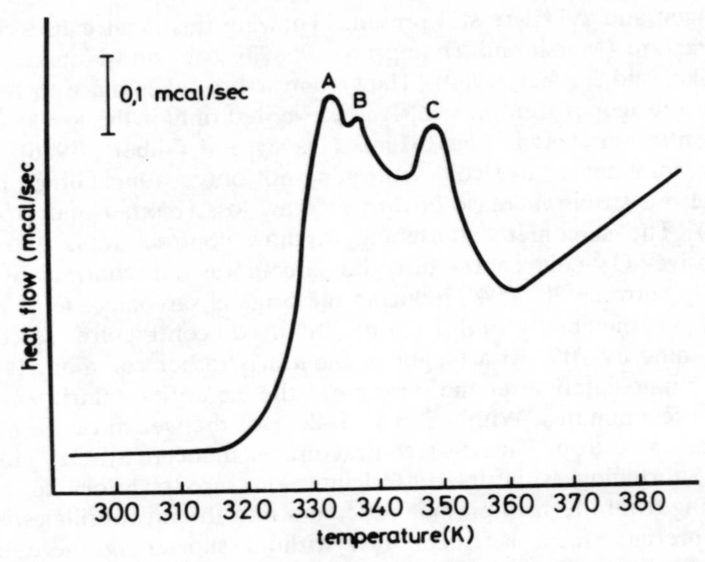

FIG. 10. Differential scanning calorimetry thermogramme of postrigor bovine muscle.

about 60°C, followed by sarcoplasmic proteins and collagen denaturation (peak B). Finally the myofibrillar protein actin or actomyosin (peak C) denatures (75–80°C). These changes are characterized by a rearrangement of the tertiary structures of the proteins, accompanied by an association into aggregates (Hamm, 1977). The heat-induced structural changes apparently lead to a reduced immobilization of water as with progressing heating time and with increasing temperature of heating the cooking loss increases as shown in Fig. 11. The cooking loss can be as high as 40% of the raw muscle weight on heating to 95°C. Slow heating rates to a certain final temperature results in a higher cooking loss (Fig. 12) than fast heating. As observed in all aspects of WHC also the cooking loss depends on the pH of the meat. The higher the pH, the lower is the cooking loss (Fig. 11). It is interesting to note as shown in Table 2 that prerigor shortening induced by storage at various temperatures has no effect on cooking loss. This fact indicates that cooking causes further reaching changes in meat than shortening or fast glycolysis do to raw meat. Drip loss never exceeds 20%, cooking may cause a loss of 45% of the original weight.

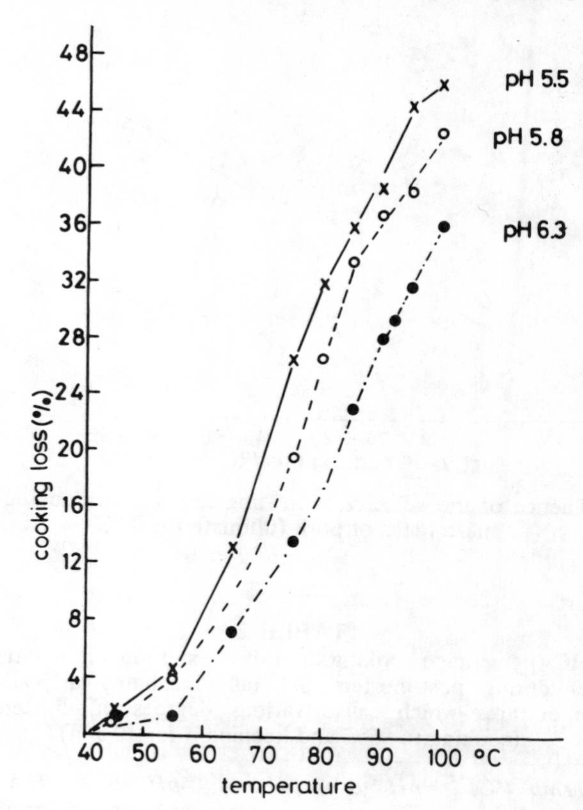

FIG. 11. Temperature of heating and cooking loss at various final temperatures of heating and the influence of the pH of the meat (heating velocity was constant 2·5°C/min).

Summary

Processing of meat is necessary to store and prepare meat. If chilling, freezing, and thawing are done in a proper way, the influence on WHC remains small; prerigor shortening must be avoided. Meat is usually consumed after heating. The meat we eat should be tender but also juicy. Juiciness is related to water retention. The demand of a high tenderness of heated meat must be combined with its WHC. We found in our own investigations that this can be done (Seuß and Honikel, 1987).

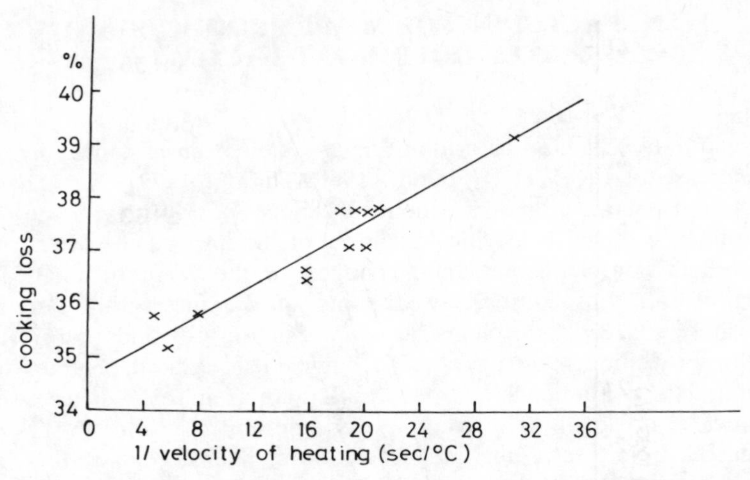

FIG. 12. Influence of the velocity of heating to 95°C on cooking loss of M. mastoideus of pork (ultimate pH 5·8).

TABLE 2

WHC (percentage cooking loss) of M. sternomandibularis of beef during post-mortem pH fall, incubated at various temperatures which cause various degrees of shortening (cooking to 95°C and keeping it for 10 min)

Temp. (°C)	*pH 6·8*	*pH 6·1*	*pH 5·9*	*pH 5·5*
0·5	34	42	44	—
4	34	40	41	44
5	36	41	42	45
7·5	37	40	42	45
10	27	38	40	45
14	33	39	41	44
17	32	40	43	44
20	37	42	43	44
20	35	39	41	43
23	33	40	42	44
24	37	44	46	53
27	37	39	40	42
30	35	41	42	45
$\bar{x}$	34·4	40·4	42·1	44·8
SD	±2·8	±1·6	±1·7	±2·7

MANUFACTURING OF MEAT PRODUCTS AND WATER-HOLDING CAPACITY

Salting

The addition of sodium chloride to meat systems causes a swelling and an increase of the WHC (Hamm, 1960; Hamm, 1972). Besides the effect of pH (see section on biochemical changes *post mortem*, page 283) the influence of salt is a typical example of the importance of charged amino acid side chains and their changes for the WHC of meat. The effect of salt ions causes a weakening of the interaction between oppositely charged side chains as shown in Fig. 13 and therefore a swelling. Swelling means the space between associated proteins like myofibrils and filaments increases and in this enlarged space water molecules move in. The water gets immobilized and the WHC increases. Infinite swelling means solution. Therefore, swelling and dissolving of myofibrillar proteins happen at the same time. Meat tissue with its ordered state of myofibrils, and after the onset of rigor also crosslinks of actomyosin, has only a limited swelling capacity. In order to increase it, meat pieces must be mechanically massaged or tumbled as is done on preparing cooked ham. Much more effective is mincing and comminuting because with knives the well ordered structures of the myofibrils are cut into small pieces and an increased swelling and dissolving can occur. In meat products a measurable

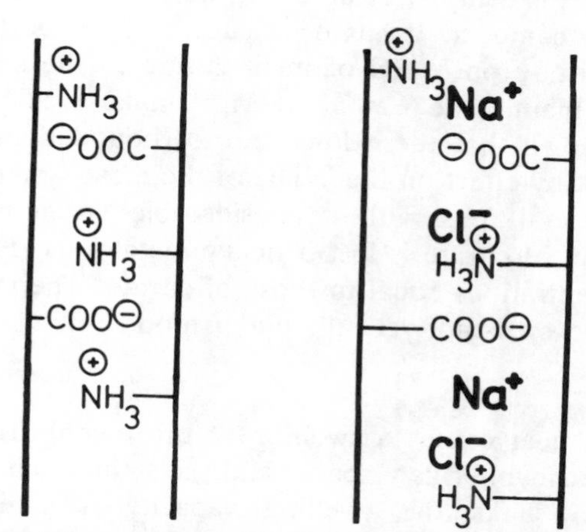

FIG. 13. Scheme of the action of salt ions on swelling of proteins due to the interaction of salt ions with charged protein side chains.

increase in WHC is observed by the addition of more than 1% salt equal to 0·17 M NaCl. The maximum swelling is achieved with about 0·85 M (5%) NaCl. Above 5% salt the myofibrillar proteins are salted out and a decrease in WHC is observed (Hamm, 1957). On ageing the crosslinks disintegrate and the swelling capacity increases with salt (Hamm, 1972).

Meat salted in the perigor state i.e. without actomyosin crosslinks swells more intensively which is also due to its higher pH. But this prerigor effect of salt remains. Even after the pH has fallen to its final value this salted meat keeps its high WHC. Therefore prerigor salting of meat is recommended to increase the WHC of sausages prepared from this presalted homogenate. This is one aspect of the process of hot deboning of carcasses.

Addition of Additives

Di- and polyphosphates are used worldwide to improve the WHC of salted meat products. As we and others (Yasui *et al.*, 1964; Bendall, 1974) have shown diphosphate acts as an ATP-analogous substance causing the dissociation of actomyosin in postrigor meat and thus increasing the swelling capacity of the myofibrillar system in the presence of salt.

Sodium citrate also improves the WHC of meat systems but to a lesser extent than diphosphate (Hamm, 1972). The mode of action of citrate, however, is completely different, because citrate does not split the actomyosin complex. It has been suggested in the literature that the effect of citrate on WHC of meat is due to the sequestering of bivalent ions, mainly the Ca^{2+} and Mg^{2+} ions. Denk and Honikel (1987) have shown that these ions are not bound tightly enough to citrate to have any effect on the WHC of the meat system. But citrate binds heavy metal ions with a considerable strength. The WHC increasing effect of citrate is lost if heavy metal ions (Fe^{2+}/Fe^{3+}) are added to concentrations equal to those of citrate. The mode of action of citrate, however, is not yet fully understood.

Summary

Salt added to meat causes a swelling of the myofibrils. More water moves in the myofibrillar space and gets immobilized. Due to actomyosin crosslinks the swelling capacity is limited. Cutting, grinding, and comminution breaks part of the crosslinks; ageing has a similar effect.

The addition of additives especially diphosphate as an ATP-analogous substance causes a dissociation of the actomyosin complex, also enhancing the swelling capacity of meat as it occurs in prerigor meat, when ATP is still present.

CONCLUDING REMARKS

It has been shown that the WHC of meat represents an important quality aspect of this food. The general term WHC can express a number of different things like drip loss in raw meat, cooking loss in heated meat and the WHC of salted products. The WHC of meat in these different states depends on a number of factors. A common factor for all the states of meat is its pH value.

The pH which changes *post mortem* from about 7 to about 5·5 lowers the WHC in any case. The rate of pH fall influences the WHC if it is too fast and denaturation of proteins and membrane disorders take place at prevailing high temperatures above 35°C at pH values below 5·8. The chilling rate may also cause prerigor contraction; contraction increases drip loss but not cooking loss. The onset of rigor mortis with the formation of actomyosin limits the swelling capacity of meat after salt addition.

Ageing, mechanical disruption by mincing and comminuting or addition of diphosphate break off the highly ordered myofibrillar postrigor system and increases the WHC by a facilitated swelling. Freezing and thawing influences WHC in a rather complicated way, leading always to an increase in thaw loss. In the same way heating and heating rate change the WHC.

The water of meat is more or less limited in movement by immobilization within the myofilamental space, within the cell surrounded by membrane and within the capillaries of the extracellular space. Movements between these spaces change the WHC. Nearly all processes during the post-mortem changes, during processing and manufacturing cause a movement of water between the different spaces. The knowledge about these movements and their causes will help to improve the eating quality of meat and initiate innovations in meat industry in developing tasty and juicy new meat products.

REFERENCES

Anon., M. C. and Calvelo, A. (1980). Freezing rate effects on the drip loss of frozen beef. *Meat Sci.*, **4**, 1–14.

Bendall, J. R. (1974). The snags and snares of freezing rapidly after slaughter. In: *Meat Freezing—Why and How*? MRI Symp. No. 3, C. L. Cutting (Ed.), MRI, Langford, UK, pp. 7.1–7·8.

Bendall, J. R. and Restall, D. J. (1983). The cooking of single myofibres, small myofibre bundles, and muscle strips of beef M. psoas and M. sternomandibularis of varying heating rates and temperatures. *Meat Sci.*, **8,** 93–117.

Bevilacqua, A. E. and Zaritzky, N. E. (1980). Ice morphology in frozen beef. *J. Food Technol.*, **15,** 589–97.

Bevilacqua, A. E., Zaritzky, N. E. and Calvelo, A. (1979). Histological measurements of ice in frozen beef. *J. Food Technol.*, **14,** 237–51.

Davey, C. L. and Gilbert, K. V. (1976). Thaw contracture and disappearance of adenosinetriphosphate in frozen lamb. *J. Sci. Food Agric.*, **27,** 1085–92.

Denk, G. and Honikel, K. O. (1987). *About the Influence of Metal Ions on the Stability of Frankfurter Type Sausages and the Effect of Citrate,* Vol. II, Proc. 33rd ICoMST, Helsinki, pp. 247–250.

Fennema, O. R. (1977). Water and protein hydration. In: *Food Proteins,* J. R. Whitaker and S. R. Tannenbaum (Eds), Avi Publ. Co., Westport, CN, pp. 50–90.

Fischer, Chr. and Honikel, K. O. (1980). Schnelles Auftauen von schlachtfrisch gefrorenem Rindfleisch: Biochemische Veränderungen, Taukontraktur und Wasserbindungsvermögen. *Fleischwirtschaft,* **60,** 1703–7.

Fischer, Chr., Hofmann, K. and Hamm, R. (1976). Erfahrungen mit der Kapillarvolumetermethode nach Hofmann zur Bestimmung des Wasserbindungsvermögens von Fleisch. *Fleischwirtschaft,* **56,** 91–5.

Grau, R. and Hamm, R. (1957). Über das Wasserbindungsvermögen des Säugetiermuskels. *Z. Lebensm. Unters. Forsch.*, **105,** 446–60.

Grau, R., Hamm, R. and Baumann, A. (1953). Über das Wasserbindungsvermögen des toten Säugetiermuskels, I. *Biochem. J.*, **325,** 1–11.

Hamm, R. (1957). Über das Wasserbindungsvermögen des Säugetiermuskels, Die Wirkung von Neutralsalzen. *Z. Lebensm. Unters. Forsch.*, **106,** 281–97.

Hamm, R. (1960). Biochemistry of meat hydration. *Adv. Food Res.*, **10,** 355–463.

Hamm, R. (1972). *Kolloidchemie des Fleisches,* Parey Verlag, Berlin.

Hamm, R. (1975). Water-holding capacity of meat. In: *Meat,* D. J. A. Cole and R. A. Lawrie (Eds), Butterworth, London, pp. 321–38.

Hamm, R. (1977). Postmortem breakdown of ATP and glycogen in ground muscle. *Meat Sci.*, **1,** 15–39.

Hamm, R. (1985). The effect of water on the quality of meat and meat products: problems and research needs. In: *3rd International Symposium on Properties of Water in Relation to Food Quality and Stability,* D. Simatos and J. L. Multon (Eds), M. Nijhoff Publ., Dordrecht, Netherlands, pp. 591–602.

Hamm, R. (1986). Functional properties of the myofibrillar system and their measurement. In: *Muscle as Food,* P. J. Bechtel (Ed.), Academic Press, New York pp. 135–99.

Hofmann, K. and Hamm, R. (1978). Sulfhydryl and disulfide groups in meat. *Adv. Food Res.*, **24,** 1–11.

Honikel, K. O. (1987*a*). How to measure the water-holding capacity of meat? Recommendation of standardized methods. In: *Evaluation and Control of Meat Quality in Pigs,* P. V. Tarrant, G. Eikelenboom and G. Monin (Eds), M. Nijhoff Publ., Dordrecht, Netherlands, pp. 129–42.

Honikel, K. O. (1987*b*). Influence of chilling on meat quality attributes of fast glycolysing pork muscles. In: *Evaluation and Control of Meat Quality in Pigs,* P. V. Tarrant, G. Eikelenboom and G. Monin (Eds), M. Nijhoff Publ., Dordrecht, Netherlands, pp. 273–83.

Honikel, K. O. and Fischer, Chr. (1980). Einfluß der Auftaugeschwindigkeit auf das Wasserbindungsvermögen von schlachtfrisch gefrorenem Rindfleisch. *Fleischwirtschaft,* **60,** 1709–14.

Honikel, K. O. and Hamm, R. (1987). Critical evaluation of methods detecting effects of processing on meat protein characteristics. In: *Chemical Changes During Food Processing, Vol. II,* S. Bermell (Ed.), Proc. IUFoST Symp., Consejo Superior de Investigationes Cientificas, Valencia, Spain, pp. 64–82.

Honikel, K. O. and Kim, C. J. (1985). Über die Ursachen der Entstehung von PSE-Schweinefleisch. *Fleischwirtschaft,* **65,** 1125–31.

Honikel, K. O., Roncales, P. and Hamm, R. (1983). The influence of temperature on shortening and rigor onset in beef muscle. *Meat Sci.,* **8,** 221–41.

Honikel, K. O., Kim, C. J., Hamm, R. and Roncales, P. (1986). Sarcomere shortening and their influence on drip loss. *Meat Sci.,* **16,** 267–82.

Howard, A. and Lawrie, R. A. (1956). *Studies on Beef Quality, Part I–III,* Dept. of Scient. and Industr. Research, Food Investigation, Her Majesty's Stationery Office, London.

Locker, R. H. and Daines, G. J. (1973). The effect of repeated freeze–thaw cycles on tenderness and cooking loss in beef. *J. Sci. Food Agric.,* **24,** 1273–5.

Marsh, B. B. and Thompson, J. F. (1958). Thaw rigor and the delta state of muscle. *J. Sci. Food Agric.,* **9,** 417–21.

Offer, G. (1984). *Progress in the Biochemistry, Physiology and Structure of Meat,* Proc. 30th Europ. Meeting of Meat Research Workers, Bristol, pp. 87–94.

Penny, J. F. (1977). The effect of temperature on drip, denaturation, and extracellular space of pork longissimus dorsi muscle. *J. Sci. Food Agric.,* **28,** 329–38.

Riedel, L. (1961). Zum Problem des gebundenen Wassers im Fleisch. *Kältetechnik,* **13,** 122–8.

Robson, R. M. and Huiatt, T. W. (1983). *Roles of Cytoskeletal Proteins, Desmin, Titin and Nebulin in Muscle,* Proc. 36th Annual Recipr. Meat Conference, National Live Stock and Meat Board, Chicago, USA, pp. 116–24.

Seuß, I. and Honikel, K. O. (1987) Einfluß der Erhitzung auf das Wasserbindungsvermögen von Fleisch verschiedener Qualitäten. *Ernährungsumschau,* **34,** 343–7.

Tsai, T. C. and Ockerman, H. W. (1981). Water binding measurement of meat. *J. Food Sci.*, **46,** 697–707.

Wierbicki, E. and Deatherage, F. F. (1958). Determination of water holding capacity in fresh meats. *J. Agric. Food Chem.*, **6,** 387–92.

Wright, D. J. Lead, J. B. and Wilding, P. (1977). Differential scanning calorimetry, studies of muscle and its constituent proteins. *J. Sci. Food Agric.*, **28,** 557–64.

Yasui, T., Fukazawa, T., Takahashi, K., Sakanishi, M. and Hashimoto, Y. (1964). Specific interaction of inorganic polyphosphates with myosin B. *J. Agric. Food Chem.*, **12,** 399–404.

Chapter 9

APPLICATIONS TO CONFECTIONERY PRODUCTS

B. Brockway

Department of Food Science and Technology,
University of Reading, UK

INTRODUCTION

The skilled confectioner will say that the relationship between water and confectionery quality is very much one of common sense. The confectioner will probably also stress the importance of getting the correct amount of water into a confectionery formula; if too much water is present then the product is likely to be sticky and if too little water is available then the product may be chewy, brittle or dry. A confectioner might also suggest that it is a good idea to protect sweets from atmospheric water by wrapping them or sealing them before storage.

Bakery-confectioners regularly use flour which must be of a suitable quality. Water can affect the quality of the raw materials, thus it will probably be stated that flours which have detrimentally high enzyme activity (amylases and proteases) are often the result of milling grain that has been stored wet (Manley, 1983).

Grain and flour are sold by weight and so there is a great temptation to add, rather than remove, water. Unfortunately when grain or flour is stored at too high a moisture level they can be spoilt by microbial growth, especially moulds (Manley, 1983).

The skilled confectioner therefore knows that water can affect the quality of his product by influencing its characteristics, both in production and storage, and that it can affect the quality and costs of raw materials.

It is necessary to have a thorough understanding of the role of water in confectionery in order to prepare reproducibly high quality products and to be able quickly to identify problems as they arise during production.

This information is also invaluable to those who are interested in improving confectionery packaging, in developing new confectionery lines or those who simply wish to replace traditional confectionery constituents with novel ingredients.

Sucrose is obviously the most important component of sugar confectionery, but it is the way that sucrose interacts with water that dictates the major attributes of the products. It could therefore be said that water is the second most important ingredient in sugar confectionery (Cakebread, 1975).

In a similar manner the rheology of flour dough affects the characteristics of the baked product and the necessary visco-elastic behaviour of dough is also as a result of how carbohydrates and in this case protein also interact with water (Manley, 1983).

It is important to remember that water is a dynamic food ingredient which affects the consistency of all foods during both their manufacture and their storage (Morley and Grove 1987). This is particularly true for confectionery products where the ratio of water to sucrose, or water to flour, is deliberately manipulated to achieve the desired final texture (Cakebread, 1975; Manley, 1983). However, not everyone involved in confectionery is thought to appreciate this and it has been said that production managers tend to view water as just another ingredient (Morley and Groves, 1987).

When preparing confectionery formulas, for sugar- or flour-based products, it is important to calculate how much water is already present in the ingredients before adding water. This prevents unnecessary energy losses and costs at the cooking or baking stage, in addition to preventing irreversible changes that may occur during a subsequent mixing or resting stage (Alikonis, 1979; Manley, 1983).

Flour which has been milled and stored correctly should contain $13{\cdot}5\% \pm 0{\cdot}4\%$ moisture. Sugar can be purchased as a range of different products which have different amounts of water present e.g. liquid sugar which contains normally $32{\cdot}9\% \pm 0{\cdot}5\%$ water and also crystal sugar which should contain less than 0·04% moisture (Manley, 1983). Most ingredients will contain some water and this may affect how we perceive quality. The higher the viscosity of ingredients such as honey, molasses, maple syrup, glucose syrup, and liquid sugars the 'richer' the ingredient is thought to be. A change of 1% can cause a perceptible change in viscosity and also the cost effectiveness of a process (Matz, 1965).

SWEET-MAKING

The technology, or perhaps more correctly the craft, of sweet-making is based firmly on the unique properties of sucrose and how sucrose behaves in the presence of different amounts of water. Ingredients such as flavours, nuts, fruits and chocolate have only a secondary influence on the final attributes that characterize the major sweet types (Cakebread, 1975).

There are a large number of different confectioneries available throughout the world which all have sucrose as their main ingredient. The large variety of confectionery is only possible because sucrose can exist in different amorphous and crystalline states. The state of the sucrose depends on the carbohydrate concentration and therefore the amount of water present. The state of the sugar can further be influenced by the rate at which the solutions are cooled and also by the addition of small amounts of water during cooling (Brook, 1971; Alikonis, 1979).

A saturated sucrose solution contains 66·4 parts of sucrose and 33·6 parts of water at 20°C. If the temperature of the sucrose solution is raised then more sucrose can be added and the proportion of water present is consequently reduced, e.g. up to 76·4 parts of sucrose can be dissolved in 23·6 parts of water at 70°C (Brook, 1971). Increasing amounts of sucrose can be added to and dissolved in heated solutions. The solutions will become concentrated as water is lost through evaporation. If the solutions are heated too much then the sucrose will eventually caramelize; however at temperatures below the carameliza- tion point concentrated sucrose solutions can easily be produced.

Interestingly the temperature at which the saturated sucrose solution boils is related to the concentration of dissolved sucrose and the atmospheric pressure. The more concentrated the sucrose solution is the higher the temperature it boils at and when the atmospheric pressure is reduced the boiling point of the solution is lowered proportionally. A thermometer is all a confectioner needs to judge when the correct amount of water is present during sweet manufac- ture. It is common practice for confectioners to manipulate the con- centration of sucrose and therefore the amount of water present in the final product, by boiling their formulas to specific temperatures. If any of the ingredients are heat labile the procedure is simply carried out under reduced pressures (Brook, 1971; Alikonis, 1979; Lees, 1980).

Having prepared saturated sucrose solutions at specific concentrations the next stage in candy production is to make the solutions supersaturated.

Supersaturated solutions are formed by carefully cooling the saturated solutions. These solutions are unstable and eventually become solid. The temporary instability of supersaturated solutions is exploited by confectioners to form many of the various types of sweets. This is done by a variety of different means such as by careful seeding using sucrose crystals to produce a product which has sucrose present both in the amorphous and the crystalline state. These types of candies which contain sugar crystals are described by confectioners as grained.

Other techniques are used to influence the final candy texture. A combination of aeration and cooling encourages the sugars to crystallize; this results in opaque fondants suitable for peppermint creams or marshmallows. Fudges are made by adding small amounts of cold water to the concentrated sucrose solution during the cooling procedure, this again enables the sucrose to crystallize and results in a solid but short textured product (Brook, 1971; Lees, 1980).

Smooth, clear, hard, boiled sweets contain approximately 96–98% carbohydrate solids, most of which is sucrose. These candies are made from supersaturated solutions which on cooling form a 'glass'. The sucrose in these candies is therefore in an amorphous state and the sweet is said by confectioners to be non-grained (Brook, 1971).

When the moisture content of a hard candy is raised from 2% to 4% the product becomes noticeably softer but not chewable. The hardness improves dramatically as the moisture is reduced. A water content of 1·5% gives boiled sweets their optimum hardness. Any further loss of moisture does not affect the hardness (Matz, 1965).

If the candy is to be crystal clear it is essential that the correct quantity and quality of water are used. Soft water is normally used in combination with the purest sucrose available, which should contain little to no ash (Matz, 1965; Alikonis, 1979; Lees, 1980).

The characteristics that determine the quality and also the desired final texture of the sweet will vary between different confectionery products, however every confectioner aims to produce a product that is stable and therefore has a long shelf-life (Gilbert, 1986). Some confectionery ingredients indirectly influence the stability of the sugar within the sweet by controlling the availability and movement of water and they are the main means by which the confectioner maintains quality (Morley and Groves, 1987). These ingredients are called 'doctors'. Examples of 'doctors' are glucose syrup, sorbitol, and egg white (Matz, 1965). For example, plain hard boiled candies also contain small quantities of glucose syrup as well as sucrose. Glucose syrup is deliberately included in the formula because it competes with sucrose for the available water and thus prevents the sucrose from crystallizing (Brook, 1971).

It is important to remember that there are many different types of glucose syrup and that they are products of starch hydrolysis. The different syrups contain glucose either as the monosaccharide or as complex mixtures of mono-, di-, and oligosaccharides. The actual composition will vary according to the production method and some syrups are more suitable for this purpose than others (Kennedy *et al.*, 1987).

STORAGE

The ratio of water to sugar within the candy is readily influenced by atmospheric water and it is therefore essential that sweet manufacture, and storage, is carried out in areas which are well ventilated and definitely not damp and that the final product is protected from the moisture in the air by the correct packaging (Bush, 1928; Matz, 1965).

A quality product must have a reasonable shelf life and therefore different components such as toffee and biscuit, should be close to or at equilibrium with themselves and with the immediate environment. A stable product can therefore only be achieved when water availability is constant and movement is prevented (Morley and Groves, 1987).

One of the major problems with hard boiled candies is that they tend to 'grain-off' by picking up moisture during storage. Sucrose in hard boiled candies is in the amorphous state where all the (—H) bonds are free to interact with the (OH) of water and thus water is readily taken up from the atmosphere. Rock candies can withstand relatively high humidities. This is because sucrose is in the crystalline state and therefore the (—H) bonds are locked internally and it is difficult for the (OH) groups of water to penetrate (Richardson, 1986).

The moisture transferred from the atmosphere to the non-crystalline, amorphous sucrose enables the sucrose molecules to change into the crystalline state, resulting in the candies becoming 'grained', i.e. crystallized, opaque, granular and sticky (Alikonis,

1979). Stickiness generally reaches a maximum twice during this deterioration process. The first occurrence of stickiness happens shortly before the beginning of crystallization and the second when the crystal lattice begins to break down. (This later stage is not seen under normal conditions.)

Deteriorated candies can be coated in granulated sugar to *conceal* but not prevent the resulting undesirable textural changes (Matz, 1965). The problem can be solved by maintaining a proper sugar–glucose syrup balance and by rapid cooking to prevent the formation of significant reducing sugar (Alikonis, 1979).

When pure sucrose solutions are heated they rapidly become acidic due to sucrose 'inverting' from the non-reducing disaccharide to the reducing monosaccharides, fructose and glucose. This sucrose inversion produces an equal molar mixture of fructose and glucose which is called 'invert sugar'. Sucrose is quite stable in neutral solutions but in acidic solutions it readily hydrolyses. The reaction can be accelerated by adding small amounts of food acids, salts of food acids, fructose or invert sugar to the solution (Cakebread, 1975).

The inversion reaction can be expressed empirically as follows:

$$\underset{\text{Sucrose}}{C_{12}H_{22}O_{11}} + \underset{\text{Water}}{H_2O} \rightarrow \underbrace{\underset{\text{Glucose}}{C_6H_{12}O_6} + \underset{\text{Fructose}}{C_6H_{12}O_6}}_{\text{Invertsugar}}$$

Note that the inversion reaction requires the presence of some water to hydrolyse the glucosidic bond. It may be necessary in some very concentrated solutions to add small amounts of water to compensate for this water loss or the inversion may be deliberately employed to concentrate formulas (Pancoast and Junk, 1980).

The water incorporated into the products of hydrolysis results in an increase of approximately 5% solids as compared with sucrose. This may be economically significant to a confectioner producing sweets on a large scale.

Some of the properties of sucrose are altered in the presence of invert sugar. One important change is that sucrose is less soluble when invert sugar is present than in pure water due to a salting-out effect (Pancoast and Junk, 1980).

Sugar mixtures are usually stable. Crystallization does not occur until one or more of the sugars become supersaturated, and even then crystallization occurs slowly or only when seed crystals are present or the solution is agitated (Pancoast and Junk, 1980).

The degree to which sucrose inversion is allowed to continue will affect the characteristics of the finished sweet. As inversion of the sugar proceeds the finished product becomes less brittle and more hygroscopic.

The presence of invert sugar is not always desirable. It can be avoided by carefully adjusting the sucrose concentration and by boiling the formula under reduced pressure (Brook, 1971).

Where invert sugar is desirable the enzyme invertase can be added to the cooling sucrose solution. This is commonly carried out with fondant formulas to catalyse the conversion of some of the sucrose to invert sugar which then results in a liquid mass suitable for the centres of chocolate creams (Brook, 1971).

The chocolate coating is necessary for two reasons. Firstly to retain the liquid centre and secondly to prevent the hygroscopic filling from absorbing any atmospheric moisture and in doing so prevent water movement. Suitable water-tight barriers, such as wrappers or edible coatings are essential for all hygroscopic candies as a variation of as little as 1% moisture can change the textural qualities of the product (Matz, 1965; Brook, 1971).

The simplified explanation for why water moves at all was first given around the turn of this century (Gilbert, 1986). It is because water in its vapour state is free to be extremely active, but the water within foodstuffs can exist either as 'free' water or as 'bound' water. The free water is available to 'move' and function as a solvent for salts and sugars, as well as entering into chemical reactions or supporting microbial growth. The bound water is associated with other chemical entities such as the hydrophilic groups of polysaccharides and protein and it is no longer available for chemical reactions nor is it free to act as a solvent (Hardman, 1986; Morley and Groves, 1987).

Free water molecules migrate from areas of high activity to areas of lower activity in order to achieve equilibrium. The areas of lower activity are where there is little water present or the existing water is in the form of bound water (Morley and Groves, 1987).

The 'activity' of water is a way to describe how much free water is present in a sample. Moisture movement in materials such as foods is known to be driven by the need to equilibrate the differences between the '*water activities*' of the various components of the food with themselves and of course with the *'relative humidity'* of their environment (Morley and Groves, 1987). (The relative humidity of the environment is the ratio of the amount of water present in the air, or its vapour pressure, to the amount of water in the air if the air were saturated, or the vapour pressure of pure water at the same temperature, multiplied by 100 to give a percentage figure.)

Water activity (a_w) values are related to the behaviour of pure water under the same conditions as the food sample. Water activity is defined as the ratio of the vapour pressure of the water present in the food to the vapour pressure of pure water at the same temperature (Morley and Groves, 1987). A food with a high water activity has free water within its matrix and therefore contains water at a high vapour pressure.

The 'water activity' of a hard boiled sweet is shown below. (The ratio of the vapour pressure of the water within the sweet to the vapour pressure of pure water at the same temperature.)

$$a_w(\text{for a hard boiled sweet}) = \frac{P_1}{P_1^0}$$

Where P_1 is the partial vapour pressure of water in the sweet and P_1^0 is the saturated vapour pressure at the same temperature.

The actual values will vary among the different varieties of hard candies but the values will all be low because there is little to no free water present. The values will fall within the range 0–0·2 which is indicative that water is strongly bound (Hardman, 1986).

Water activity values between 0·2 and 0·7 indicate that all the water is not firmly bound and a significant amount is able to behave as free water.

The water activity of pure water is one because P_1 and P_1^0 are obviously the same value. Water activity values between 0·7 and 1·0 indicate that most of the water is free. When salts and sugars are dissolved in water the value of P_1 falls and consequently the water activity values reduce.

The fall in water activity is proportional to the ratio of the number of water molecules to the number of dissolved molecules. Therefore the addition of low molar mass material, such as invert sugar, is more effective at lowering water activity on a weight-for-weight basis than the addition of higher molar mass materials such as sucrose or the much larger oligosaccharides found in glucose syrups (Hardman, 1986;

Morley and Groves, 1987). A 66% sucrose solution has a water activity of 0·86 whereas a 66% solution of glucose has a water activity of 0·78 (Richardson, 1986).

The various methods for measuring the water activity in confections has recently been thoroughly reviewed by Von Elbe (1986) and therefore will not be included in this discussion. Methods for determining water activities in the range 0·80–0·99 which are commonly found in doughs, batters and baked goods is discussed by Fett (1973).

When discussing water activity data for foods it is important to remember that values will vary by ±0·02 depending upon the method used for the determination. A wider variation will be found in foods with higher water activities (Labuza *et al.*, 1976).

This concept that the equilibration of the various water activities determines the availability of water for its reactions affecting storage stability has recently been disputed (Gilbert, 1986). A number of key reactions ranging from ascorbic acid oxidation to microbial growth call this concept into question.

An alternative theory is that 'available water' is water in a thermodynamic state whose differential free energy for transfer from matrix to reaction system defines its availability (Gilbert, 1986).

Whichever theory is most correct it is certain that water plays a dominant role in controlling stability and that catastrophes will occur if a confectionery contains one layer with a greater water activity than another.

Confections such as fondant and fudge are composed of sucrose in the form of crystals suspended in a saturated sugar syrup. The solid crystals have no effect on the water activity and the phases are stable provided that the system is protected from atmospheric moisture by a suitable wrapper (Morley and Groves, 1987).

The water activity of the syrup phase will be related to both its composition and to any drying or water uptake that may have occurred in its history. Water activities of 0·73 and 0·75 have been quoted for fudge and fondant respectively.

Interestingly caramel which has more or less the same formula as fudge has a lower water activity. A water activity of 0·62 has been quoted for caramel. This is because caramel is a dispersion of fat globules in a high solids, highly supersaturated sugar matrix, which is a very viscous, single phase that cannot crystallize and therefore all the component sucrose molecules, rather than just those in a syrup phase, will influence the water activity (Jewell, 1986; Morley and Groves, 1987).

Most confectioneries are multicomponent containing layers of different sugary material such as caramel, grained fruit paste, biscuit, wafer, jelly, nougat, marshmallow, fudge, fondant cream, chocolate, etc. Each of these components has a different potential water activity. Water movement occurs as all the different water activities equilibrate with, in the case of unwrapped sweets, the surrounding atmosphere (Morley and Groves, 1987).

If a composite candy was prepared using caramel and nougat (which both have the same water activity) then the candy, if suitably protected from atmospheric moisture, would be in balance and there would be no moisture transfer. However if a third layer of marshmallow were introduced then, since marshmallow has a water activity of 0·71 (Morley and Groves, 1987), the candy would no longer be in balance and the caramel and nougat phases would take up water from the marshmallow. The marshmallow would eventually become rubbery and tough. The nougat would soften as sugar crystals dissolve and the caramel would become less viscous, eventually crystallizing to form a fudge.

Further problems could occur if this composite candy were placed on a biscuit base since biscuit must have a water activity of below 0·35 if it is to remain crisp (Cakebread, 1976; Manley, 1983).

CONFECTIONERY WRAPPINGS

A quality confectionery product must therefore contain all its 'layers' at the same water activity. Water activity can be altered by increasing the proportion of humectants such as glucose syrup, invert sugar, sorbitol, or glycerine present in the formula. Sometimes this makes the candy too soft and an alternative solution is sought. A barrier can be created around the offending layers to prevent water movement (Morley and Groves, 1987).

Similarly the balanced product must be protected from the moisture in the environment and this means completely enrobing in a barrier such as chocolate or better still, a suitable wrapping (Morley and Groves, 1987).

Chocolate makes a good but not perfect barrier. The moisture in chocolate is less than 2%. Under normal conditions the sugar and soluble materials in chocolate are completely immobilized because of the almost complete absence of water. However if the surface becomes wet due to condensation or bad storage conditions then its sugar dissolves, apparently to an appreciable depth and migrates to the surface where it is deposited as crystals when the moisture is reabsorbed by the atmosphere (Matz, 1965).

The different types of wrapping film available allow varying amounts of water vapour transmission. Nylon and cellulose acetate films allow significant vapour transmission whereas polyvinyl alcohol and foil laminates stop all transmission (Morley and Groves, 1987). Hermetically sealed containers may be ideal because the equilibration between the free water in the candy and the water in the low volume of air may rapidly establish a suitably stable atmosphere.

The ideal wrapping film may not be the one that completely protects the confectionery from the atmosphere. The types of film that totally prevent moisture movement also trap condensation. Condensation is inevitable if the confectionery is packaged before it has completely cooled to ambient temperature. It will also develop if the wrapped product is warmed and cooled at any time during transportation or storage. Often it is impossible for the manufacturer to control the temperatures that the product is subjected to after it leaves the place of production and so the quality of the confectionery is best preserved if it is wrapped in a film that allows some water movement (Morley and Groves, 1987).

Water condenses on the surface of a confectionery product in the form of pure water which is fully capable of dissolving the sugar present, regardless of the water activity of the surface layer since pure water has a water activity of 1·0. Hence even barrier layers, such as chocolate which have low water content, will be spoilt by condensation.

Chocolate develops sugar bloom. This is a fine layer of sugar crystals which as the name suggests appears on the surface of chocolate as a white bloom which spoils the finish and often makes the product unsaleable. If the level of condensation is left unchecked eventually mould will grow and the confectionery will be not only unsightly but it will also be totally ruined (Alikonis, 1979; Morley and Groves, 1987).

Most confectionery is resistant to microbial spoilage because the low levels of available moisture in candies effectively reduce the moisture available to support microbial growth. Most microorganisms require water activity values in the range of 0·8–0·85 or above (Von Elbe, 1986).

When sucrose is being used as a preservative the substrate must contain sucrose in solution at concentrations above 70% soluble solids. If the product is acid (say below pH 4·5) then the concentration of sucrose can be reduced to about 65% soluble solids (Brook, 1971).

JAMS AND JELLIES†

The manufacture and preservation of jams depends on this high sugar, high acid principle (Brook, 1971). The quality of the colour and flavour can be faithfully retained.

The mould growth that is commonly seen on the surface of jams is due to the localized reduction of soluble solids due to condensation. This is most commonly seen on home-made preserves where the sugar concentration is not so stringently controlled (Brook, 1971).

Jellies have a sucrose content of over 60% which protects them from most types of microbial growth.

Yeasts are ubiquitous food spoilage organisms and some of them will tolerate, or become adapted to tolerate, relatively high concentrations of sucrose. If sensible sterile procedures are not carried out then these yeasts will ferment the sugar and, in most cases, spoil the product.

The quality of a confectionery product will be affected by the quality of the raw materials used. It is important therefore that storage tanks of sugar syrup intended for the bulk production of confectionery should also be kept at the low moisture levels appropriate to microbial inhibition.

Fructose and glucose are, as you would expect from their size, more effective than sucrose as microbial inhibitors but the high solubility of sucrose gives practical advantages (Brook, 1971).

† In light of the possible ambiguities regarding the US/UK interpretations of these and other words in this chapter the reader is advised that the UK interpretation is intended throughout.

The jelly-like confections are blends of carbohydrates stabilized to a semi-firm consistency by a colloidal system. The gel formers are usually starch, agar–agar or pectin (Matz, 1965; Alikonis, 1979; Anon., 1986). These substances are far more complicated than sucrose or invert sugar. Under the correct conditions of hydration the polysaccharides form very high molecular weight colloidal solutions which will also reduce water activity. However, because of their extreme effects on viscosity and texture, they can be only used in low and therefore less significant quantities (Richardson, 1986).

There are basically two types of starch used by confectioners. These are thin boiling starch and thick boiling starch. Thin boiling starch is most suitable for jelly work because it does not become too viscous to handle for most uses, however, slab jellies and Turkish Delight are made with thick boiling starch. This effect is called syneresis.

Starch gel formulas contain higher amounts of water than most other confectionery formulas but this is necessary to fully gelatinize the starch. Fifteen parts of starch are usually mixed with 80 parts of water. The sugar is dissolved in the remaining 20 parts of water in a separate pan and added along with the glucose syrup. The cooked starch is added to the boiling syrup and the mixture is cooked until a spatula full of syrup pours as a string which is judged scientifically as when a refractometer indicates that the syrup contains 78% total solids (Lees, 1980).

Starch gels sweat, i.e. water appears on the jelly's surface if all the excess water is not removed at the boiling stage and if the ingredients have not been sufficiently cooked.

Pectin is another type of gelling agent commonly used in confectionery. It is the partially methoxylated ester of polygalacturonic acid. Within its molecular construction are both acid and ester groups. It therefore behaves as an acid and an ester and it only forms gels in acid solutions. If the solution changes to alkaline the gel breaks down.

High levels of carbohydrates also impair the solubility of pectin. Pectin is water soluble, but it must be dispersed evenly in the dry ingredients before it is hydrated or it will clump. Pectin gels are therefore first made by mixing powdered pectin with some of the sugar in a dry container.

Water is then added and the mixture is slowly stirred and brought to the boil. Glucose syrup is added with the remaining sugar while the mixture is still boiling. The sugar is completely dissolved and the batch is heated to 115°C (this results in a solution of 82·5% solids).

The mixture is cooled and a small amount of citric acid is dissolved in the minimum amount of water and added to the mixture. The batch is then flavoured, coloured and mixed carefully without touching the sides of the container.

The cooked gel is then deposited in starch moulds and left in a warm place to set (Cakebread, 1975; Alikonis, 1979).

The traditional Jelly Baby is moulded from a gel made from gelatine. Gelatine jellies are more resilient than starch gels and are also somewhat more rubbery (Lees, 1980).

The gelatine has to be soaked in water before it can be added to boiled sugars. It is added while the sugar is still hot. Citric acid, flavours and colouring are usually added after the gelatine, when the syrup is a little cooler. The gel is then poured into suitable starch moulds (Cakebread, 1975; Alikonis, 1979).

Another traditional sweet that contains gelatine is marshmallow. Originally marshmallow came from a plant of the same name, however today it is a man-made confectionery which is sometimes referred to as a stabilized sugar foam. The dispersed air in the candy is stabilized by proteins such as gelatine, egg albumen, vegetable albumen or soyprotein. These give body to marshmallow. If gelatine is not included in the formula the sugar syrup will not whip to the required light consistency. The gelatine is an example of a 'whipping' protein that forms a fibril structure throughout the batch of marshmallow and it retains the moisture (Cakebread, 1975; Alikonis, 1979).

Fast setting gelatines are preferred to other slower setting gelatines. Egg albumen is another 'whipping' protein that sometimes is included in marshmallow formulas but only in small amounts.

Syneresis and discoloration are prevented by adjusting the pH correctly. Sugar dominates the flavour of marshmallow however vanilla is occasionally included (Cakebread, 1975; Alikonis, 1979).

The quality of a marshmallow is judged by the uniformity of the dispersed air and the thin wall structure. To achieve the best results the rate of whipping and the temperature are important. So too is the moisture since moisture is lost throughout the preparation. It is important therefore to initially add as much water as possible to compensate for these inevitable losses. Higher moisture levels improve tenderness and eating qualities, however too much water results in syneresis and too little results in a tough sweet.

Those more familiar with microbiology than food science may be surprised to discover that agar–agar is used in candies as well as for microbiological plates. On its own it produces jellies with a not-too-pleasing texture. These gels also exhibit syneresis so agar gels are usually combined with pectin, starch, gelatine or gums for stability (Cakebread, 1975; Alikonis, 1979).

Agar is a carbohydrate which is extracted from certain species of seaweed. Agar forms a thermally reversible double helix. The helix is occasionally interrupted by a 'kinking residue'. The kinking residue is a slightly different sugar to the other sugars in the repeating sequence. They are too bulky to fit comfortably into the helix and they cause the helix to kink.

The strands of agar associate to form double helices with the occasional kink. These are called soluble domains. The soluble domains then associate together either through the bridging action of divalent ions, which bridge like charges, or if the molecules are uncharged the domains associate through hydrogen bonds.

Agarose will form gels in the presence of relatively high amounts of water. As little as 0·2% agarose will form a gel and 0·5% agarose sets quite firmly (Cakebread, 1975).

The properties of agar vary according to its source. It is therefore important to always check with the suppliers specification before deciding how much water to add to a confectionery formula requiring agar.

Agar gels are prepared by simply dissolving a solid strip or agar powder in warm water. The water is then boiled and cooled and heat labile ingredients can be added during the cooling phase.

Gums are also commonly used in confections. Gum arabic is added to sweets such as fruit gums and is also used as a glaze.

Gum arabic comes from the Acacia tree. Most of the requirements of the confectionery industry are supplied from Africa. The gum is collected from the tree by cutting a 'V'-shaped incision into the bark and the gum slowly oozes out. It is then left to dry in the sun and is eventually cut away as a solid.

The colour varies from straw to darker reds. The darker gums are inferior because they contain tannins which have unpleasant flavours.

Water affects the yield of gum which improves with increased rainfall or irrigation and the unique thing about gum arabic is its high water solubility. It is capable of achieving concentrations of over 50%

whereas other gums cannot form concentrations above 5%.

A concentration of 40% gum arabic at 24°C gives a highly viscous gel similar to those achieved with high starch concentration. This gel holds over a wide pH range and it is often used to stabilize emulsions and to hold solids in a paste form e.g. in a lozenge.

The molecular structure of gum arabic is very complex. It contains calcium, magnesium and potassium salts of glycuronic acid, with galactose and arabinose arranged between the simple sugars and the hemicelluloses.

Recently gum arabic has become scarce and expensive so other gums have been substituted. One such gum is tragacanth which comes from the thorny *Artagaluse* shrub which grows in Turkey, Persia, Syria, and India.

This gum has a high resistance to acid hydrolysis, which is important in formulas containing fruit acids. The gum swells in water and if the water is warmed the gum will take up more water thus forming a more viscous solution.

Unfortunately this gum forms unstable dispersions and in time they separate into a sol and a sludge. This occurs even at its most stable pH (pH 5–6) and gelatine is usually added to compensate for this problem. Gum arabic is *never* added because this precipitates tragacanth gum.

Another gum which is commonly included in confectionery formulas is Locust Bean gum. This comes from the seeds of the *Carob* or *Cerotonia,* the Locust bean tree. This gum is unusual because it thickens rather than breaking down on standing. So it is added to confectioneries to stabilize them, i.e. to prevent syneresis, shrinkage, cracking and separation. (It is especially used with agar gels.)

The final gum of importance to the role of water in confectionery is guar gum. This gum is unusual because it is soluble in cold water and it is used when formulas cannot be heated. As with the other gums discussed it resists syneresis when added to starch gels. It is a good example of a thixotropic material, that is to say if it is stirred the gel returns to a fluid state.

Guar gum, locust bean gum or xanthan gum have been added, with mixed success, to cereal flours with poor bread-making qualities to improve their bread making performance (Hoseney, 1986; Aybendajo, 1987). These additives 'improve' the doughs by increasing the number of water binding sites. Hydroxypropylmethyl cellulose is another such compound that has been successfully used with rice flour to make bread suitable for people suffering with coeliac disease (Kent, 1984).

BAKED DOUGH PRODUCTS

Bread is basically a baked dough which is prepared from a mixture of flour and water and salt. Other refinements can be made such as producing leavened bread from yeasted doughs or by adding sugars and flavours (Hoseney, 1986).

The amount of water added is critical. Too much water and the flour forms a batter or the dough becomes sticky and difficult to handle. The bread from this dough is wet, soggy and more susceptible to microbial damage. If there is too little water in the dough then the gluten will not develop and the dough will be tough without the necessary visco-elastic characteristics. This type of dough is more suited to biscuit production than to bread manufacture. Doughs which are formed with only just below the optimum water ferment badly and stale faster than normal (Elie and LeGry, 1987).

The baker would like the flour to hold as much water as possible since bread is sold by weight. Several factors govern the actual optimum amount of water that can actually be added to a bread or biscuit dough.

One factor is the variety of flour. Other factors are the soil and climatic conditions that prevailed when it was growing and the effects of milling.

The major function of water in flour is to bind the flour and other ingredients together. Bakers always use tap water because it contains the essential minerals required by the yeast and these minerals also help to toughen the gluten. Starch cannot gelatinize in the absence of water. Also water vapour contributes to the initial volume increase that occurs when the dough first enters the oven (the 'oven-spring') (Hoseney, 1986).

There are several different ways in which bread is produced commercially. The simplest bread-making procedure is a 'straight-dough system' (this is the method which is also commonly carried out at home). All the ingredients are mixed with warm water (about 27°C) and developed into a yeasted dough. The dough is allowed to ferment for about 100 min. After the fermentation the dough is reworked or 'knocked-back' in order to evenly distribute the gas pockets. The dough is then allowed to ferment for a further 55 min before being divided into the desired weight pieces, moulded and placed into the baking pan. The dough is then given an additional fermentation or 'proof' until it reaches the desired size. The dough is then baked.

This is a very lengthy procedure and is rarely used for bulk fermentation. An alternative method commonly used in the USA is the 'sponge-and-dough' method. A loose dough is prepared by mixing approximately two-thirds of the flour with part of the water and the yeast. This 'sponge' dough is fermented for up to 5 h. It is then combined with the remaining ingredients and left to prove for a relatively short time before being divided, moulded and proved as before. Bread produced by this method is soft and finely structured and retains more moisture than the loaves prepared by the previous method (Kent, 1984; Hoseney, 1986).

In the UK there is no legal standard for the moisture content of bread, however there are strict standards governing the weight of the bread as manufactured and so a further method, the 'Chorleywood process' is popular. This method actually incorporates slightly more water into the final product than the long straight dough procedure method without noticeably altering the texture.

The Chorleywood process mixes all the ingredients with a considerable amount of work (11 Watt-hours/kg) within 5 min. Ascorbic acid is included in the ingredients along with some additional fat and 3·5% more water (Kent, 1984). Once the dough is developed it is allowed to ferment, mould and prove before being divided in much the same manner as discussed earlier.

Moisture is lost from the bread during the baking stage; some processes inject steam into the oven to try and counteract some of this loss. The bread produced in this way and bread which has been cooled incorrectly may have softer leathery crusts because of the effects of surface water.

Flour doughs with very low water contents are used to make biscuits. Generally biscuits are made from varieties of wheats which produce flours with weak gluten. The formulas are high in sugar and fat as well as being relatively low in water. The ingredients are mixed but the gluten is not fully developed. It is developed just enough to hold the dough together during sheeting (Manley, 1983).

Biscuits are usually not yeasted but contain sodium bicarbonate which among other things keeps the pH alkaline and slows gluten hydration.

The doughs are commercially cooked in long tunnel ovens. When the biscuits first enter the oven the fats melt and the dough is free-flowing. The sugar then dissolves in the small amount of water present, increasing the volume and making the dough sticky. The chemical leavening agents then become active and the dough expands in all directions. Biscuits do not reach the temperature when starch gelatinizes (at temperatures above 115°C) because the high sugar/low water that is present raises the gelatinization temperature of starch. Finally moisture is lost from the system and the sugar concentrates and crystallizes. The final moisture content of biscuits is between 2–5%. If the moisture is higher then the biscuit will appear soggy (Manley, 1983).

Water is a unique non-food material which is an essential ingredient of all confectionery. It allows changes to occur in the other ingredients which could not occur in its absence. The amount of water available affects the state and quality of the product, therefore too much water or too little water is worse than no water at all!

REFERENCES

Anon. (1986). Susswaren transportieren and lagern. Das Prinzip der Frisch- und Weichhaltung. (Transport and storage of confectionery. Principles of retention of freshness and softness). In: *Zucker- und Susswarenwirtschaft,* **39** (4), 95–7.

Alikonis, J. J., (1979). *Candy Technology,* Avi Publishing Co. Inc., Westport, Connecticut, USA.

Aybendajo, T. (1987). Thesis. Reading University, UK.

Brook, M. (1971). Sucrose and the food manufacturer. In: *Sugar: The chemical, biological and nutritional aspects of sucrose,* J. Yudkin, J. Edelmam and L. Hough (Eds). Butterworth, UK, pp. 32–46.

Bush, J. W. (1928). *Skuse's complete confectioner,* 12th Edn, W. J. Bush and Company, Ltd, UK.

Cakebread, S. H. (1975). *Sugar and Chocolate Technology,* Oxford Univ. Press, UK.

Cakebread, S. H. (1976). Ingredient migration in composite products. I. *Confect. Prod.,* **42** (4), 172–3.

Elie, U. and LeGry, G. A. (1987). *Proceedings of the 7th World Food Congress,* (in press).

Fett, H. M. (1973). Water activity determination in foods in the range 0·80–0·99. *J. Food Sci.,* **38,** 1097–8.

Gilbert, G. S. (1986). New concepts on water activity and storage stability. In: *The Shelf Life of Foods and Beverages,* G. Charalambous (Ed.), Elsevier Applied Science, 791–802.

Hardman, T. M. (1986). Interaction of water with food components. In: *Interactions of Food Components,* G. G. Birch and M. G. Lindley (Eds), Elsevier Applied Science, London, pp. 19–33.

Hoseney, R. C. (1986). *Principles of Cereal Science and Technology,* American Association of Cereal Chemists, Inc., Minnesota, USA, 232 pp.

Jewell, G. G. (1986). Interactions of confectionery components., In: *Interactions of Food Components,* G. G. Birch and M. G. Lindley (Eds), Elsevier Applied Science, London, pp. 277–97.

Kennedy, J. F., Cabral, J. M. S., Sa-Correia, I. and White, C. A. (1987). Starch biomass: A chemical feedstock for enzyme and fermentation processes. In: *Starch: Properties and Potential,* Vol. 13, T. Galliard (Ed.), John Wiley and Sons, UK, pp. 115–48.

Kent, N. L. (1984). *Technology of Cereals,* 3rd Edn, Pergamon Press Ltd, UK.

Labuza, T. P., Acott, K., Tatini, S. R. and Lee, R. Y. (1976). Water activity determination: A collaborative study of different methods. *J. Food Sci.,* **41,** 910–17.

Lees, R. (1980). *A Basic Course in Confectionery,* Specialised Publications Ltd (Books), Surbiton, UK.

Manley, D. J. R. (1983). *Technology of Biscuits, Crackers and Cookies,* Ellis Horwood Ltd, UK.

Matz, S. A. (1965). *Water in Foods,* Avi Publishing Co. Inc., Westport, Connecticut, USA.

Morley, G. M. and Groves, R. (1987). Study of water activity holds the key to freshness. *Candy Industry,* **152** (1), 44–8.

Pancoast, H. M. and Junk, W. R. (1980). *Handbook of Sugars,* 2nd Edn, Avi Publishing Co. Inc., Westport, Connecticut, USA.

Richardson, T. (1986). ERH of confectionery food products. *Manuf. Confect.,* **66** (12), 85–9.

Von Elbe, J. H. (1986). Measurement of water activity in confections. *Manuf. Confect.,* **66** (11), 51–6.

Chapter 10

THE EFFECT OF WATER ACTIVITY ON THE STABILITY OF VITAMINS

JANICE RYLEY

Procter Department of Food Science, University of Leeds, UK

SYMBOLS AND ABBREVIATIONS

a_e	External water activity (outside packaged food)
a_i	'In packet' water activity
A	Package area
BET	Brunauer–Emmett–Teller
C	Concentration at time $t = t$
C_0	Concentration at time $t = 0$
D	Quality parameter index at time t
DHA	Dehydroascorbic acid
D_0	Quality parameter index at time $t = 0$
EDTA	Ethylene diamine tetraacetic acid
Ext	Extent of oxidation μl oxygen absorbed per g food at standard temperature and pressure
E_a	Activation energy
FAD	Flavin adenine dinucleotide
FMN	Flavin mononucleotide
HPLC	High performance liquid chromatography
IMF	Intermediate moisture food
k	Rate constant
K_{H_2O}	Water vapour permeability of packaging material
m	Moisture gain g/100 g
P_0	Water vapour pressure at storage temperature
Q_{10}	(Rate at $T + 10$)/(Rate at T)
RAA	Reduced ascorbic acid
RH	Relative Humidity
t	Time

T	Temperature in degrees K
TAA	Total ascorbic acid
TDT	Thermal death time
X	Thickness of packaging material

INTRODUCTION

Vitamins are a structurally disparate group of substances and there is no chemical reason for grouping them except in so far as their behaviour can be fitted into the general picture of chemical reactivity in relation to water activity summarised by Labuza (1980*b*) and Rockland and Nishi (1980).

The effect of processes which reduce water activity on various aspects of nutrient stability has been discussed by Labuza (1972); Bluestein and Labuza (1975, 1987); Bender (1978); Kirk (1981) and Leung (1987). In 1972 Labuza, and in 1975 Bluestein and Labuza commented that most data in this area had been obtained by 'end point' analysis and that very little work had been done in which temperature and water activity had been systematically varied. They made the same point in 1987, as did Leung. In recent years attention has been given to conditions which might be encountered during extrusion processing (Laing *et al.*, 1978), in intermediate moisture foods (Neale *et al.*, 1978; Chirife *et al.*, 1980) and in connection with the conditions which would arise using a combination of mild processes (Fox *et al.*, 1982; Fernandez *et al.*, 1986). Nutritional labelling regulations have given impetus to shelf life studies (Labuza, 1972) some of which have considered moisture changes which occur due to permeable packaging and to moisture migration which arises as a result of fluctuating storage temperature.

The value of studies at constant temperature and humidity lies in the insight they give into the mechanisms of change although, as pointed out by Lund (1983) most reactions in foods appear to be described by zero order or first order kinetics, particularly if one of the reactants is present in excess. The kinetic data derived find a practical application in the prediction and subsequent optimisation of processing and storage conditions. However the pursuit of systematic studies for this purpose may have been hampered by the fact that doubts have been cast on the validity of applying data obtained under isothermal conditions to non-isothermal processes (Thompson, 1982). Additionally the enormous expansion in computer software has made it feasible to carry out curve fitting exercises on data obtained under dynamic conditions without the need for prior mechanistic information (Mizrahi and Karel, 1977).

Biological activity is usually found in a group of chemically related substances of varying potency so that, in considering the effects of processes on nutrients, the relevance of the assay method must be considered (Gregory III, 1983). For example, beta-carotene is relatively well studied in relation to the effect of water activity due to its importance as a food colour, but because less biologically active carotenoids are equally important colour contributors, many studies of carotenoids primarily monitor colour (Arya *et al.*, 1983) and do not necessarily reflect changes in biological potency. In addition, during processing, interconversion of biologically active forms with varying potencies and stabilities may occur, as exemplified by vitamin B_6 in a model breakfast food (Gregory III and Kirk, 1978).

Undoubtedly chemical assay techniques in general have been revolutionised by the increased possibilities for separation and consequent specificity afforded by high performance liquid chromatography, but for nutrients in complex mixtures method evaluation is still in a relatively early stage with most success having been achieved in the area of the fat soluble vitamins (Brubacher *et al.*, 1985). Ligand-binding assays have great potential for the future due to their inherent specificity. The assay of folic acid and its derivatives in foods still presents a major challenge to analysts and studies of its kinetic behaviour are at an early stage (Ruddick *et al.*, 1980).

Frequently kinetic studies have been carried out on systems which have been spiked with a pharmaceutical form of the vitamin, either to simulate fortification or to raise the natural level for ease of monitoring. However the vitamins contained within the food structure have often been found to be more stable than the same entity in a buffer or a fortified food and the stability may vary between sites and between chemical forms of the vitamin. For example, beta-carotene on the surface of dehydrated sweet potato flakes was destroyed 100 times more rapidly than bound beta-carotene (Walter and Purcell, 1974) during autoxidation. Cocarboxylase (thiamin pyrophosphate) is less stable than free thiamin when added to farina, wheat or oatmeal (Farrer, 1955).

WATER SOLUBLE VITAMINS

Vitamin B_1

Biological activity is attributed to free thiamin, thiamin pyrophosphate (cocarboxylase) and protein-bound thiamin, each of which has similar biological potency but different chemical stability characteristics (Farrer, 1955). Kinetic studies in foods have often been carried out on systems fortified with thiamin salts (Hollenbeck and Obermeyer, 1952; Mulley *et al.*, 1975*a,b*).

The thiochrome procedure for the assay of thiamin, based on the original procedure of Strohecker and Henning (1963) and recently assessed by Brubacher *et al.* (1985) has stood the test of time and has been widely used in studies related to water activity. The extraction and purification methods are documented for a wide range of foods so that probably comparability of studies is more likely to be possible for thiamin than for other vitamins. Kamman *et al.* (1981) and Dennison *et al.* (1977) used a modification involving a continuous flow method (Kirk, 1974).

However thiamin, whatever its origin, is usually monitored as total thiamin using extraction and cleanup procedures which hydrolyse the pyrophosphate and protein-bound forms.

The findings on thiamin stability related to water activity can be summarised as:

1. The degradation rate increases as moisture content and water activity increase and can be described by first order kinetics. These findings have emerged in studies on fortified flour (Hollenbeck and Obermeyer, 1952); fortified corn–soy–milk powder (Bookwalter *et al.*, 1968); simulated fortified breakfast food (Dennison *et al.*, 1977) and enriched pasta (Kamman *et al.*, 1981). The change in the degradation rate is continuous above and below the BET monolayer up to $0{\cdot}65a_w$. The effect of water activity and temperature on the half life of fortified pasta and simulated breakfast cereal is shown in Table 1 from which it can be seen that a water activity increase of 0·2 units is approximately equivalent to an increase of 10°C in its degradation effect.

2. The Arrhenius activation energy for degradation in pasta (Kamman *et al.*, 1981) is in the range 110–130 kJ/mol (Table 2) which is in the same range as found by Mulley *et al.* (1975*a,b*) for thiamin chloride hydrochloride added to phosphate buffers, peas-in-brine purée, pea purée and beef purée; for fortified beef baby food by

TABLE 1
Effect of water activity and temperature on the half-life[a] of thiamin

1. Fortified pasta (Kamman *et al.*, 1981)

a_w	*Half-life (days)*			
	25°C	*35°C*	*45°C*	*55°C*
0·44	3480	1070	313	44
0·54	3120	730	180	34
0·65	1130	380	110	27

2. Simulated breakfast food stored at 45°C (Kirk, 1981)

a_w	*Half-life (days)*	
	B_1-fortified	*AB_1C-fortified*
0·10	1050	4950
0·24	761	1050
0·40	103	107
0·50	63	75
0·65	80	73

[a] Calculated from the first order rate constant.

Lappo (1987); for meat loaves cooked in convection ovens (Skjöldebrand *et al.*, 1983); for chemical systems whose water activity had been lowered to 0·9–0·95a_w by a range of different solutes (Fernandez *et al.*, 1986) or at different pH (Fox *et al.*, 1982).

3. In pasta the Arrhenius activation energy falls with increasing water activity in the range 0·44–0·65 (Kamman *et al.*, 1981) (Table 2). This phenomenon has only been observed in one other study and concerned ascorbic acid in dehydrated tomato juice (Riemer and Karel, 1978; Table 6). In model systems studied only at 0·65a_w and above, activation energies were at the lower end of the quoted range as would be expected if extrapolating from work at lower water activity (Lund, 1983; Fernandez *et al.*, 1986).

4. In simulated breakfast food the degradation rate is dramatically increased at 45°C above 0·24a_w (Table 1) (Dennison *et al.*, 1977). This has been attributed to the role played by thiamin in Maillard reactions

TABLE 2
Effect of water activity on the apparent activation energy for thiamin

Product	a_w	*Activation energy (kJ/mol)*		*Source*
1. Enriched pasta	0·44	127		Kamman *et al.* (1981)
	0·54	123		
	0·65	110		
2. Model systems with differing solutes	0·90–0·95			Fernandez *et al.* (1986)
NaCl		114		
KCl		106		
Glycerol		97		
Na_2SO_4		98		
		IMF3	IMF4	
3. Model systems[a]	0·65	106	95	Arabshahi and Lund (1985)
	0·75	111	101	
	0·85	115	106	
4. Model systems of glycerol/buffer	0·90	111		Fox *et al.* (1982)
	0·93	111		
	1·00	113		
		114		

[a] Microcrystalline cellulose, propylene glycol, sodium propionate, thiamin, riboflavin, niacin, and water.

(Van der Poel, 1956). Since Maillard reactions have activation energies in the range 100–200 kJ/mol (Thijssen and Kerkhof, 1977), Maillard reactions involving thiamin would accelerate more than other modes of thiamin degradation as temperature rises. Indeed, in a simulated breakfast food, thiamin degradation was found to be greatest above 45°C at 0·3–0·4a_w where browning was most pronounced (Dennison *et al.*, 1977).

5. At high water activity (>0·90) such as might be used in combination processes, the rate of thiamin hydrochloride degradation in a buffered solution of glycerol at 85–95°C is independent of water activity but highly dependent on pH (Fox *et al.*, 1982). Similar experiments were carried out by Fernandez *et al.* (1986) at lower temperatures using a range of solutes in which the type of water activity lowering solute was found to be very important. Thiamin

hydrochloride was less stable in solutions of univalent ions than in glycerol or divalent ions (Table 3) at the same water activity. This is in line with the findings of Chen and Karmas (1980) who found that 5% (0·86 M) NaCl had a greater water activity lowering effect than 15% (1·36 M) $CaCl_2$ which was explained as a greater net structuring effect by sodium ions. In Fernandez' experiments where systems of different ions were brought to the same water activity the univalent ion systems would be less concentrated than the divalent ions on both a weight and molar basis suggesting that at high water activity a concentration effect operates on thiamin stability which is distinct from the water activity effect *per se*.

Thus, within the range of studies available the effect of water activity on thiamin stability shows remarkably similar rate constants when the range of solutes in the system is large and the pH range is limited as occurs in real and model food systems. However in chemical systems where one solute predominates a specific solute effect is displayed at high water activities but has not been reported at lower water activities. The relatively narrow range of Arrhenius activation

TABLE 3
Effect of solute on the half-life of thiamin

System 1	a_w	*Half-life (days)*			
		35°C	*45°C*	*55°C*	*65°C*
NaCl	0·90–0·95	28·9	7·3	1·9	—
KCl		—	11·1	3·9	1·2
Glycerol		—	96·3	28·9	11·1
Na_2SO_4		—	144	48·1	16·0
					Fernandez *et al.* (1986)

System 2	a_w	*Half-life (days)*			
		84·2°C		*93·9°C*	
		pH = 7	*pH = 5*	*pH = 7*	*pH = 5*
Glycerol/ buffer	0·90	0·149	1·25	0·057	0·444
	0·93	0·156	1·41	0·059	0·532
	0·96	0·158	1·71	0·059	0·545
	1·00	0·154	2·03	0·061	0·656
					Fox *et al.* (1982)

energies reported and their similarity to values reported for a wide range of conditions infers that the mechanism by which the methylene bridge is cleaved is similar throughout. Studies have not involved conditions in which thiamin would be expected to be extremely unstable, as in the presence of nucleophiles such as sulphite which can be inadvertently present via ingredients such as dried fruit and could migrate during storage.

Vitamin B_2

Vitamin B_2 activity resides in foods as FMN (flavin mononucleotide, riboflavin monophosphate) and as FAD (flavin adenine dinucleotide, in which riboflavin and adenosine are joined through pyrophosphate). The free form is rarely found in foods but it is the form typically used by food manufacturers. Riboflavin is entering a new phase of importance as a food colour as a result of the current consumer pressure to replace synthetic colours by those of natural origin (Kearsley and Rodriguez, 1981). Chemical methods of riboflavin assay mostly utilise the natural fluorescence of the flavin entity which is unaffected by the phosphate moiety although suitable extraction methods must be used to liberate the protein-bound material. The efficacy of chemical methods is very dependent on successful removal of interfering fluorescence so that microbiological assay, which is both specific and sensitive, is still often preferred for natural levels in food. HPLC has recently been classed as only tentative for riboflavin (Brubacher *et al.*, 1985).

The stability of riboflavin in food processing has been briefly reviewed by Archer and Tannenbaum (1979) and it is well known for its stability during processing as long as it is kept in the dark and is not alkaline. For example, after the sterilisation process in the canning of soya beans, the total riboflavin content of the beans plus the can liquor was >99% of the preprocessing value (Abdel-Kader, 1985). In light conditions, riboflavin is subject to degradation the rate of which is dependent on pH and light intensity (Singh *et al.*, 1975; Kearsley and Rodriguez, 1981) and follows first order kinetics (Allen and Parks, 1979).

Few studies have been undertaken of the effect of water activity on riboflavin. The main findings may be summarised as:

1. When a simulated breakfast cereal product was stored in the dark a water activity dependent increase in the degradation rate occurred above 37°C (Table 4) and was faster in a paperboard package than in a can. This result was attributed to the oxygen

TABLE 4
Effect of storage conditions on the half life of riboflavin

Product	a_w	*Half-life (days)*	*Source*
1. Storage at 4°C exposed to 150 foot candles			
Elbow macaroni	—	14·8 (1st phase) 630 (2nd phase)	Furuya *et al.* (1984)
NFDM	—	25·3 (1st phase) 193 (2nd phase)	
Skim milk (fluid)	—	0·58	
2. Storage in TDT cans at 37°C			
Simulated breakfast food in TDT cans at 37°C	0·1	3013	Dennison and Kirk (1977)
	0·24	369	
	0·40	264	
	0·50	169	
	0·65	138	
In paperboard boxes at 30°C	% RH of surroundings		
	10	158	
	40	161	
	85	161	

permeability of the package (Dennison *et al.*, 1977). Riboflavin has not formerly been regarded as sensitive to oxygen. However it is sensitive to pH, so that if other changes such as the concurrent loss of ascorbic acid caused an increase in pH the apparent oxygen sensitivity could be explained.

2. The degradation rate of riboflavin in macaroni is significantly faster in very low light exposures (25 foot candles) than in the dark. At higher levels of light intensity, changes in the range 100–450 foot candles which have dramatic effects on riboflavin degradation in liquid milk do not alter the degradation rate in enriched macaroni (Woodcock *et al.*, 1982).
3. During exposure of macaroni at 75 foot candles the loss rate increased slightly as water activity increased in the range 0·11–0·52 but above 0·11a_w the values were not significantly different from each other after two days (Furuya *et al.*, 1984).
4. Light induced loss in fluid milk gives a single phase in the first order rate plot. In pasta a second slower phase of loss develops after 2–3 days (Furuya *et al.*, 1984).
5. The accumulation of lumichrome, the major product under acid or neutral conditions, only accounted for 25% of the lost riboflavin during the storage of pasta in the light (Furuya *et al.*,

1984). Thus riboflavin, when exposed to light in low water activity products, as would happen in the storage of pasta in transparent wraps, is less stable than thiamin as can be seen from the half lives in Tables 1 and 4, but is quite similar to thiamin when stored in the absence of light.

Although small changes in water activity in a solid have only a small effect on riboflavin stability the rate of the light catalysed reaction is greatly reduced relative to the liquid state, as exhibited by the behaviour of milk, and may be a consequence of the relative importance of the lumichrome and lumiflavin pathways.

Vitamin C

In common with most studies of the nutrient effects of food processing, data on ascorbic acid (where applicable) far outweigh data for other nutrients. This can be directly attributed to its value as a non-subjective, relatively easily measured criterion of food quality, along with non-enzymic browning, as well as to its well earned reputation as the most labile nutrient under many conditions met within food processing. These properties make it a useful parameter for monitoring processes and testing the applicability of mathematical models.

Of the many studies involving vitamin C relatively few have systematically varied moisture content or water activity and calculated both rate constants and activation energies. The foods which have been so studied include seaweed (Jensen, 1969; Ogawa *et al.*, 1984); dehydrated tomato juice (Riemer and Karel, 1978); concentrated grapefruit juice (Saguy *et al.*, 1978); intermediate moisture apple (Singh *et al.*, 1983); and a variety of model systems designed to simulate 'dry' or intermediate moisture food (IMF) formulations (Vojnovich and Pfeifer, 1970; Lee and Labuza, 1975; Waletzko and Labuza, 1976; Kirk *et al.*, 1977; Dennison and Kirk, 1978; Laing *et al.*, 1978). Thompson (1982) has summarised kinetic studies and, in some cases, calculated rate constants and activation energies when they had not been reported, if sufficient data were available. In other studies, designed to study storage effects, observations have been carried out at one temperature only (orange juice crystals, Karel and Nickerson (1964); sweet potatoes, Haralampu and Karel (1983)) or insufficient time intervals monitored for rate constants to be evaluated (tomato juice powder, Wong *et al.* (1956)).

The estimation of vitamin C by chemical means has been extensively reviewed by Cooke (1981). In many, particularly the earlier, of the above studies samples were assayed using various modifications of the 2:6 dichlorophenol indophenol technique which only monitor reduced ascorbic acid (RAA). The first stage of the aerobic degradation is the formation of dehydroascorbic (DHA) which has similar biological activity to the reduced form (Linksweiler, 1958). Riemer and Karel (1978) included a reduction stage to include DHA. DHA can make an important contribution to the total vitamin C content particularly after prolonged storage (Ogawa *et al.*, 1983; Attwood, 1987). Studies which have utilised the more recent fluorimetric procedure using *o*-phenylene diamine based on the original procedure of Deutsch and Weeks (1965) have reported either total ascorbic acid (TAA = RAA + DHA) (Purwadaria *et al.*, 1980; Ogawa *et al.*, 1983) or total ascorbic acid and dehydroascorbic acid from which reduced ascorbic acid can be calculated (Kirk *et al.*, 1977; Dennison and Kirk, 1978, 1982; Riemer and Karel, 1978; Ogawa *et al.*, 1983). Some studies have used an HPLC technique, which would be expected to be more specific (Singh *et al.*, 1983; Mishkin *et al.*, 1984*b*).

The results of studies of the effect of varying the water activity or moisture content on ascorbic acid can be summarised as:

1. First order reactions with respect to 'ascorbic acid' (however measured) have been reported by many workers over a wide range of water activities and systems. Exceptionally Laing *et al.* (1978) working with a beef/soya flour system between 0·69 and 0·90a_w over the range 61–105°C (in relation to extrusion processing conditions) found that degradation was adequately described by zero order kinetics. However, as the authors pointed out their studies did not exceed 50% loss before which it is difficult to distinguish between zero and first order kinetics.

2. The rate of loss increases as water activity increases and as temperature increases. An exception was reported by Ogawa *et al.* (1983), who found that dried seaweed, laver 'nori', was more stable at 0·05a_w than in the 'dry' state. The magnitude of the first order rate constants at specified temperatures is system dependent and values have been reviewed by Thompson (1982). Table 5 shows the effect of water activity and temperature on the half life of ascorbic acid in dried seaweed (laver 'nori', Ogawa *et al.* (1984)) and in dehydrated tomato juice (Riemer and Karel, 1978). The difference in the half life values

TABLE 5
The effect of water activity and temperature on the half-life[a] of (reduced) ascorbic acid

1. Seaweed (laver 'nori') (Ogawa *et al.*, 1984)

a_w	*Half-life (days)*		
	10°C	*20°C*	*30°C*
'0'	184	87	51
0·05	281	130	70
0·1	129	67	33
0·2	54	26	14
0·4	21	8·6	3·2
0·6	7·2	2·4	0·8

2. Dehydrated tomato juice (Riemer and Karel, 1978)

a_w	*Half-life (days)*		
	20°C	*37°C*	*51°C*
'0'	533	141	23·1
0·11	117	34·7	6·9
0·32	69	9·9	2·5
0·57	13·9	3·5	0·99
0·75	6·3	1·3	0·59

3. Intermediate moisture apple (Singh *et al.*, 1983)

a_w	*Half-life (days)*			
	25°C	*35°C*	*45°C*	*55°C*
0·62–0·64	32	7·5	4·8	1·8
0·77–0·80	13	5·3	2·3	1·1
0·83–0·85	8·5	3·8	1·8	0·6
0·88–0·89	6·2	2·7	1·1	0·6

4. 'Hemmican' fortified (Waletzko and Labuza, 1976)

a_w	*Half-life (days)*		
	25°C	*35°C*	*45°C*
0·85 packaged in air	10	5	2
Protected from air	69	21	6

[a] Calculated from first order rate constants.

for the two products reflects the lack of sensitivity of tomato juice powder to oxygen reported by Riemer and Karel. The values are significantly lower than half lives already described for thiamin and riboflavin in Tables 1 and 4 and as described by Lueng (1987), Kirk (1981) and Thompson (1982).

For a cellulose/glycerol/corn oil system, Lee and Labuza (1975) showed a factor increase of 69 and 55 in the 45°C rate constant for the desorption and adsorption processes respectively over the water activity range 0·32–0·93. The rate constant for samples prepared by desorption was always greater than that for samples prepared by adsorption in keeping with the concept of increased molecular mobility due to decreased viscosity at higher moisture contents.

Although Mishkin *et al.* (1984b) working on the dehydration of potato slices in a laboratory drier concluded that there was a maximum degradation rate which occurred at a moisture content of 2·5 g per g of dry solids the authors attributed the effect to physical rather than chemical phenomena by postulating a loss of membrane integrity.

An exception to the general findings of increased rate with increasing temperature was reported by Laing *et al.* (1978). They found, in the same beef/soya IMF system that although the rate increased with increasing water activity, a reduction in the rate occurred above 92°C. They attributed this to the decrease in oxygen solubility with temperature and a possible shift to the much slower anaerobic mechanism. However Mohr (1980) interpreted the observed effect in terms of the heterogeneous nature of the system and the reduction in the partial pressure of oxygen of the product environment due to the increased partial pressure of water as the temperature rose. Assuming that the solubility of oxygen in the food system was the same as in pure water, Mohr pointed out that the amount of oxygen in the liquid phase was very much less than that required to react and had therefore to be continuously replenished from the gas phase. He showed that the results of Laing *et al.* (1978) could be predicted by postulating a mechanism in which the oxygen reacts rapidly in a thin film at the interface with the gas phase if the rate was proportional both to the concentration of RAA and to the concentration of oxygen. It was not necessary to involve the anaerobic mechanism in order to explain the findings.

3. The dependence of the Arrhenius activation energy on water activity (and thus the evidence for a change of mechanism) is unclear (Table 6) since linear but divergent Arrhenius plots have been found by several workers.

TABLE 6
Effect of water activity on the apparent activation energy of ascorbic acid

Product	*Entity*	a_w *(or % water or °Brix)*	*Activation energy (KJ/mol)*	*Source*
1. Seaweed in airtight tins	RAA(?)	10%	32	Jensen (1969)
		20%	55	
		25%	127	
2. Seaweed packaged in air (laver 'nori')	RAA	'0'	45·5	Ogawa *et al.* (1984)
		0·05	49·5	
		0·1	48·9	
		0·2	48·1	
		0·4	67·9	
		0·6	77·6	
3. Dehydrated tomato juice (stored 0–21% O_2)	RAA	'0'	101	Riemer and Karel (1978)
		0·11	92	
		0·32	83	
		0·57	74	
		0·75	67	
4. Model breakfast food	TAA	0·1	44·1	Kirk (1981)
		0·4	65·9	
		0·65	75·4	
A303 cans (excess O_2)	RAA	0·1	44·1	
		0·4	64·3	
		0·65	70·0	
TDT cans (limited O_2)	TAA	0·1	33·4	
		0·24	65·5	
		0·40	72·6	
		0·50	79·1	
		0·65	79·1	
	RAA	0·10	29·2	
		0·24	58·5	
		0·40	73·3	
		0·50	74·6	
		0·65	79·5	
5. Grapefruit juice concentrate	RAA	11·2°	20·5	Saguy *et al.* (1978)
		31·2°	21·6	
		47·1°	27·6	
		55·0°	35·4	

In spite of an apparent trend to lower values of the apparent activation energy at high water activity in a glycerol/microcrystalline cellulose/corn oil system in the range 0·32–0·93a_w, Lee and Labuza (1975) concluded that, because of scatter in the results the differences were not significant. Riemer and Karel (1978) who also found smaller values at higher water activity concluded that in dehydrated tomato juice the variation in the activation energy fitted the term:

$$E_a = \exp(a - ba_w)$$

where a and b are constants with the values $a = 9{\cdot}9990$ and $b = 0{\cdot}470$.

With reference to their findings of increasing activation energy with water activity in the range 0·24–0·65 for a simulated breakfast cereal (Table 6) Kirk (1981) has invoked a thermodynamic explanation based on the idea that the higher activation energy is offset by a reduction in the system entropy which occurs as a result of the more ordered structure at higher moisture contents, from which it can be shown that the Gibbs free energy of activation remains constant. Therefore, Kirk concluded that in the system studied there was no evidence for a change in mechanism. Nevertheless the range of values reported for the apparent activation energy of ascorbic acid is considerably wider than that reported for thiamin.

Additionally, Mishkin *et al.* (1984*a,b*) from data collected under dynamic conditions concluded that the apparent activation energy for ascorbic acid degradation in the dehydration of potato slices was moisture dependent and increased continuously from 0 to 4·0 g water/g dry solids.

4. The role of oxygen

The degradation pathways of ascorbic acid have been summarised by Archer and Tannenbaum (1979). The metal catalysed and the non-catalysed aerobic reactions both give dehydroascorbic acid as a breakdown product, but the very much slower anaerobic reaction does not. The degradation rate of ascorbic acid in dehydrated tomato juice was not reduced by packaging in zero oxygen relative to 20% oxygen (Riemer and Karel, 1978). In the processing of concentrated grapefruit juice Saguy *et al.* (1978) postulated an anaerobic mechanism based on the system's lack of response to de-aeration and nitrogen flushing. However, while Saguy *et al.* found an activation energy of only 20·6–46·6 kJ/mol, Riemer and Karel found values in the range 65·9–98·9 kJ/mol which is in the same range as found for systems believed to be aerobic.

Kirk *et al.* (1977) working with a simulated breakfast food packed into thermal death time (TDT) cans where the ratio of mol oxygen/mol ascorbic acid was 10, found that at $0{\cdot}11a_w$ (below the BET monolayer) the slope of the plot of ln k against $1/T$ degrees K was less steep than the rest of the range studied (0.24–$0{\cdot}65a_w$) indicating a dramatic reduction in the activation energy. The authors suggest that at this very low water activity, although excess oxygen was available in the package headspace the amount available in solution was limiting because of the low moisture content. The activation energy found at $0{\cdot}11a_w$ (33·4 kJ/mol for TAA and 29·3 kJ/mol for RAA) was closer to the value of 13·6 kJ/mol reported for canned tomato juice by Lee and Labuza (1975) for the anaerobic reaction than the values of 58·5–79·5 kJ/mol found in the range $0{\cdot}24$–$0{\cdot}65a_w$.

5. The role of dehydroascorbic acid

Since dehydroascorbic acid is the product of the oxidation of reduced ascorbic acid but itself retains biological value its stability is of interest, but has been studied very little. The main conclusions have been that total and reduced ascorbic acid degrade at the same rate in dehydrated tomato juice (Riemer and Karel, 1978) with or without oxygen. In simulated breakfast food, however, Kirk *et al.* (1977) calculated rate constants for both total and reduced ascorbic acid and found that at the water activity levels studied the rate constants were larger for reduced ascorbic acid and therefore concluded that the degradation of reduced ascorbic acid is the determining factor in total ascorbic acid loss. However at low water activity (0·1) and low temperature (10 and 20°C) dehydroascorbic acid increased during the first few days and thus influenced total ascorbic acid loss. In a mineral (iron or copper) fortified system similar results were observed. The destruction rates of TAA and RAA were very similar (Dennison and Kirk, 1982).

Ogawa *et al.* (1983) working with dried laver 'nori' (Table 7), also reported that the proportion of DHA increased initially during storage above $0{\cdot}3a_w$ and reached 85% of TAA at $0{\cdot}8a_w$. Attwood (1987) working with an IMF-fortified chocolate coated low energy meal replacement product at $0{\cdot}6a_w$ (Table 7) found that the proportion of DHA was between 60 and 95% Thus it appears that, in some instances the contribution of DHA to the vitamin C content may have been neglected, particularly in studies at ambient temperatures.

6. Metal salts

Dennison and Kirk (1982) found that Fe^{2+} and Zn^{2+} salts in a simulated breakfast food did not significantly affect the degradation

TABLE 7
Effect of storage on the ratio DHA/RAA

	a_w	*DHA/RAA after storage (weeks)*				
		1	*4*	*8*	*10*	*18*
1. Dried seaweed (laver 'nori') 13–15°C	0·2	0·13	0·22	0·31	0·30	—
Ogawa *et al.*	0·4	0·06	0·11	0·37	0·29	—
(1983)	0·6	0·05	0·54	2·84	2·02	—
	0·8	0·06	1·06	3·28	5·04	—
2. Fortified chocolate coated low energy meal replacement product						
Type 1	≃0·6	2·7	2·0	2·5	—	1·5
Type 2	≃0·6	12·5	3·7	4·6	—	1·1
~20°C Attwood (1987)						

rates of TAA or RAA at 0·1 or 0·4a_w but a small acceleration was observed in the presence of copper. At 0·65a_w the degradation rates were increased relative to unfortified systems and was attributed to increased mobility of the metal ions in the capillary absorption region. This effect is apparent in the protection for beta-carotene exerted by copper and ascorbic acid in autoxidising systems (Kanner *et al.*, 1977).

FAT SOLUBLE VITAMINS

Work on fat soluble vitamins is mainly confined to retinol and its derivatives and to beta-carotene In his brief update on vitamins in relation to the effect of water activity Leung (1987) was unable to report any work on tocopherols other than that of Widicus *et al.* (1980) who found that in a model system containing no fat, α-tocopherol degradation rate increased with increasing water activity and temperature in the ranges studied (0·1–0·65a_w and 20–37°C). The degradation followed first order kinetics and activation energies were in the range 36–54 kJ/mol.

Work on fat soluble vitamins tends to get classified either with fats because the mechanisms of oxidative degradation both proceed via the formation of hydroperoxides, or, in the case of beta-carotene, with pigments. Because of the importance of beta-carotene as a food colour and its vulnerability to oxidation and consequent loss of colour it has been far more extensively studied than retinol and its derivatives.

Although many studies have monitored loss over time under different water activity conditions and reported first or second order plots few have reported both rate constants and activation energies.

Studies fall into two groups—those which have reported that stability increases with increasing water activity (i.e. water exhibits a protective effect at all water activity levels) and those which report water activity values which exhibit maximum stability at or near the BET monolayer.

In studies of synthetic chemical systems (beta-carotene adsorbed on solid supports) the continuous protective effect from the 'dry' state through values of 0·11, 0·23, 0·52, and 0·75a_w on a range of adsorbents whose BET monolayer water activity value was from 0·18 (cellulose) to 0·28 (starch), was shown by Ramakrishnan and Francis (1979*a*) and by Arya *et al.* (1979*a*). Goldman *et al.* (1983) found that stability was greater at 0·33 and 0·84a_w than in the 'dry' state. In contrast, Chou and Breene (1972) and Baloch *et al.* (1977) showed that the degradation rate was lower at the BET monolayer than in the 'dry' state or at higher water activity. Despite the differences in the findings the methods of sample preparation, treatment and analysis were essentially the same. The beta-carotene was applied to the dry support in an organic solvent which was then removed by evaporation under vacuum. The required water activity was achieved by equilibrating the sample in a desiccator containing a saturated solution of the appropriate salt. All operations were carried out in the dark under nitrogen.

Similarly both types of result have been found in food systems and in extracts from foods. Premavalli and Arya (1985) working with carotenoids extracted from watermelon and suspended on microcrystalline cellulose, carboxymethyl cellulose, pectin, sugar, starch, or gelatin found the protective effect of water to be continuous.

Other studies have shown that a water activity close to the monolayer exerts maximum protection. Chen and Gutmanis (1968) showed maximum stability for pigments extracted from chilli pepper at around 9% moisture, but pigments from paprika required 14%

moisture ($0{\cdot}64a_w$) for maximum storage stability. Papaya and carrot carotenoids have been found to be most stable at $0{\cdot}43a_w$ on microcrystalline cellulose (Arya *et al.*, 1979*b*, 1983). Haralampu and Karel (1983) found that the first order rate constants for the degradation of beta-carotene in dehydrated sweet potato were inversely proportional to the water activity so that the maximum effect on the rate constant was observed below $0{\cdot}3$–$0{\cdot}4a_w$. Thereafter the rate changed little.

Carotenoids mostly occur in the lipid component of foods. Two types of evidence on the relationship between carotenoid oxidation and fat composition exist in the literature. The school represented by Bickoff *et al.* (1955) suggests that carotenoids are protected by unsaturated fats. The alternative view represented by Budowski and Bondi (1960) indicates that unsaturation accelerates degradation. Ramakrishnan and Francis (1979*b*) made a significant contribution to the understanding of the coupled oxidation of beta-carotene and lipids, by producing both effects in chemical systems (Table 8) and showing that relative to hexane unsaturated bonds accelerate beta-carotene destruction, but relative to methyl stearate unsaturated bonds protect beta-carotene. The extreme sensitivity of beta-carotene to oxidation in methyl stearate was explained by considering the characteristics of the systems in relation to the most likely free radical reactions. A saturated system cannot react by addition so that hydrogen abstraction was considered to be the route of propagation of free radicals. However saturated peroxy radicals cannot be stabilised by resonance and would therefore react rapidly with the most likely

TABLE 8
Half-life of β-carotene in methyl esters of fatty acids

Medium (% methyl linoleate in hexane)	*Half-life (hours)*	*Medium (methyl ester)*	*Half-life (hours)*
0	87·7	Stearate	3·2
0·5	61·2	Oleate	8·7
1·0	47·2	Linoleate	21·5
5·0	33·4	Linolenate	44·3
100·0	22·5		

From: Ramakrishnan and Francis (1979*b*).

hydrogen donor. In the presence of beta-carotene abstraction of hydrogen from the 4 and 4′ positions is seen as energetically favoured. However in the presence of unsaturated esters better sites for hydrogen abstraction are postulated. In addition, unsaturated peroxy radicals are stabilised by resonance and addition offers an alternative reaction pathway. Extreme stability in saturated hydrocarbons is attributed to the lack of free radical formation.

Model systems which offer an explanation of the occurrence of a water activity zone of maximum stability have been devised by Kanner *et al.* (1977) and Kanner and Budowski (1978). Pigments extracted from red pepper fruits were suspended on microcrystalline cellulose with linoleate or oleoresin and ascorbic acid and copper. Both antioxidant and pro-oxidant effects were observed depending on the water activity and the level of ascorbic acid. Ascorbic acid had no significant pro- or antioxidant effect at low water activity (0·01). The antioxidant effect observed at high levels of ascorbic acid was attributed to chelate formation between ascorbate and copper. At low ascorbic acid levels a pro-oxidant effect was observed which increased with increasing water activity. As increasing water activity also accelerates ascorbate destruction the combined effect would show as a peak on a water activity/stability curve of variable position depending on the exact composition of the system.

Systems in which retinol and its derivatives have been studied in relation to the specific effect of water activity are very limited. In the most comprehensive review of the effect of water activity on vitamin stability (Kirk, 1981) vitamin A stability is presented as having features in common with lipids with which it is normally associated in foods. Thus there is a water activity of maximum stability below which antioxidant mechanisms attributed to water do not function (Labuza, 1975). This water activity represents the zone of greatest stability because at higher water levels antioxidant effects due to hydration of trace metals, hydrogen bonding of hydroperoxides, promotion of radical combination and dilution of metal concentration may be offset by increased mobility, dissolution of precipitated catalysts and the increased availability of catalytic surfaces due to swelling of solid matrices. As would be expected in view of the findings of Ramakrishnan and Francis (1979*a*) with a similar system, Guevara (1985) working with a model system of retinol on microcrystalline cellulose carefully freed from metal contamination by treatment with EDTA, showed that the second order rate constant decreased below the BET monolayer. Thus in the absence of metal ions the pro-oxidant effect attributed to the unhydrated ions does not function.

First order kinetics have been reported for beta-carotene adsorbed on solid supports or model systems by Chou and Breene (1972); Baloch *et al.* (1977); Ramakrishnan and Francis (1979*a,b*) and Teixeira-Neto *et al.* (1981). Complex kinetics were obtained by Goldman *et al.* (1983). They showed that the reaction had the characteristic sigmoidal shape of a chain reaction. As it was catalysed by a free radical initiator they concluded that it was a free radical reaction.

In foods and food extracts Chen and Gutmanis (1968) found second order kinetics for ground chilli pepper.

Values in excess of one for the ratio of mol of oxygen consumed per mol of polyene lost have been observed in low moisture foods containing retinoid polyenes by several workers. Walter and Purcell (1974) working with dehydrated sweet potato concluded that more oxygen was absorbed than could be accounted for by peroxide formation. Teixeira-Neto *et al.* (1981) in a model system of milk powder on a microcrystalline cellulose support containing beta-carotene found a ratio of 7 mol of oxygen per mol of beta-carotene decolourised.

Goldman *et al.* (1983) in a similar system found 8 mol of oxygen were required per mol of beta-carotene decolourised. No such reports have been made for retinol or its derivatives—reflecting perhaps the greater importance of beta-carotene as a component of plant foods and its role as a food colour. However oxygen uptake by both beta-carotene and retinyl acetate has been investigated in solutions and in thin films during extensive mechanistic studies. Gargarina *et al.* (1970) observed that, in the dark oxidation of beta–carotene in benzene and xylene, the maximum rate of absorption of oxygen exceeded the rate of carotene consumption by a factor of 3·5 and that the absorption of oxygen continues to take place for a long time after the disappearance of the original polyene. Under conditions in which the only reaction of retinol was oxidation of the alcohol group the ratio of mol of oxygen consumed per mol of retinol consumed was 0·5 (Karrer and Hess, 1957). Pekkarinen and Montonen (1973), working with retinol, retinyl acetate and anhydro-retinol in liquid paraffin or acetic acid found ratios in the range 2·2–2·5 mol of oxygen per mol of polyene consumed.

Finkel'shtein *et al.* (1981) using retinyl acetate in chlorobenzene extrapolated that 3 mol of oxygen per mol of retinyl acetate were consumed at infinite dilution but that this tended to a limiting value of one at high concentration. The products of oxidation of retinyl acetate and beta-carotene differ; in beta-carotene oxidation peroxides and the 5–6 epoxides have been identified but in retinyl acetate, peroxides and the 11–12 epoxide have been found (Ogata, 1970). According to Finkel'shtein et al. (1978), the most energetically favoured position for hydrogen abstraction from retinol and retinyl acetate is from C_{14} resulting in the 11–12 epoxide as opposed to C_4 in beta-carotene giving the 5–6 epoxide.

Evteeva and Gargarina (1983) studied the consumption of unsaturated bonds during the oxidation of retinyl acetate and examined some simple models of the sequence:

$$A_n \rightarrow P_{n-1} \rightarrow P_{n-2} \rightarrow P_0$$

In the first of these models all reactive centres and the free radicals formed from them are identically active. In the second model some of the products of the sequence are so active they are consumed at the same time as the original polyene. In model three, all products P are very reactive. Experiment showed that the primary products of transformation of retinyl acetate contain polyene fragments with a higher reactive capacity with respect to free radicals than the system of conjugated bonds. The authors concluded that the rate of consumption of unsaturated bonds can only be greater than the rate of consumption of retinyl acetate if this is the case. It was suggested that the ratio of oxygen uptake/double bond loss is less than one when some double bonds are lost in cyclisation reactions without further uptake of oxygen.

Finkel'shtein *et al.* (1981) working with retinyl acetate in chlorobenzene suggested that the high ratio of oxygen consumed per mol of retinyl acetate lost depends on the rate constant for the dissociation of the radical $RO_2\cdot$ being small compared with the rate constant for its isomerisation, where RH is the original polyene.

Thus the high rate of the chain propagation reaction during the oxidation of retinyl acetate is secured by the high rate of isomerisation of the $RO_2\cdot$ radical which competes with its dissociation reaction.

Since carotenoids and retinoids in low water activity systems have been found to take up many moles of oxygen per mol of retinol lost they do not have the characteristics of a concentrated solution for which this ratio tends to one at high concentration (Finkel'stein *et al.*,

1981). In a dilute solution intramolecular reactions become more important and it has been shown by the same workers by extrapolation of results from solutions of retinyl acetate in chlorobenzene that at infinite dilution the mole ratio of oxygen absorbed to polyene lost tends to a value of three. In solid films molecular mobility affects the rate of oxidation (Finkel'shtein *et al.*, 1970) as shown by the high activation energy for the oxidation of thin films of retinyl acetate between 10 and 25°C. Reduced mobility in the solid state might also be expected to increase the relative importance of intramolecular reactions and hence the uptake of oxygen. Teixeira-Neto *et al.* (1981) working in model food systems containing microcrystalline cellulose showed uptake by beta-carotene of 6–7 mol of oxygen per mole of beta-carotene decolourised.

Thus it may be tentatively proposed that due to reduced molecular mobility in low water activity systems the uptake of oxygen by retinoids shows characteristics in common with mechanisms undergone by solid films rather than by solutions.

PREDICTING VITAMIN LOSSES DURING STORAGE WHEN WATER ACTIVITY IS CHANGING

Approaches to and progress in the modelling of food processes in order to predict quality changes have been discussed by Labuza (1972); Saguy and Karel (1980); Thompson (1982) and Lund (1983). Impetus for the generation of kinetic data on nutrient stability, particularly stability during storage, has come in part from the need to predict shelf life in order to meet the requirements of nutritional labelling regulations in the US (Labuza, 1972). Additionally nutrient stability, particularly ascorbic acid, is often used as a relatively easily measured quality monitoring index (along with non-enzymic browning) to test the validity of mathematical models of food processing and storage. Thiamin has not been as much used as ascorbic acid to test predictive and optimisation models for processing and storage in the field of changing water activity as it has been in thermal sterilisation (Teixeira *et al.*, 1969, 1975; Saguy and Karel, 1980). At first sight this seems strange because systems in which both temperature and moisture content are changing are mathematically more complex to model than thermal sterilisation in closed containers where the moisture content is effectively constant. Also ascorbic acid is undoubtedly a more difficult entity to deal with than thiamin because of the greater complexity of its degradation pathways. However, predictive studies are more satisfactory to verify experimentally if the entity in question has the potential to be seriously degraded within a satisfactorily short period of time thus avoiding unnecessarily lengthy storage periods. Rate constants for the degradation of ascorbic acid are large compared with thiamin (Tables 1 and 4) and therefore ascorbic acid offers advantages in modelling the effects of storage, and processes in which the temperature changes are relatively small.

The ‘classical’ approach is to perform a ‘balanced’ matrix of studies in which rate constants are measured in a series of isothermal systems—a time consuming operation. The use of kinetic data obtained under isothermal conditions in unsteady-state heating processes is mathematically simpler if other parameters remain constant, so that the majority of modelling exercises have involved canning processes where the total water content does not change. Thompson (1982) pointed out that, although information obtained on both foods and model systems under isothermal conditions gives insight into degradation mechanisms, such information is not essential in mathematical modelling processes. Whilst knowledge that a degradation process follows first order kinetics is helpful in choosing a mathematical model, the great expansion in computer software means that curve fitting can be carried out to find a fit for experimental data without the need for prior mechanistic information. Also doubts have been cast on the validity of transferring data obtained under isothermal conditions to unsteady-state conditions (Labuza and Saltmarsh, 1982).

Thus there has been considerable interest by food engineers in finding predictive models for processes in which the water content is changing using data obtained under dynamic conditions. The ultimate objective of such studies is to optimise processing and storage parameters (Mishkin *et al.*, 1982). The indices utilised for these predictions are required to be both rapid and simple to measure and to be relatively labile so that significant changes occur within a reasonable experimental timespan. Both ascorbic acid and non-enzymic browning fall into this category.

Storage studies related to water activity effects can conveniently be discussed in four sections

Studies Assuming Isothermal Storage Conditions and Constant Water Activity

Early studies undertaken to simulate storage simplified the mathematical treatment required by assuming isothermal storage in which the rate of ascorbic acid loss was dependent on water content but water content itself did not change and was treated as a reactant (Wanninger, Jr, 1972):

$$\ln C/C_0 = -k[H_2O]t$$

where C = concentration at time t; C_0 = initial concentration; k = first order rate constant at a specified temperature; and t = time.

Using this equation Wanninger predicted the results of experiments of Vojnovich and Pfeifer (1970) which showed that ascorbic acid degradation rate increased with increasing water content between 26 and 45°C over the ranges 12–14% for wheat flour, 8–11% for corn/soy/milk powder and 5–10% for mixed cereal.

Riemer and Karel (1978) developed an equation which related the retention of ascorbic acid during the isothermal storage of dehydrated tomato juice to the temperature and the water activity. The relationship was developed from the Arrhenius equation:

$$k = k_0 \exp(-E_a/(RT)) \tag{1}$$

where k = rate constant; k_0 = frequency factor; E_a = activation energy; R = gas constant; and T = temperature in degrees Kelvin, and the first order rate equation:

$$\frac{dc}{dt} = -kc \tag{2}$$

where c = nutrient concentration; t = time; and k = rate constant.

Using percentage retention RTN = $100C/C_0$ and substituting for k into the integrated form of eqn (2):

$$\mathrm{RTN} = \exp(4{\cdot}6051 - k_0 \exp(-E_a/(RT))) \tag{3}$$

The unknowns in the above equation were found by considering the behaviour of k_0, and E_a with water activity from a bank of experimental data which were shown by regression analysis to have the forms:

$$k_0 = \exp(a - ba_w)10^{10}$$
$$E_a = \exp(p - qa_w)$$

where a, b, p and q are constants. k_0 and E_a were then substituted into (3):

$$\mathrm{RTN} = (\exp 4{\cdot}6051 - \exp(a - ba_w)\, 10^{10} \exp(-(p - qa_w)/(RT)))$$

The values of the constants $a = 8{\cdot}100$, $b = 8{\cdot}962$, $p = 9{\cdot}999$ and $q = 0{\cdot}470$ were determined by a curve fitting exercise using a data base of experimentally determined values covering the temperature range 20–51°C and water activity from 0–0·75. The authors point out that as degradation of ascorbic acid in tomato juice is not affected by oxygen the above equation is a general model for dehydrated tomato juice.

Saguy *et al.* (1985) predicted decolouration in a model system of beta-carotene on microcrystalline cellulose under both static and dynamic oxygen conditions and incorporated water activity in the mathematical model. The oxygen consumption model was based on the free radical chain mechanism of Émanuel *et al.* (1967):

$$2RH \rightarrow 2R\cdot \quad (1)$$
$$R\cdot + O_2 \rightleftarrows RO_2\cdot \quad (2)$$
$$RH + RO_2\cdot \rightarrow R\cdot + ROOH \quad (3)$$
$$R\cdot + R^1H \rightarrow R^1\cdot + RH \quad (4)$$
$$RO_2\cdot + RO_2\cdot \rightarrow \text{products} \quad (5)$$
$$R\cdot + R^1\cdot \rightarrow \text{products} \quad (6)$$

Using a second order chain termination they derived a linear plot from which an effective rate constant (σ) could be determined.

$$\ln[(1 + \sqrt{1 - C/C_0})/(1 - \sqrt{1 - C/C_0})] = \sigma t$$

where σ was a function of the rate constants of reactions 3 and 5, the equilibrium constant of reaction 2, the partial pressure of oxygen and the solubility coefficient of oxygen in beta-carotene. C and C_0 are the concentrations at time $= t$ and time $= 0$ respectively.

From data collected under constant conditions of water activity (dry to 0·84) and oxygen partial pressure (2–20·9%) corrections were made to σ to incorporate both oxygen partial pressure and water activity.

$$\sigma = [2{\cdot}982*10^{-2} + 1{\cdot}656*10^{-2}P_{O_2}] \exp(-0{\cdot}4833a_w)$$

When applying this empirical model to dynamic oxygen conditions it was found necessary to incorporate a correction factor to accommodate physical diffusion processes.

Studies Assuming Isothermal Conditions in which the Water Activity Changes

More sophisticated approaches to isothermal storage modelling incorporated water activity changes in the product caused by permeable packaging. Mizrahi and Karel (1977, 1978) utilised a 'no model' concept (no prior knowledge of the kinetic model of the effect of moisture on the degradation rate) to develop a rapid test for determining the storage life of a packaged product in terms of any chosen quality index. They obtained good agreement with experiment for ascorbic acid in tomato powder and browning in dehydrated cabbage. The accelerated technique was based on 'programming' rapid moisture change in the product by packaging in a highly permeable material and assuming constant external water activity.

The time required to reach any given moisture content is inversely proportional to the permeability.

$$dm/dt = K(a_e - a_i)$$

where $K = 100(K_{H_2O}P_0A(a_e - a_i)/W_sX)$; K_{H_2O} = permeability of the packaging material; a_e = external water activity; a_i = 'in packet' water activity in equilibrium with contents; P_0 = water vapour pressure at storage temperature; A = package area; X = thickness of packaging material; W_s = weight of solids per package; and dm/dt = rate of moisture gain (g/100 g solid/day).

If the change in the quality parameter index is

$$(D - D_0) = f(m)t$$

where D_0 = quality parameter index at time $t = 0$ and D = quality parameter index at time $t = t$.

When moisture is introduced into the food at constant rate

$$m = m_0 + bt$$

where m_0 = the initial moisture content. dt can be replaced by dm/b

$$(D - D_0) = 1/b \int_{m_0}^{m} f(m)\, dm$$

$$= (f(m) - f(m_0))/b$$

For cases in which b, m_0 and m are kept constant, the change in the index of deterioration is inversely proportional to b, so

$$(D - D_0)_2/(D - D_0)_1 = b_1/b_2$$

Although the moisture regain is not constant, it can be broken into constant rate periods for mathematical treatment. Since $b_1/b_2 = K_1/K_2$, the time taken for a specific quality change to be achieved in the highly permeable packaging can be extrapolated to the less permeable material for which a shelf life study would be unreasonably long. Using this technique, ascorbic acid loss in dehydrated tomato after 14–19 days could be predicted within a 2–3 day storage period.

Quast and Karel (1970*a,b*) developed a model for the oxidation of potato chips (crisps), measured as μl O_2/g/h at STP as a function of oxygen pressure, extent of oxidation and relative humidity, which had the form:

Rate of oxidation

$$= [\text{Ext} + (P_1/\text{RH}^{1/2}) + (P_2\,\text{Ext}/\text{RH}^{1/2})][P_{O_2}/(P_3 + P_4 P_{O_2})]$$

where Ext = extent of oxidation in μl O_2 absorbed/g at STP; RH = relative humidity; P_{O_2} = oxygen partial pressure in atmospheres; and P_1, P_2, P_3 and P_4 are constants.

This model incorporated knowledge of lipid oxidation kinetics and was further developed by Quast, Karel and Rand (1972), to incorporate the packaging variables oxygen and water vapour permeability and the oxygen and water activity gradient across the packaging film. The outcome was that by specifying maximum acceptable storage levels for the extent of oxidation and water activity it was possible to choose between different packaging systems.

Studies Assuming Fluctuating Storage Temperatures

The most sophisticated storage models simulate fluctuating temperatures. In reality storage is not isothermal and there is evidence that the effect of fluctuating temperature is greater than that predicted from knowledge of the mean temperature (Labuza, 1979).

The simplest technique for dealing with fluctuating temperature is that illustrated by Labuza and Saltmarsh (1982) who used the Hicks–Schwimmer–Labuza model (Labuza, 1979) for predicting available lysine loss and browning in whey powder at 0·33, 0·44 and 0·65a_w, stored in desiccators and alternated every five days between 25°C and 45°C.

The effective rate constant is given by:

$$k_{\text{effec}} = \tfrac{1}{2} k_{T_m}[Q_{10}^{[(\theta/10)(T_m+10)/(T_m+\theta)]} + Q_{10}^{-[(\theta/10)(T_m+10)/(T_m-\theta)]}]$$

where θ(°C) = the amplitude of the square wave fluctuation; k_{T_m} = the rate constant at the mean temperature; and Q_{10} = (Rate at T + 10)/(Rate at T).

Browning increased, and available lysine (measured as dinitrophenyl lysine and as relative nutritive value) fell more rapidly at $0{\cdot}44a_w$ than at $0{\cdot}33a_w$ or $0{\cdot}66a_w$ and the rates were greater under fluctuating storage conditions than at the mean temperature.

Models in which both Water Activity and Storage Temperature Change Simultaneously

Singh *et al.* (1984) predicted a temperature profile, a water activity profile and browning and ascorbic acid retention in 3-cm slices of IMF apple, packaged in aluminium pouches at 25°C and subjected to a square wave fluctuating temperature of ±15°C cycling over 12 h. The results were compared with actual storage experiments. Ascorbic acid was significantly more stable at $0{\cdot}6a_w$ compared with $0{\cdot}84a_w$. There was an induction period of seven days for non-enzymic browning during which time 25% of ascorbic acid had been lost.

The basis of this method was that one-dimensional heat and moisture transfer equations were used to predict the temperature, moisture content and water activity at the surface and at the centre after each temperature change.

The value of techniques in this group, which predict water activity changes lies in their power to predict microbial stability in enclosed environments where the water activity increases due to the permeability of the packaging or to moisture migration due to fluctuating temperature.

CONCLUSION

The lack of rapid, accurate and specific methods of assay and the work content and prolonged time associated with the generation of data banks which cover an adequate range of conditions have largely restricted studies to the most labile and the most easily monitored vitamins in 'chemical' systems and in a relatively small number of foods. However, the fact that storage conditions inevitably involve some degree of temperature fluctuation or cycling which cause local changes in water activity reduce the efficacy of data collected under steady state conditions for predictive purposes.

REFERENCES

Abdel-Kader, Z. M. (1985). Diffusion of riboflavin under sterilisation conditions. PhD thesis, University of Leeds, UK.

Allen, C. and Parks, O. W. (1979). Photodegradation in milks exposed to fluorescent light. *J. Dairy Sci.,* **62,** 1377–9.

Arabshahi, A. and Lund, D. B. (1985). Considerations in calculating kinetic parameters from experimental data. *J. Fd Proc. Eng,* **7,** 239–52.

Archer, M. C. and Tannenbaum, S. R. (1979). Vitamins. In: *Nutritional and Safety Aspects of Food Processing,* S. R. Tannenbaum (Ed.), Marcel Dekker Inc., New York and Basel, pp. 47–95.

Arya, S. S., Natesan, V., Parihar, D. B. and Vijayaraghavan, P. K. (1979*a*). Stability of beta-carotene in isolated systems. *J. Fd Technol.,* **14**(6), 571–8.

Arya, S. S., Natesan, V., Parihar, D. B. and Vijayaraghavan, P. K. (1979*b*). Stability of carotenoids in dehydrated carrots. *J. Fd Technol.,* **14,** 579–86.

Arya, S. S., Natesan, V. and Vijayaraghavan, P. K. (1983). Stability of carotenoids in freeze dried papaya. *J. Fd Technol.,* **18**(2), 177–81.

Attwood, J. S. (1987). An investigation of the storage stability of riboflavin and ascorbic acid in an intermediate moisture food. MSc. thesis, University of Leeds.

Baloch, A. K., Buckle, K. A. and Edwards, R. A. (1977). Stability of beta-carotene in model systems containing sulphite. *J. Fd Technol.,* **12**(3), 309–16.

Bender, A. E. (1978). *Food Processing and Nutrition,* Food Science and Technology Monographs, Academic Press, London.

Bickoff, E. M., Thompson, C. R., Livingston, A. L., van Atta, G. R., and Guggolz, J. (1955). Effect of animal fats and vegetable oils on the stability of carotene in dehydrated alfalfa meal. *J. Agric. Fd Chem.,* **3,** 67–9.

Bluestein, P. M. and Labuza, T. P. (1975). Effects of moisture removal on nutrients. In: *Nutritional Evaluation of Food Processing,* 2nd Edn, R. S. Harris and E. Karmas (Eds), AVI Publishing Co. Inc., Westport, Connecticut, pp. 289–324.

Bluestein, P. M. and Labuza, T. P. (1987). Effects of moisture removal on nutrients. In: *Nutritional Evaluation of Food Processing,* E. Karmas and R. S. Harris (Eds), AVI, New York, pp. 393–422.

Bookwalter, G. N., Moser, H. A., Pfeifer, V. F. and Griffin, E. L., Jr (1968). Storage stability of blended food products, formula no. 2: a corn–soy–milk food supplement. *Fd Technol.,* **22,** 1581–4.

Brubacher, G., Müller-Mulot, W. and Southgate, D. A. T. (Eds) (1985). *Methods for the Determination of Vitamins in Food. Recommended by COST 91,* Elsevier Applied Science Publishers, London and New York, pp. 119–28.

Budowski, P. and Bondi, A. (1960). Autoxidation of carotene and vitamin A. Influence of fat and antioxidants. *Archs Biochem. Biophys.,* **89,** 66–73.

Chen, A. C. C. and Karmas, E. (1980). Solute activity effect on water activity. *Lebensm-Wiss. u.-Technol.,* **13,** 101–4.

Chen, S. L. and Gutmanis, F. (1968). Auto-oxidation of extractable color pigments in chili pepper with special reference to ethoxyquin treatment. *J. Fd Sci.,* **33,** 274–80.

Chirife, J., Iglesias, H. A. and Boquet, R. (1980). Retention of available lysine after long term storage of intermediate moisture beef formulated with various humectants. *Lebensm-Wiss. u.-Technol.,* **13**(1), 44–5.

Chou, H. and Breene, W. M. (1972). Oxidative decoloration of beta-carotene in low moisture model systems. *J. Fd Sci.*, **37**(1), 66–8.

Cooke, J. R. and Moxon, R. E. D. (1981). The detection and measurement of vitamin C. In: *Vitamin C,* J. N. Counsell and D. H. Horning (Eds), Applied Science Publishers, London, pp. 167–98.

Dennison, D. B. and Kirk, J. R. (1978). Oxygen effect on the degradation of ascorbic acid in a dehydrated food system. *J. Fd Sci.*, **43**(2), 609–12, 618.

Dennison, D. B. and Kirk, J. R. (1982). Effect of trace mineral fortification on the storage stability of ascorbic acid in a dehydrated model food system. *J. Fd Sci.*, **47**(4), 1198–1200, 1217.

Dennison, D., Kirk, J., Bach, J., Kokoczka, P. and Heldman, D. (1977). Storage stability of thiamin and riboflavin in a dehydrated food system. *J. Fd Proc. Pres.*, **1**(1), 43–54.

Deutsch, M. J. and Weeks, C. E. (1965). Microfluorimetric assay for vitamin C. *J. Ass. Off. Anal. Chem.*, **48**(6), 1248–56.

Émanuel, N. M., Denisov, E. T. and Maizius, Z. K. (1967). *Liquid Phase Oxidation of Hydrocarbons,* Plenum Press, New York.

Evteeva, N. M. and Gargarina, A. B. (1983). Consumption of unsaturated bonds during oxidation of retinyl acetate. *Dokl. Akad. Nauk SSSR* **269**(1), 135–40. Translation in: *Dokl. Phys. Chem.*, (1983). **269**(1–3), 117–21.

Farrer, K. T. H. (1955). The thermal destruction of vitamin B in foods. *Adv. Fd Res.*, **6,** 257–311.

Fernandez, B., Mauri, L. M., Resnik, S. L. and Tomio, J. M. (1986). Effect of adjusting the water activity to 0·95 with different solutes on the kinetics of thiamin loss in a model system. *J. Fd Sci.*, **51**(4), 1100–1.

Finkel'shtein, E. I., Koslov, E. I. and Samokhvalov, G. I. (1970). Kinetic laws of the oxidation and stabilisation of polyunsaturated compounds. II. Autoxidation of vitamin A acetate in the solid state. *Kinet. Katal.*, **11**(1), 71–4.

Finkel'shtein, E. I., Dolotov, S. M. and Koslov, E. I. (1978). The mechanism of the oxidation of retinyl polyenes by molecular oxygen. The possible role of reversibility in the reaction $R\cdot + O_2 \rightarrow RO_2\cdot$ *Zh. Org. Khim.*, **14**(3), 525–34.

Finkel'shtein, E. I., Mednikova, N. A. and Koslov, E. I. (1981). Kinetic relationships in the liquid phase oxidation of retinyl acetate *Zh. Org. Khim.*, **17**(5), 929–33.

Fox, M., Loncin, M. and Weiss, M. (1982). Investigations into the influence of water activity, pH, and heat treatment on the breakdown of thiamine in foods. *J. Fd Qual.*, **5**(3), 161–2.

Furuya, E. M. Warthesen, J. J. and Labuza, T. (1984). Effects of water activity, light intensity, and physical structure of food on the kinetics of riboflavin photodegradation. *J. Fd Sci.*, **49**(2), 525–8.

Gargarina, A. B., Kasaikina, O. T. and Émanuel, A. N. M. (1970). Kinetics of autoxidation of polyene hydrocarbons. *Dokl. Akad. Nauk SSSR,* **195**(2), 387–90.

Goldman, M., Horev, B. and Saguy, I. (1983). Decolorization of beta-carotene in model systems simulating dehydrated foods. *J. Fd Sci.*, **48**(3), 751–4.

Gregory III, J. F. (1983). Methods of vitamin assay for nutritional evaluation of food processing. *Fd Technol.*, **37,** 75–80.

Gregory, J. F. and Kirk, J. R. (1978). Assessment of storage effects on vitamin B_6 stability and bioavailability in dehydrated food systems. *J. Fd Sci.*, **43**(6), 1801–15.

Guevara, L. G. (1985). The stability and oxygen consumption of retinol at low water activity. PhD thesis, University of Leeds, UK.

Haralampu, S. G. and Karel, M. (1983). Kinetic models for moisture dependence of ascorbic acid and beta-carotene degradation in dehydrated sweet potato. *J. Fd Sci.*, **48**(6), 1872–3.

Hollenbeck, C. M. and Obermeyer, H. G. (1952). Relative stability of thiamin mononitrate and thiamin chloride hydrochloride in enriched flour. *Cereal Chem.*, ·**29,** 82–7.

Jensen, A. (1969). Tocopherol content of seaweed and seaweed meal. 3. Influence of processing and storage on the content of tocopherol, carotenoids and ascorbic acid in seaweed meal. *J. Sci. Fd Agric.*, **20,** 622–6.

Kamman, J. F., Labuza, T. P. and Warthesen, J. J. (1981). Kinetics of thiamin and riboflavin loss in pasta as a function of constant and variable storage conditions. *J. Fd Sci.*, **46**(5), 1457–61.

Kanner, J. and Budowski, P. (1978). Carotene oxidising factors in red pepper fruits (*Capsicum annum* L.): Effect of ascorbic acid and copper in a beta-carotene–linoleic acid solid model. *J. Fd Sci.*, **43**(2), 524–6.

Kanner, J., Mendel, H. and Budowski, P. (1977). Pro-oxidant and anti-oxidant effects of ascorbic acid and metal salts in a beta-carotene–linoleate model system. *J. Fd Sci.*, **42,** 60–4.

Karel, M. and Nickerson, J. T. R. (1964). Effects of relative humidity, air and vacuum in browning of dehydrated orange juice. *Fd Technol.*, **18**(8), 1214–18.

Karrer, von, P. and Hess, W. (1957). Uber die katalytische oxydation von vitamin A mit sauerstoff und platin zu vitamin A aldehyd (eine neue methode). *Helv. Chim. Acta,* **40**(34), 265–6.

Kearsley, M. W. and Rodriguez, N. (1981). The stability and use of natural colours in foods: anthocyanin, beta-carotene and riboflavin. *J. Fd Technol.*, **16,** 421–31.

Kirk, J. R. (1974). Automated method for the analysis of thiamine in milk, with applications to other selected foods. *J. Ass. Off. Anal. Chem.*, **57**(5), 1081–4.

Kirk, J. R. (1981). Influence of water activity on the stability of vitamins in dehydrated foods. In: *Water Activity: Influences on Food Quality,* L. B. Rockland and G. F. Stewart (Eds), Academic Press, pp. 531–66.

Kirk, J. R., Dennison, D. B., Kokoczka, P. and Heldman D. R. (1977). Degradation of ascorbic acid in a dehydrated food stystem. *J. Fd Sci.*, **42**(5), 1274–9.

Labuza, T. P. (1972). Nutrient losses during drying and storage of dehydrated foods. *CRC Crit. Rev. Fd Technol.*, **3**(2), 217–40.

Labuza, T. P. (1975). Oxidative changes in foods at low and intermediate moisture levels. In: *Water Relations of Foods,* R. B. Duckworth (Ed.), Academic Press, London, pp. 455–76.

Labuza, T. P. (1979). A theoretical comparison of loss in foods under fluctuating temperature sequences. *J. Fd Sci.,* **44,** 1162–8.

Labuza, T. P. (1980*a*). Enthalpy/entropy compensation in food reactions. *Fd Technol.,* **34,** 67–77.

Labuza, T. P. (1980*b*). The effect of water activity on reaction kinetics of food deterioration. *Fd Technol.,* **34,** 36–41, 59.

Labuza, T. P. and Saltmarsh, M. (1982). Kinetics of browning and protein quality loss in whey powders during steady state and non-steady state storage conditions. *J. Fd. Sci.,* **47,** 92–6, 113.

Laing, B. M., Schlueter, D. L. and Labuza, T. P. (1978). Degradation kinetics of ascorbic acid at high temperature and water activity. *J. Fd. Sci.,* **43**(5), 1440–3.

Lappo, B. P. (1987). Operational and nutritional optimisation in the thermal sterilisation of canned food. PhD thesis, University of Leeds, UK.

Lee, S. H. and Labuza, T. P. (1975). Destruction of ascorbic acid as a function of water activity. *J. Fd Sci.,* **40,** 370–3.

Leung, H. K. (1987). Influence of water activity on chemical reactivity. In: *Water Activity: Theory and Applications to Food,* L. B. Rockland and L. R. Beuchat (Eds), Basic Symposium Series, Marcel Dekker Inc., New York and Basel, pp. 27–54.

Linksweiler, H. (1958). The effect of the ingestion of ascorbic acid and dehydroascorbic acid upon the blood levels of these two components in human subjects. *J. Nutr.,* **64,** 43–54.

Lund, D. B. (1983). Considerations in modeling food processes. *Fd Technol.,* **37,** 92–4.

Mishkin, M., Karel, M. and Saguy, I. (1982). Applications of optimisation in food dehydration. *Fd Technol.,* **36,** 101–9.

Mishkin, M., Saguy, I. and Karel, M. (1984*a*). Optimisation of nutrient retention during processing: ascorbic acid in potato dehydration. *J. Fd Sci.,* **49,** 1262–6.

Mishkin, M., Saguy, I. and Karel, M. (1984*b*). A dynamic test for kinetic models of chemical changes during processing: ascorbic acid. *J. Fd Sci.,* **49,** 1267–70, 1274.

Mizrahi, S. and Karel, M. (1977). Accelerated stability tests of moisture sensitive products in permeable packages by programming rate of moisture content increase. *J. Fd Sci.,* **42,** 958.

Mizrahi, S. and Karel, M. (1978). Evaluation of kinetic model for reactions in moisture-sensitive products using dynamic storage conditions. *J. Fd Sci.,* **43,** 750–3.

Mohr, Jr, D. H. (1980). Oxygen mass transfer effects on the degradation of vitamin C in foods. *J. Fd Sci.,* **45,** 1432–3.

Mulley, E. A., Stumbo, C. R. and Hunting, W. M. (1975*a*) Kinetics of thiamine degradation by heat. A new method for studying reaction rates in model systems and food products at high temperature. *J. Fd Sci.,* **40,** 985–8.

Mulley, E. A., Stumbo, C. R. and Hunting, W. M. (1975*b*) Kinetics of thiamine degradation by heat. Effect of pH and form of the vitamin on its rate of destruction. *J. Fd Sci.,* **40,** 989–92.

Neale, R. J., Obanu, Z. A., Biggin, R. J., Ledward, D. A. and Lawrie, R. A. (1978). Protein quality and iron availability of intermediate moisture beef stored at 38°C. *Ann. Nutr. Alim.,* **32,** 587–96.

Ogata, Y., Kosugi, Y. and Tomizawa, K. (1970). Kinetics of autoxidation of vitamin A catalysed by cobalt ion. *Tetrahedron,* **26**(24), 5939–44.

Ogawa, H., Araki, S., Oohusa, T. and Kayama, M. (1983). Studies on the quality preservation of dried laver nori, *Porohyra yezoensis.* IV. The relationship between water and ascorbic acid in a dried laver nori during storage. *Nippon Suisan Gakkai-shi,* **49**(7), 1143–7.

Ogawa, H., Araki, S., Oohusa, T. and Kayama, M. (1984). Studies on the quality preservation of dried laver hoshi-nori, *Porphyra yezoensis.* VI. The cause of ascorbic acid destruction in hoshi-nori (dried laver) during storage. *Nippon Suisan Gakkai-shi,* **50**(12), 2085–9.

Pekkarinen, L. and Montonen, M. (1973). Uptake of oxygen in the autoxidation of retinol and its derivatives. *Suomen Kemistilehti, Series B,* **46**(3), 130–2.

Premavalli, K. S. and Arya, S. S. (1985). Stability of watermelon carotenoid extract in isolated model systems. *J. Fd Technol.,* **20**(3), 359–66.

Purwadaria, H. K., Heldman, D. R. and Kirk, J. R. (1980). Computer simulation of vitamin degradation in a dry model food system during storage. *J. Fd Proc. Eng,* **3**(1), 7–28.

Quast, D. G. and Karel, M. (1970*a*). Computer simulation of storage life of foods undergoing spoilage by two interacting mechanisms. *J. Fd. Sci.,* **37,** 679–83.

Quast, D. G. and Karel, M. (1970*b*). Effects of environmental factors on the oxidation of potato chips. *J. Fd Sci.,* **37,** 584–8.

Quast, D. C., Karel, M. and Rand, W. M. (1972). Development of a mathematical model for oxidation of potato chips as a function of oxygen pressure, extent of oxidation and equilibrium relative humidity. *J. Fd Sci.,* **37**(5), 673–8.

Ramakrishnan, T. V. and Francis, F. J. (1979*a*). Stability of carotenoids in model aqueous systems. *J. Fd Qual.,* **2**(3), 177–89.

Ramakrishnan, T. V. and Francis, F. J. (1979*b*). Coupled oxidation of carotenoids in fatty acid esters of varying unsaturation. *J. Fd Qual.,* **2**(4), 277–87.

Riemer, J. and Karel, M. (1978). Shelf life studies of vitamin C during food storage: prediction of L-ascorbic acid retention in dehydrated tomato juice. *J. Fd Proc. Pres.,* **1**(4), 293–312.

Rockland, L. B. and Nishi, S. K. (1980). Influence of water activity on food product quality and stability. *Fd Technol.,* **34,** 42–50, 51, 59.

Ruddick, J. E., Vanderstoep, J. and Richards, J. F. (1980). Kinetics of thermal degradation of methyl tetrahydrofolate. *J. Fd Sci.,* **45,** 1019.

Saguy, I. and Karel, M. (1980). Optimal temperature profile in optimising thiamine retention in conduction type heating of canned foods. *J. Fd Sci.,* **44,** 1485–90.

Saguy, I. and Karel, M. (1980). Modelling of quality deterioration during food processing and storage. *Fd Technol.*, **34**(2), 78–88.
Saguy, I., Kopelman, I. J. and Mizrahi, S. (1978). Simulation of ascorbic acid stability during heat processing and concentration of grapefruit juice. *J. Fd Proc. Eng*, **2**(3), 213–25.
Saguy, I., Mizrahi, S., Villotta, R. and Karel, M. (1978). Accelerated method for determining the kinetic model of ascorbic acid loss during dehydration. *J. Fd Sci.*, **43**(6), 1861–4.
Saguy, I., Goldman, M. and Karel, M. (1985). Prediction of beta-carotene decolorization in model systems under static and dynamic conditions of reduced oxygen environment. *J. Fd Sci.*, **50**(2), 526–30.
Singh, R. P., Heldman, D. R. and Kirk, J. R. (1975). Kinetic analysis of light-induced riboflavin loss in whole milk. *J. Fd Sci.*, **40,** 164–7.
Singh, R. K., Lund, D. B. and Buelow, F. H. (1983). Storage stability of intermediate moisture apples. Kinetics of quality change. *J. Fd Sci.*, **49,** 939–44.
Singh, R. K., Lund, D. B. and Buelow, F. H. (1984). Computer simulation of storage stability in intermediate moisture apples. *J. Fd Sci.*, **49,** 759–64.
Skjöldebrand, C., Anas, A., Oste, R. and Sjodin, P. (1983). Prediction of thiamine content in convective heated meat products. *J. Fd Technol.*, **18,** 61–73.
Strohecker, R. and Henning, H. M. (1963). *Vitamin Determination,* Verlag Chemie. Weinheim/Bergstr.
Teixeira, A. A., Dixon, J. R., Zahradnik, J. W. and Zinsmeister, G. E. (1969). Computer optimisation of nutrient retention in the thermal processing of conduction heated foods. *Fd Technol.*, **23,** 845–50.
Teixeira, A. A., Zinsmeister, G. E. and Zahradnik, J. W. (1975). Computer simulation of variable retort control and container geometry as a possible means of improving thiamine retention in thermally processed foods. *J. Fd Sci.*, **40,** 656–9.
Teixeira-Neto, R. O., Karel, M., Saguy, I. and Mizrahi, S. (1981). Oxygen uptake and beta-carotene decoloration in a dehydrated food model. *J. Fd Sci.*, **46**(3), 655–9, 667.
Thijssen, H. A. C. and Kerkhof, P. J. A. M. (1977). Effect of temperature and water concentration during processing on food quality. *Fd Proc. Eng*, **1**(3), 129–47.
Thompson, D. R. (1982). The challenge in predicting nutrient changes during food processing. *Fd Technol.*, **36,** 97–108, 115.
Van der Poel, G. H. (1956). Participation of B vitamins in non-enzymic browning reactions. *Voeding*, **14,** 452–5.
Vojnovich, C. and Pfeifer, V. F. (1970). Stability of ascorbic acid in blends with wheat flour, CSM and infant cereals. *Cereal Sci. Today*, **15**(9), 317–22.
Waletzko, P. and Labuza, T. P. (1976). Accelerated shelf-life testing of an intermediate moisture food in air and in an oxygen free atmosphere. *J. Fd Sci.*, **41**(6), 1338–44.
Walter, W. M. and Purcell, A. E. (1974). Lipid autoxidation in precooked dehydrated sweet potato flakes stored in air. *J. Agric. Fd Chem.* **22**(2), 298–302.

Wanninger, L. A. Jr (1972). Mathematical model predicts stability of ascorbic acid in food products. *Fd Technol.*, **26,** 42–5.
Widicus. A., Kirk, J. R. and Gregory, J. F. (1980). Storage stability of alpha-tocopherol in a dehydrated model food system containing no fat. *J. Fd Sci.*, **45,** 1015–18.
Wong, F. F., Dietrich, W. C., Harris, J. G. and Lindquist, F. E. (1956). Effect of temperature and moisture on storage stability of vacuum-dried tomato juice powder. *Fd Technol.*, **10**(2), 96–100.
Woodcock, E. A., Warthesen, J. J. and Labuza, T. P. (1982). Riboflavin photochemical degradation in pasta measured by high performance liquid chromatography. *J. Fd Sci.*, **47,** 545–9, 555.

INDEX